SUSTAINABLE POULTRY PRODUCTION IN EUROPE

Poultry Science Symposium Series

Executive Editor (Volumes 1–18): B.M. Freeman

*Out of print
Volumes 1–24 were not published by CAB International. Those still in print may
be ordered from:

Carfax Publishing Company
PO Box 25, Abingdon, Oxfordshire OX14 3UE, UK

Sustainable Poultry Production in Europe

Poultry Science Symposium Series
Volume Thirty One

Edited by

Emily Burton
School of Animal, Rural and Environmental Sciences, Nottingham, UK

Joanne Gatcliffe
AB Agri, Peterborough, UK

Helen Masey O'Neill
AB Vista Headquarters, Wiltshire, UK

and

Dawn Scholey
School of Animal, Rural and Environmental Sciences, Nottingham, UK

www.cabi.org

CABI is a trading name of CAB International

CABI	CABI
Nosworthy Way	745 Atlantic Avenue
Wallingford	8th Floor
Oxfordshire OX10 8DE	Boston, MA 02111
UK	USA
Tel: +44 (0)1491 832111	T: +1 (617) 682 9015
Fax: +44 (0)1491 833508	E-mail: cabi-nao@cabi.org
E-mail: info@cabi.org	
Website: www.cabi.org	

A catalogue record for this book is available from the British Library, London, UK.

ISBN-13: 978 1 78064 530 8

Commissioning editor: Caroline Makepeace
Editorial assistant: Emma McCann
Production editor: Shankari Wilford

Typeset by AMA DataSet, Preston, UK.
Printed and bound in the UK by CPI Group (UK) Ltd, Croydon, CR0 4YY, UK.

CONTENTS

Contributors

A. Amerah, *Danisco Animal Nutrition, Marlborough, Wiltshire SN8 1XN, UK; e-mail: Ahmed.amerah@dupont.com*

R. Andersson, *University of Applied Sciences, Am Krümpel 31, 49090 Osnabrueck, Germany; e-mail: robby.andersson@uni-osnabrueck.de*

M.C. Appleby, *World Animal Protection, 5th floor, 222 Gray's Inn Road, London WC1X 8HB, UK; e-mail: michaelappleby@worldanimalprotection.org*

S. Avendaño, *Aviagen Limited, Newbridge, Midlothian, UK; e-mail: s.avendano@aviagen.com*

P.E. Baker, *University of Bristol, School of Veterinary Sciences, Langford House, Langford, Bristol, BS40 5DU, UK*

E. Ball, *Agri-Food and Biosciences Institute, 18a Newforge Lane, Belfast, BT9 5PX, UK; e-mail: elizabeth.ball@afbini.gov.uk*

M.R. Bedford, *ABVista, Marlborough, UK; e-mail: mike.bedford@abvista.com*

E. Blesbois, *INRA, UMR-PRC, 37380 Nouzilly, France; e-mail: Elisabeth. Blesbois@tours.inra.fr*

E. Burton, *School of Animal, Rural and Environmental Sciences, Nottingham Trent University, Brackenhurst Campus, Southwell, Nottinghamshire NG25 0AF, UK; e-mail: emily.burton@ntu.ac.uk*

L.F. Caron, *Department of Basic Pathology, Universidade Federal do Paraná, Setor de Ciências Biológicas, Centro Politécnico, PO Box 19031, PR 81531-990 Curitiba, Brazil; e-mail: caronvet@uol.com.br*

M. Charismiadou, *Department of Animal Science and Aquaculture, Agricultural University of Athens, 75, Iera Odos, Athens, Attiki, Greece; e-mail: ahus7ham@ aua.gr*

P. Clarke, *Poultry World, Quadrant House, The Quadrant, Sutton, Surrey SM2 5AS, UK; e-mail: philip.clarke@rbi.co.uk*

J. De Gussem, *Poulpharm, Prins Albertlaan 111, B-8870 Izegem, Flanders, Belgium; e-mail: jeroendegussem@poulpharm.be*

M. De Gussem, *Vetworks, Knokstraat 38, B-9880 Poeke, Flanders, Belgium; e-mail: maarten.de-gussem@alpharma.com*

S. De Smet, *Laboratory for Animal Nutrition and Animal Product Quality, Department of Animal Production, Ghent University, Melle, Belgium; e-mail: Stefaan.DeSmedt@UGent.be*

S. Deligeorgis, *Department of Animal Science and Aquaculture, Agricultural University of Athens, 75, Iera Odos, Athens, Attiki, Greece; e-mail: sdel@aua.gr*

R. Devaney, *Queen's University, 22 Bristol Park, Newtownards, Belfast, BT23 4RH, UK; e-mail: rdevaney03@qub.ac.uk*

S. Dhaouadi, *Impextraco nv, wiekevorstsesteenweg 38, Heist-op-den-berg, Belgium; e-mail: sidki@impextraco.be*

F.E. Dias, *Vetworks, 129 Cherry Hinton Road, Cambridge, CB1 7BX, UK; e-mail: francisco.dias@vetworks.eu*

N. Dierick, *Laboratory for Animal Nutrition and Animal Product Quality, Department of Animal Production, Ghent University, Melle, Belgium; e-mail: noel.dierick@ugent.be*

J. Doppenberg, *Schothorst Feed Research B.V., Lelystad, the Netherlands; e-mail: Jdoppenberg@schothorst.nl*

S. Ellis, *2 Sisters Food Group, 1 Colmore Row, Birmingham, B3 2BJ, UK; e-mail: stephen.p.ellis@btinternet.com*

D. Filmer, *FLOCKMAN, Wascelyn, 48 Brent Street, Brent Knoll, Somerset TA9 4DT, UK; e-mail: david@flockman.com*

C. Fisher, *EFG Software, 9 New Mart Square, Edinburgh, EH14 1TJ, UK e-mail: cfisher345@gmail.com*

P. Garland, *Premier Nutrition, Brereton Business Park, The Levels, Rugeley, Staffordshire WS15 1RD, UK; e-mail: patrick.garland@premiernutrition.co.uk*

M. Goliomytis, *Department of Animal Science and Aquaculture, Agricultural University of Athens, 75, Iera Odos, Athens, Attiki, Greece; e-mail: mgolio@aua.gr*

F. Goodarzi Boroojeni, *Institute of Animal Nutrition, Department of Veterinary Medicine, Freie Universität Berlin, Germany; e-mail: farshad.goodarzi@fu-berlin.de*

H.M. Hafez, *Institute of Poultry Diseases, Department of Veterinary Medicine, Freie Universität Berlin, Germany; e-mail: hafez@vetmed.fu-berlin.de*

W.G. Hill, *Institute of Evolutionary Biology, University of Edinburgh, West Mains Road, Edinburgh, EH9 3JT, UK; e-mail: w.g.hill@ed.ac.uk*

G. Hogarth, *Aviagen Limited, Newbridge, Midlothian, UK; e-mail: ghogarth@aviagen.com*

R.A. Holser, *Russell Research Center, USDA, 950 College Station Rd, Athens, GA 30605, USA; e-mail: ronald.holser@ars.usda.gov*

H. Jewhurst, *AFBI, 18a Newforge Lane, Belfast, BT9 5PX, UK; e-mail: h.jewhurst@qub.ac.uk*

C. Kahuwai, *Department of Animal Science, Ahmadu Bello University, Zaria, Nigeria; e-mail: ckahuwai@abu.edu.ng*

F. Karadas, *Yuzuncu Yil University, Van, Turkey; e-mail: fkaradas@yyu.edu.tr*

M. Karakeçili, *Yuzuncu Yil University, Van, Turkey; e-mail: mkarakecili@yyu.edu.tr*

N. Kartsonas, *Department of Animal Science and Aquaculture, Agricultural University of Athens, 75, Iera Odos, Athens, Attiki, Greece; e-mail c/o: acom@aua.gr*

F. Kaufmann, *University of Applied Sciences, Am Krümpel 31, 49090 Osnabrueck, Germany; e-mail: f.kaufmann@hs-osnabrueck.de*

S. Keller, *Novus Deutschland GmbH, Gudensberg, Germany; e-mail c/o: info@novusint.com*

R. Kempsey, *Stonegate Ltd, Sutton Farm, Claverley, Wolverhampton, WV5 7DD, UK; e-mail: richard.kempsey@btconnect.com*

A. Kominakis, *Department of Animal Science and Aquaculture, Agricultural University of Athens, 75, Iera Odos, Athens, Attiki, Greece; e-mail: acom@aua.gr*

C. Kwakernaak, *Schothorst Feed Research B.V., Lelystad, the Netherlands; e-mail: ceesc.kwakernaak@wur.nl*

M. Laget, *Taminco BVBA, Axxes Business Park, Building H, Guldensporenpark 74, 9820 Merelbeke, Belgium; e-mail: mia.laget@taminco.com*

A. Lauwaerts, *Taminco BVBA, Ghent, Belgium; e-mail: angelo.lauwaerts@taminco.com*

I. Leinonen, *School of Agriculture, Food and Rural Development, Newcastle University, Kings Road, Newcastle upon Tyne, NE1 7RU, UK; e-mail: ilkka.leinonen@ncl.ac.uk*

A. Lovegrove, *Rothamsted Research, West Common, Harpenden, AL5 2JQ, UK; e-mail: alison.lovegrove@rothamsted.ac.uk*

P.C. Machado Júnior, *Impextraco Latin America, Rua Eng. Sady Souza, 650, PR 81290-020 Curitiba, Brazil; e-mail: pedro@impextraco.com*

A.M. Mackenzie, *National Institute for Poultry Husbandry, Harper Adams University, Edmond, Newport, Shropshire, UK; e-mail: ammackenzie@harper-adams.ac.uk*

K. Männer, *Institute of Animal Nutrition, Department of Veterinary Medicine, Freie Universität Berlin, Germany; e-mail: katmaenn@zedat.fu-berlin.de*

M. Martínez-Cummer, *Elanco Animal Health, Lily House, Priestly Road, Basingstoke, Hampshire RG24 9NL, UK; e-mail: martinez_marco_antonio@elanco.com*

M.J. McGrew, *Division of Developmental Biology, The Roslin Institute and Royal (Dick) School of Veterinary Studies, Easter Bush Campus, University of Edinburgh, Roslin, Midlothian EH25 9PS, UK; e-mail: mike.mcgrew@roslin.ed.ac.uk*

J. Michiels, *Department of Applied Biosciences, Ghent University, Valentin Vaerwyckweg 1, 9000 Ghent, Belgium; e-mail: joris.michiels@ugent.be*

N.K. Morgan, *Nottingham Trent University, Brackenhurst, Southwell, Notts, UK; e-mail: nat.morgan@ntu.ac.uk*

N.-D. Mulder, *Rabobank International, PO Box 17100, 3500 HG Utrecht, the Netherlands; e-mail: Nan-Dirk.Mulder@rabobank.com*

A.M. Neeteson-van Nieuwenhoven, *Aviagen Group, Newbridge, Midlothian, UK; e-mail: aneeteson@aviagen.com*

N.P. O'Sullivan, *Hy-Line International, Dallas Center, Iowa, USA; e-mail: NOSullivan@hyline.com*

J.J. Omage, *Department of Animal Science, Ahmadu Bello University, 80001 Zaria, Nigeria; e-mail: jjomage@abu.edu.ng*

P.A. Onimisi, *Department of Animal Science, Ahmadu Bello University, 80001 Zaria, Nigeria; e-mail: onimisiphil@gmail.com*

A. Ovyn, *Laboratory for Animal Nutrition and Animal Product Quality, Department of Animal Production, Ghent University, Melle, Belgium*

P. Parrott, *Harper Adams University, Edgmond, Newport, Shropshire TF10 8NB, UK; e-mail: tparrott@harper-adams.ac.uk*

T. Pellny, *Rothamsted Research, West Common, Harpenden, AL5 2JQ, UK; e-mail: till.pellny@rothamsted.ac.uk*

V. Pirgozliev, *National Institute for Poultry Husbandry, Harper Adams University, Edgmond, Shropshire TF10 8NB, UK; e-mail: vpirgozliev@harper-adams.ac.uk*

K. Poulsen, *Elanco Animal Health, Greenfield, Indiana, USA; e-mail: karl.poulsen@elanco.com*

S. Pritchard, *Premier Nutrition, The Levels, Rugeley, WS15 1RD, UK; e-mail: steve.pritchard@premiernutrition.co.uk*

T. Rogge, *Proviron, Oudenburgsesteenweg 100, 8400, Oostende, Flanders, Belgium; e-mail c/o: info@proviron.com*

B. Roosendaal, *RCL Foods, 1 The Boulevard, Westway Office Park, Westville 3629, KwaZulu-Natal, South Africa; e-mail: brett.roosendaal@rclfoods.com*

S.P. Rose, *National Institute for Poultry Husbandry, Harper Adams University, Shropshire TF10 8NB, UK; e-mail: sprose@harper-adams.ac.uk*

C. Sanni, *Nottingham Trent University, Brackenhurst, Southwell, Notts, UK; e-mail: colin.sanni@ntu.ac.uk*

D. Scholey, *School of Animal, Rural and Environmental Sciences, Nottingham Trent University, Brackenhurst, Southwell, Notts NG25 0QF, UK; e-mail: dawn.scholey@ntu.ac.uk*

A. Schwarz, *Vetworks, Knokstraat 38, B-9880 Poeke, Flanders, Belgium; e-mail c/o: info@vetworks.eu*

A. Scrase, *University of Bristol, School of Veterinary Sciences, Langford House, Langford, Bristol, BS40 5DU, UK; e-mail: as12141@bristol.ac.uk*

M. Shahbaz Yousaf, *University of Veterinary & Animal Sciences, Lahore, Pakistan; e-mail: drmshahbaz@uvas.edu.pk*

P.R. Shewry, *Rothamsted Research, West Common, Harpenden, AL5 2JQ, UK; e-mail: peter.shewry@rothamsted.ac.uk*

P. Simitzis, *Department of Animal Science and Aquaculture, Agricultural University of Athens, 75, Iera Odos, Athens, Attiki, Greece; e-mail: pansimitzis@aua.gr*

V.J. Smyth, *AFBI, AFBI-Stormont, Belfast BT4 3SD, UK; e-mail: victoria.smyth@afbini.gov.uk*

D. Speller, *Applied Group, Incl Geotek Heating Lower Farm, Eastmoor, Derbyshire S42 7DH, UK; e-mail: david@appliedgroup.org.uk*

J. Trudgett, *AFBI, 18a Newforge Lane, Belfast, BT9 5PX, UK; e-mail: James.Trudgett@afbini.gov.uk*

H. Ur-Rehman, *University of Veterinary & Animal Sciences, Lahore, Pakistan; e-mail: habibrehman@uvas.edu.pk*

W. Vahjen, *Institute of Animal Nutrition, Department of Veterinary Medicine, Freie Universität Berlin, Germany; e-mail: wilfried.vahjen@fu-berlin.de*

P.J. van der Aar, *Schothorst Feed Research B.V., Meerkoetenweg 26, 8218NA Lelystad, the Netherlands; e-mail: pvdaar@schothorst.nl*

V. Van Hamme, *Impextraco nv, wiekevorstsesteenweg 38, 2220 Heist-op-den-berg, Belgium; e-mail: valentine@impextraco.com*

H. Van Meirhaeghe, *Vetworks, Knokstraat 38, B-9880 Poeke, Flanders, Belgium; e-mail c/o: info@vetworks.eu*

A. Wahlstrom, *Zinpro Corporation, Akkerdistel 2E, 5831 Boxmeer, the Netherlands; e-mail: awahlstrom@zinpro.com*

C.L. Walk, *ABVista, Marlborough, UK; e-mail: Carrie.Walk@abvista.com*

A. Walker, *Slate Hall Veterinary Services, 27 Cow Lane, Rampton, Cambridgeshire CB24 8QG, UK; e-mail: andrew@slatehall.co.uk*

P. Wall, *University College Dublin, Belfield Campus, Dublin 4, Ireland; e-mail: patrick.wall@ucd.ie*

K. Walley, *Harper Adams University, Newport, Shropshire, UK; e-mail: kwalley@harper-adams.ac.uk*

J. Walton, *University of Bristol, School of Veterinary Sciences, Langford House, Langford, Bristol, BS40 5DU, UK; e-mail: j.walton@bristol.ac.uk*

C.A. Weeks, *University of Bristol, School of Veterinary Sciences, Langford House, Langford, Bristol, BS40 5DU, UK; e-mail: claire.weeks@bristol.ac.uk*

I. Whiting, *National Institute for Poultry Husbandry, Harper Adams University, Edgmond, Shropshire TF10 8NB, UK; e-mail: iwhiting@harper-adams.ac.uk*

C.T. Whittemore, *British Society of Animal Science, 17 Fergusson View, West Linton, Peeblesshire EH46 7DJ, UK; e-mail: colin.whittemore@btinternet.com*

J. Whyte, *Division of Developmental Biology, The Roslin Institute and Royal (Dick) School of Veterinary Studies, University of Edinburgh, Roslin, Midlothian EH25 9PS, UK; e-mail: jemimawhyte@hotmail.com*

A.G. Williams, *School of Energy, Environment and Agrifood, Cranfield University, Bedford, MK43 0AL, UK; e-mail: adrian.williams@cranfield.ac.uk*

P. Williams, *AG-Bio, Silver Street, Brixworth, NN6 9BY, UK; e-mail: peterevwilliams@gmail.com*

A. Wolc, *Department of Animal Science, Iowa State University, Ames, Iowa, and Hy-Line International, Dallas Center, Iowa, USA; e-mail: awolc@iastate.edu*

Z. Yu, *University College Dublin, Ireland; e-mail: yuzhongyi509@qq.com*

J. Zentek, *Institute of Animal Nutrition, Department of Veterinary Medicine, Freie Universität Berlin, Germany; e-mail: juergen.zentek@fu-berlin.de*

PREFACE

'Sustainable' has evolved from a term commonly describing the securing of environmental resources for future generations to a much wider and more complex meaning. Sustainability of European poultry production encompasses all the factors needed to create a durable industry. This symposium aimed to explore and ultimately define sustainability in the context of poultry production in Europe and to answer the questions around how we achieve a sustainable poultry industry in Europe.

The symposium, the 31st in the Poultry Science Symposium Series, was held on 8–10 September 2014, at the Queen Hotel, Chester, UK, and featured global leaders sharing their expertise across the full spectrum of sustainable poultry production in Europe. This book contains the manuscripts produced by the presenting authors and their supporting co-authors, or in some cases, a modified transcript of their talk. Sadly, family circumstances prevented one speaker, Professor Patrick Wall, from presenting at the symposium, but the contents of the talk are now offered as a chapter.

This symposium identified both the resilience and evolutionary factors needed to create a durable industry capable of thriving from tomorrow to 2050. Talks examined the role of cutting edge technologies and how other new (and not so new) approaches relate to the three pillars of sustainability; Environmental, Social and Economic. In addition to the papers of invited speakers, this book also contains abstracts from posters on original scientific communications relating to the four major themes of the symposium: Resources – securing material supplies and maintaining a skilled workforce; Market – strengthening positive links to end users; Risk management – identifying and containing threats from disease and economic fluctuations; and Green credentials – maximizing our contribution to waste management and food production and minimizing our use of global resources.

Following the challenge laid down by Dr John Hodges during his keynote address at the 23rd World's Poultry Congress in Brisbane on 30 June 2008,[1] the WPSA has begun opening a path towards sustainable production. While 'Small is Beautiful' is not a bedfellow of European poultry production, this symposium shows that the sustainability concepts espoused by economist E.F. Schumacher[2] regarding environmental resources as capital and maximizing well-being for minimum consumption are becoming embedded in our approach to poultry production. Similarly, while criticism may be laid at our door for limiting the scope of our symposium to Europe, the global diversity of sustainability-related challenges led us to put our own house in order first.

I am greatly indebted to my colleagues on the organizing committee who were fellow editors and gave their time into this symposium and book despite packed schedules in their regular jobs. I also feel immense gratitude to those with experience in the mires of symposium organization who stepped in with advice along the way, such as Paul Hocking, Vicky Sandilands, Howard Birley and Patrick Garland. Many others have contributed in ways as diverse as reconstruction of graphics, transcribing documents, proof reading, legal advice, logo design and delegate bag stuffing, for which I offer unending thanks. The scientific committee must be credited with the breadth of the programme, and their generosity and honesty with their thoughts and time.

E.J. Burton
Nottingham Trent University
Nottingham, UK
July 2015

REFERENCES

[1]Hodges, J. (2009) Emerging boundaries for poultry production: Challenges, opportunities and dangers. *World's Poultry Science Journal* 65, 5–21.

[2]Schumacher, E.F. (1973) *Small Is Beautiful: A Study of Economics As If People Mattered*. Blond and Briggs Ltd, London. ISBN 9780060916305.

ACKNOWLEDGEMENTS

Special thanks to those involved in organizing the Symposium:

Organizing Committee: Emily Burton, Joanne Gatcliffe and Helen Masey O'Neill
Scientific Committee: Emily Burton, Joanne Gatcliffe, Helen Masey O'Neill, Steve Wilson, Paul Hocking, Anne-Marie Neeteson, Colin Fisher, Steve Lister, Andrew Joret, Patrick Garland and Dawn Scholey

Support for this symposium is gratefully acknowledged from the following organizations:

Platinum
Zoetis

Gold
P.D. Hook (Hatcheries) Ltd
DSM Nutrition

Silver
Nutriad Animal Feed Additives
Huvepharma
Moypark Ltd
Evonik Industries
Elanco Animal Health
Aviagen
AB Vista Feed Ingredients

Bronze
DuPont Danisco Animal Nutrition
BOCM Pauls, now ForFarmers
ABN
Oxford University Press
Premier Nutrition
Alltech
Flockman Advanced Poultry Technology

PART I

Creating a Resilient Industry

CHAPTER 1

Making a Resilient Poultry Industry in Europe

Anne-Marie Neeteson-van Nieuwenhoven,[1]*
Michael C. Appleby[2] and George Hogarth[1]

[1]*Aviagen Group, Newbridge, Midlothian, UK;* [2]*World Animal Protection, London, UK*

INTRODUCTION

This chapter is the first of a series on the subject of 'Sustainable poultry production in Europe' written after the UK World Poultry Science Association conference on the same topic. Its aim is to give an overview of the factors that can contribute to a resilient poultry industry, and which factors may be a threat. Subsequent chapters will highlight some of these in more detail. This chapter discusses the factors that may contribute to a resilient poultry industry in Europe taking into account the global scale, and which factors may be a threat. From 2011 to 2050 available land resources will decline from 0.7 to 0.5 ha/person. More people will eat poultry products. Markets will move from producer to consumer markets. Food safety and animal welfare will become more important to the affluent in society. Although without import levies European producers may have difficulty surviving in the short term, in the longer term the need for global food security will also influence European food production. Keeping a poultry career interesting for young people will require dedicated efforts. The poultry sector has ample opportunities to improve its performance in a sustainable way balancing environmental, societal and economic aspects. Transparency and communication will be important to bridge the gap between the public and food producers, improve the sector's image and maintain its licence to produce.

Definitions

If we look at sustainability, defined in the Brundtland report as 'meeting the needs of the present without compromising the ability of future generations to meet their own needs' (World Commission on Environment and Development,

*Corresponding author: aneeteson@aviagen.com

1987, p. 15), there are opportunities and challenges to the future of the poultry industry, which lead us to consider its resilience.

According to the Intergovernmental Panel on Climate Change (2014, p. 5), resilience is:

> the capacity of social, economic, and environmental systems to cope with a hazardous event or trend or disturbance, responding or reorganizing in ways that maintain their essential function, identity, and structure, while also maintaining the capacity for adaptation, learning, and transformation.

Since the domestication of poultry millennia ago (estimates for chickens range from 6000 to 2000 BCE, see Miao *et al.*, 2013), mankind has developed the skills needed to breed and manage poultry. At first, animals that were easy to handle and to reproduce and that survived harsh seasons were those that were kept as farmed poultry. After World War II change accelerated, and a combination of management, housing, feeding and breeding changes have made poultry the most affordable source of animal protein and enabled less-thriving agricultural regions to develop into professional areas, helping to maintain employment in the countryside. The actors in the poultry production chain, from suppliers to farmers to processors (but not including retailers), together form the 'poultry industry'.

Poultry production in Europe

In 2012 total poultry meat production in the EU-27 was around 13 million t. The majority of poultry meat is broiler meat with a total in 2012 of 9.9 million t. This came from about 7700 million broilers. Five countries in the EU produce more than 0.6 million t of broiler meat each: Poland produces the most, followed by the UK, France, Spain, Germany, the Netherlands and Italy. In recent years, total EU poultry meat production has been growing slightly.

Europe produced around 181 billion eggs (10.6 million t) in 2012. Eight countries produced more than 5 billion each: France most, followed by Italy, Germany, Spain, the UK, the Netherlands, Poland and Romania (FAOSTAT, undated; Hiemstra and Ten Napel, 2013; Van Horne and Bondt, 2013; British Poultry Council, 2014).

Global outlook

The world's human population is expected to grow to 9 billion by 2050, with people in the emerging economies adding a lot of animal produce to their diets (subject to availability), producing what is often referred to as the demand-driven livestock revolution (Delgado *et al.*, 1999). As fewer and fewer people are being employed in agriculture, and because 'cities seldom contribute to the production of their food . . . generally, they simply consume it' (Ghirotti, 1999), fewer rural people will be producing more food for many more urban people (Neeteson-van

Nieuwenhoven *et al.*, 2013). As a consequence they will have to produce more food per worker (from animals and/or crops) from a decreasing amount of land. 'Sustainable intensification' of agriculture is a phrase that has been used to describe this (Foresight, 2011).

A consumer market

These developments also put poultry production at a greater distance from citizens. Expectations of poultry production change: while most citizens are familiar with pets and horses, most are less familiar with farm animals, except perhaps for those on a 'city farm', and emotion adds to the debate on how they should be kept. People have the right to know how their food is produced and feel good about it. With farmers in the minority, and both voting and buying power with the consumer, the poultry landscape in the affluent European countries has changed from a 'producer' to a 'consumer' market where 'the different attitudes and needs of different stakeholders, including producers, retailers, consumers and governments, with welfare and environmental considerations [are] playing an important role. Political and legal decisions both affect and are affected by the attitudes of people to poultry and their management' (Appleby, 2012, p. 53).

In this chapter the main drivers influencing the resilience of the poultry industry both in Europe itself and in its global context will be discussed, and how they may contribute to or limit the resilience of European poultry production. These will be considered in the three categories of environmental, social and economic aspects, corresponding to the key aspects of sustainability according to the United Nations (UN) General Assembly (2005).

ENVIRONMENTAL ASPECTS

Resources

A number of developments will influence the possibility of the globe to provide for our nutrition. First, population and agricultural land (including pasture) are not equally divided across the globe: compared with the relative scarcity of agricultural land in Europe and Asia (0.6 and 0.4 ha/person, respectively), North and Latin America have more space: 1.4 and 1.2 ha/person (2011 data from FAOSTAT, undated). If there is no alteration in available land, Africa, with a fast growing population, will change from 1.1 ha/person in 2011 to 0.5 in 2050. Globally, the area for producing food per person will decrease from 0.7 to 0.5 ha/person (Fig. 1.1). Furthermore, the need for sustainable water supplies is one of humanity's most critical resource needs (United Nations Environment Programme (UNEP), 2012). In 2050, Europe, although with little change in land per person, will probably have less opportunity to import both food and feed, or will have to pay high prices. Therefore, the future of poultry production in Europe will be even more dependent on knowledge and skills to manage poultry in the short

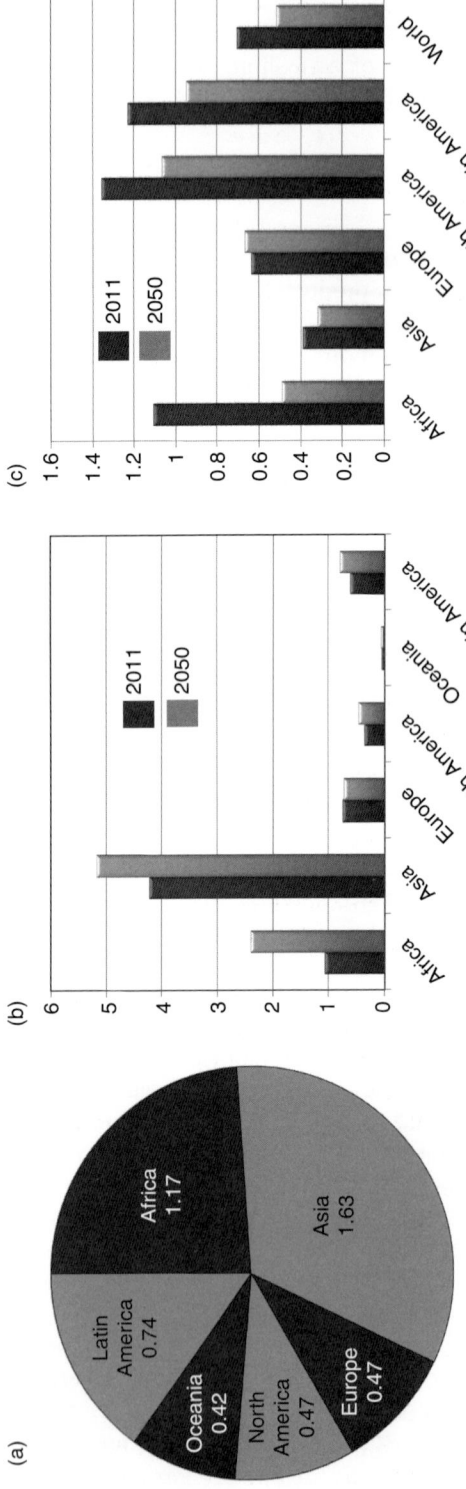

Fig. 1.1. Population and agricultural land in the regions of the world: (a) agricultural land 2011 (billion ha); (b) population size (billion) 2011 and 2050; and (c) agricultural land (ha/person). Assuming available land will not change, 2011 areas are divided by 2050 population projections (FAOSTAT, undated).

and medium term, and on development of efficient European protein-containing crops that poultry can digest well, as in the long term we will probably have to become more self-sufficient in European produce.

Second, 'burgeoning populations and growing economies' will increase competition for available resources between feed, food, fibre and fuel (UNEP, 2012, p. 26). For the EU to produce 5.75% of its fuel as biofuel (the (not achieved) goal for 2010), it would need 13–15% of the EU agricultural area. Goals for 2015 and 2030 are 8% and 25%, respectively (Plakké *et al.*, 2008; European Commission, 2013).

Third, with societies becoming more affluent, people are tending to eat more animal protein, starting with poultry, and this trend is expected to continue. Generally, meat consumption patterns are shifting towards poultry – currently 70% of global meat consumption is pork and poultry, of which 50% is poultry, and its share is growing (Delgado *et al.*, 1999; Neeteson-van Nieuwenhoven *et al.*, 2013). Some policy makers (e.g. UNEP, 2012) call for moderation of meat consumption, but globally an increase is likely to precede any possible decline in animal protein consumption.

Fourth, climate change will increase insecurity in harvesting and food availability. Food security in the widest sense will become more of an issue, with problems in volatility of feed prices and food availability. This will have most impact on less-wealthy regions and people.

Life cycle Assessment

For these reasons, efforts to increase efficiency in utilization of natural resources, including water, energy and land, while maintaining or improving other requirements of food (e.g. safety, availability, animal welfare), will become ever more important. One way to measure the environmental footprint of whole production systems is Life Cycle Assessment (LCA), which has advanced considerably during the last decade. Compared to other livestock systems, poultry has a low environmental impact. The global warming potential (GWP/kg CO_2-eq. produced per kg of product) of beef, sheep meat, pork and poultry meat are 16, 17, 6.4 and 4.6, respectively (Williams *et al.*, 2006) and the GWP for milk is 10.6 kg CO_2-eq. produced per 10 l (about 1 kg dry matter), and for eggs 5.5 kg CO_2-eq. produced per 20 eggs (Williams *et al.*, 2006). Aquaculture LCAs show values in the same range as poultry (e.g. Vandenburg *et al.*, 2012). Comparisons of conventional, free range and organic poultry systems by Williams *et al.* (2006) and Leinonen *et al.* (2012) show relative environmental advantage for conventional over free range over organic broiler systems for primary energy used, GWP, eutrophication and acidification potential, abiotic resource use and land use. LCA assessments give indications where and to what extent improvements in the environmental impact of systems can be made, provided the model contains sufficient data input and few extrapolations. To improve comparability of LCA exercises globally, the Food and Agriculture Organization (FAO) is currently developing global LCA guidelines for feed, poultry and other species (Livestock Environmental Assessment and Performance Partnership, 2014a, b).

Breeding and management

Much of the favourable LCA of poultry is due to a combination of genetic selection and improved management. Over the last 30 years, average feed conversion ratio (FCR) achieved for broilers in practice has changed by –0.015 kg/kg/year (Neeteson-van Nieuwenhoven *et al.*, 2013). For a specified quantity of production, this will have reduced land use considerably. As well as land use, there is also a relative decline in the use of water, and of materials and space for farm buildings.

Over the last three decades the breeding goals for poultry have widened (Laughlin, 2007; Farm Animal Welfare Committee (FAWC), 2012; Hiemstra and Ten Napel, 2013) to include dozens of traits, with the target of improving the traits simultaneously and gradually. This requires large populations to choose from, powerful data-gathering and management systems, and careful specification of new and improved traits for welfare, health, robustness, animal behaviour and production. Changes produced by selection are cumulative and permanent, but at any specific point in time 70% or more of the variation in final performance in animal production comes from management factors: the genetic potential is realized via management, health care and nutrition to suit the changing birds. Providers of feed and hatchery equipment, breeders and other actors in the poultry chain continually develop customized poultry management information.

Waste and losses

Waste and losses along the food chain form a large part of its inefficiency, including waste on the farm, during transport, and by retail, restaurants and consumers at home. Globally, 10–40% of agricultural production is wasted and a further 30% for human consumption (UNEP, 2012). Europe wastes 90 million t of food annually. Whereas 40% of food losses occur after harvest and during processing in developing countries, in industrialized countries over 40% is at retail and consumer level (European Commission Health and Consumers, 2014). Apart from further decreasing losses in the poultry chain, then, Europe must improve end-user awareness of this problem and actions to reduce it. An example is the call for redistribution of unsold food made by the EU Committee of the UK Parliament's House of Lords (European Union Committee, 2014).

Some countries, such as the Netherlands, use by-products of human foods as feed, and organizing production systems to increase this practice might be a way to reduce the poultry footprint further. Unfortunately, many biofuel by-products (from grain or sugarbeet) are not very useable as poultry feed, but some plant oil by-products (from soya and cole/rape seed) can be usefully included (Plakké *et al.*, 2008). Another opportunity to reduce wastage of valuable animal resources would be the reintroduction into feed of processed animal proteins (PAPs) that are currently not allowed in the EU, such as meat and bone meal from animals other than birds. This is under discussion: since 2013, the EU have allowed PAPs in fish feed. However, not all European member states are expected to adopt this practice, including the UK.

Other environmental impacts

Regional variation in use of feed resources might be a possible means to increase the resilience of food production and to reduce environmental impact, for example because of (real or perceived) lower energy costs for transport. Whole chain impacts including scale effects should be calculated to quantify the validity of these assumptions.

The resilience of poultry production in general may also benefit from diverse production systems and breeds being available. At the Convention on Biological Diversity in 1989, most countries agreed that they are responsible for the diversity in their constituency. European governments assist in the preservation and evaluation of indigenous breeds, through Animal Genetic Resource evaluations, committees, research projects, subsidies, and support for rare breeds and speciality produce systems. Thus the Dutch Rare Breeds Society received €500,000 funding from government to organize its professional office, and the French rare breeds office is also government funded. Diverse systems and breeds are perceived well in European society (for example by press, chefs, and citizen polls) but market shares are small (<5%), except in France (Hiemstra and Ten Napel, 2013). Management of diversity within and between commercial lines is under the custody of breeding organizations, each maintaining or having access to dozens of lines and managing inbreeding below the FAO inbreeding guideline of 1% (Department for Environment, Food and Rural Affairs (DEFRA), 2010; Hiemstra and Ten Napel, 2013).

SOCIAL ASPECTS: PEOPLE AND ANIMALS

Poultry production affects many groups of stakeholders: the animals that are farmed, the farmers managing the birds and the people providing support and processing, the organizations marketing the animals and products, and the consumers who eat poultry meat and eggs. The animals should be looked after well with care, knowledge and skill. The people managing poultry production perform a role in society of providing food, and are due an income from that. The members of society who consume poultry products want a safe and secure supply. Additionally, society includes some people who do not eat poultry produce.

Laws

With fewer people producing food, and more animals managed per person, familiarity of consumers and citizens with food production has decreased. Yet as Appleby (2012, pp. 53, 54, 56) explains:

> Political and legal decisions [see Table 1.1] both affect and are affected by the attitudes of people to poultry and their management . . . Concern has largely developed in urban people whose involvement with animals differed from that in rural areas . . .

Table 1.1. Laws and codes of good practice concerning poultry (as per 2014).

Council of Europe convention	EU laws and regulations	OIE guidelines	European codes of good practice and quality schemes	National codes of good practice and quality schemes in Europe (excerpt)
1976 convention of the protection of animals kept for farming purposes[a]	1988/166/EEC[d] minimum standards protection laying hens battery cages	Slaughter	avec code of good turkey farm management practice –	Beter leven (layer, broiler, NL) www.beterleven. dierenbescherming.nl
1995 standing committee recommendations for domestic fowl[b]	1998/58/EC[e] protection of animals kept for farming purposes	Transport	turkey welfare at the farm	Freedom food (layer, turkey, broiler, UK) www.rspca.co.uk/freedomfood
2001 standing committee recommendations for turkeys[c]	1999/74/EC[f] minimum standards for the protection of laying hens	Broiler welfare (under development)	Code – EFABAR www. responsiblebreeding.eu	Kip van morgen (broiler, NL)
	2002/4/EC[g] registration of establishments keeping laying hens			IKB braadkuiken (broiler, BE) www.belpulme.be
	2005/1/EC[h] animal transport directive			IKB kip (layer, turkey, broiler, NL) www.ikbkip.nl
	2006/778/EC[i] collection of information during inspections			Initiative Tierwohl (broiler, GE)
	2007/43/EC[j] broiler directive			QBT (turkeys, UK) www.britishpoultry.org.uk/ areas-of-work/quality-produce/ quality-british-turkey
	1234/2007/EC[k] organization of agricultural markets for agricultural products			QS (broiler, turkey, GE) www.qs.-de

589/2008/EC10[l] marketing
standards of eggs
1099/2009/EC[m] slaughter
directive
COM(2012) 6 final/2[n]
animal welfare strategy

Red tractor (broiler, UK)
www.redtractor.org.uk
Tierschutzlabel (broiler, GE)
www.tierschutzlabel.info
ZDG Eckwerte (turkey, GE)
www.zdg-online.de/
uploads/tx_userzdgdocs/
VDP_Broschuere_
EckwerteMastputen_29-04_
ohne_Unterschriffen.pdf

BE, Belgium; GE, Germany; NL, the Netherlands
[a] http://conventions.coe.int/Treaty/en/Treaties/Html/087.htm
[b] http://www.coe.int/t/e/legal_affairs/legal_co-operation/biological_safety_and_use_of_animals/farming/Rec%20fowl%20E.asp
[c] http://www.coe.int/t/e/legal_affairs/legal_co-operation/biological_safety_and_use_of_animals/farming/Rec%20Turkeys.asp
[d] http://ec.europa.eu/food/fs/aw/aw_legislation/hens/88-166-eec_en.pdf
[e] http://eur-lex.europa.eu/legal-content/EN/TXT/PDF/?uri=CELEX:31998L00058&from=EN
[f] http://eur-lex.europa.eu/LexUriServ/LexUriServ.do?uri=OJ:L:199:203:0053:0057:EN:PDF
[g] http://eur-lex.europa.eu/legal-content/EN/ALL/?uri=CELEX:32002L0004
[h] http://eur-lex.europa.eu/legal-content/EN/TXT/PDF/?uri=CELEX:32005R0001&rid=8
[i] http://eur-lex.europa.eu/legal-content/EN/ALL/;jsessionid=QyDGTzyKG0kSvpFVm1xg8gnb1vvhhCzVwmB7NGssl0J1vBnTFBNl-5853969940?uri=CELEX:32006D0778 7
[j] http://eur-lex.europa.eu/LexUriServ/LexUriServ.do?uri=OJ:L:2007:182:0019:0028:EN:PDF
[k] http://eur-lex.europa.eu/LexUriServ/LexUriServ.do?uri=OJ:L:2007:299:0001:0149:EN:PDF
[l] http://eur-lex.europa.eu/LexUriServ/LexUriServ.do?uri=OJ:L:2008:163:0006:0023:EN:PDF
[m] http://ec/europa/eu/food/animal/welfare/slaughter/regulation_1099_2009_en.pdf
[n] http://ec/europa/eu/food/animal/welfare/actionplan/docs/aw_strategy_19012012_en.pdf

Correspondingly, Northern countries [of Europe] have detailed laws, with codified lists of actions that are prohibited. Southern countries tend simply to state that animals must not be ill-treated. Legislation is also enforced more strictly in some countries than in others. Increasingly, however, legislation in European countries originates from the EU . . . Among farm animals, a considerable proportion of this attention has been paid to poultry, including by the large numbers of societies and groups that have been set up in most countries. The core staff of these organisations is generally professional, but they need to retain the support of their amateur supporters for their actions . . .

EU directives must be implemented in the member countries within 2 years. In addition to variation in enforcement, there are also EU member countries, and states within countries, that implement more stringent or detailed legal provisions.

Markets

In line with variation in legislation and its implementation, according to Neeteson *et al.* (2013, p. 20):

Within Europe, there is a clear distinction between markets where a) welfare is defined mainly in terms of production outcomes, and b) welfare is mainly defined and driven by consumer emotion and perception of animal production. The latter is newsworthy and attracts votes, and thus political interest. Socio-economic factors play an important role in how welfare is defined and how communication and transparency will be effective.

Discussions with North African turkey producers at an Aviagen Turkeys Management School in 2012 learned that responsibility and care for animals is embedded in Islam. Southern European markets will rate naturalness of production, and taste/quality higher, as well as working conditions (Neeteson *et al.*, 2013).

Professional associations

Appleby (2012, p. 55) points out that:

Politics does not just involve the actions of professional politicians but all developments in policy and public affairs. In agribusiness this includes the activities of trade associations, which recruit a high proportion of producers as members . . . While there has in the past sometimes been resistance to pressure for change from those organizations, they have become more active on animal welfare in recent years to reflect increased concern for this issue.

Professional associations increasingly organize stakeholders to develop codes of good practice and quality schemes (Table 1.1). Such cooperation is important, and in line with the recommendation of the Welfare Quality Advisory Committee (2009) for a holistic approach to welfare assessment.

Communication

Neeteson *et al.* (2013, p. 19) comment that, although:

> the agriculture industry has been historically poor at conveying a positive case for poultry production . . . in being transparent about animal food production in a proactive and honest way while engaging in continuous welfare improvement, [they] can play an important role to close the gap between welfare perception and welfare of the animal itself.

Turkey producers, at a BUT50 Aviagen Turkeys welfare workshop in 2012, indicated they saw ample opportunities to work towards better understanding of turkey production and turkey meat, e.g. turkey meat is healthy, this is a strong message; improving transparency of production such as opening doors for citizens; highlighting the role of welfare in breeding, management and transport; and last but not least contacting young people at schools or universities to learn about farming from a young age (Neeteson *et al.*, 2013).

Animal welfare

Farm animal welfare is a major item in the public debate. According to Neeteson *et al.* (2013, p. 19), at the *bird level* welfare means the bird is:

> performing well, in good health, under good conditions and in a way that the animal is able to do easily what it is good at. . . . It centres around the homeostatic balance of animals between their intrinsic potential and the production environment, which includes nutrition, housing, health, the social environment, and stockmanship. The social environment of the animal includes behaviour, the way it perceives its environment, and whether this perception is in line with the individual's intrinsic value as an animal. These [factors] are key to achieving animal production under responsible welfare conditions. Managing animals in such a way that all these conditions are met, will then lead to good animal welfare.

People do not all have the same concept of welfare, and in addition to physical aspects (such as health and growth), many people put more emphasis on mental aspects (such as suffering and animals' ability to express their preferences) and on naturalness (of environments and behaviour) (Fraser *et al.*, 1997; Appleby *et al.*, 2004). This is important because animal welfare groups and members of the public express concern not only about the physical welfare of birds (including leg health in broilers and bone strength in laying hens), but also about mental welfare (do certain treatments of birds cause pain or fear?) and the naturalness or unnaturalness of conditions in which they are kept (for example, very large group sizes) (Rollin, 1993). This diversity of approach is conveyed by FAWC's (2011) description of welfare in terms of the 'Five Freedoms': freedom from hunger and thirst, from discomfort, from pain, injury or disease, from fear and distress, and freedom to express normal behaviour. These freedoms 'list the provisions that should be made for farm animals . . . [they] define ideal states rather than standards for acceptable welfare . . . [and to achieve these provisions] stockmanship, plus the training and supervision necessary to achieve required standards, are key factors in the handling and care of livestock.' FAWC (2007) also outlines the 'Three Essentials of Stockmanship': knowledge of animal husbandry, skills in animal husbandry, and personal qualities of affinity and empathy with animals, dedication and patience.

The ethical and societal values are the part of the debate on animal welfare concerning welfare perception. Ideally, there should be a complete match between the welfare of the animal, and the perception of welfare by the consumer and the citizen. That will serve the welfare of the animal itself best (Neeteson *et al.*, 2013). More work is needed on both welfare and perception, and there are examples in both the meat and egg sectors. In laying hens, the issue of frequent bone fractures both during lay and at the end of lay (FAWC, 2010) have generated many projects in management improvement (for example, DEFRA, 2013) and layer breeding has adopted techniques to measure and select for better bone strength (Silversides *et al.*, 2012). In broilers, foot-pad dermatitis incidence during winter months is subject to monitoring (trigger system, DEFRA, 2010), management (De Jong and Van Harn, 2012) and breeding improvements (Kapell *et al.*, 2012). Acceptability to all stakeholders of production methods used is an important aspect of sustainability, and for that it is necessary for welfare to be maximized, taking into account all aspects: physical, mental and natural.

Other public awareness issues

As well as animal welfare, other issues are relevant for public awareness, such as dietary health (e.g. consumption of lean animal protein), indoor versus outdoor farming, farm size, use of machinery and technology, feed type (e.g. with or without animal protein, whether or not organic, whether or not genetically modified) and bird health (e.g. use of prophylactic or curative antibiotics). Some aspects of both human and animal health raise less attention, although 'the increase in number of free range systems also poses an increased risk of introducing avian influenza (AI) from the wild bird population into the . . . chain. This is more of a human health risk than a food safety risk' (Hiemstra and Ten Napel, 2013, p. 105). The majority of AI outbreaks in north-west Europe are related to outside poultry production: in the Netherlands, all AI outbreaks in 2011, 2012 and 2013 (except for two cases involving turkeys) were on poultry farms with outside areas.

When societies become more wealthy, requirements for food availability, safety and traceability increase. Food safety procedures in north-west Europe are strict, allowing high safety levels in poultry production. However, across the globe, food 'scandals' related to traceability (horse meat sold as beef, conventional produce sold as organic), safety (baby milk with dangerous additives) or hygiene (*Salmonella*) raise increasing press and public concern, making people insecure about the safety of their food. Even if food scandals are an exception, they may have a very high public profile, and it remains essential to address food safety both in practice and in perception.

People working with poultry

Perceptions of poultry production and the availability of people to work in the sector are related: as poultry farming has been criticized, working in poultry has become less popular. Improvements in welfare and food safety and decreasing use of antibiotics have not yet had positive results, as the poultry sector has not managed to get its improvements widely known. Dedicated, trained people are key for balanced and efficient production with acceptable environmental impact and good animal welfare. Although training of people who already work in poultry is gaining interest, keeping people in the sector and attracting new people remains a challenge. In north-west Europe the image of poultry production is not inviting for young people, and in the new EU member states burgeoning development makes the choice of employment in agriculture a less exciting option than non-food-producing opportunities. Nevertheless, occasionally education centres manage to be exceptions to this trend. Wageningen University (the Netherlands) almost doubled its number of students from 2002 to 2013, while Osnabrück University (Germany) runs a successful applied poultry management education programme. After decades, it seems that a balance has been achieved in poultry education integrating the full spectrum of management, physiology, economics, health, behaviour, breeding, reproduction, nutrition, housing and awareness of law and public perception. This is important, as the full spectrum of animal production aspects is required to educate the professional for today's and tomorrow's society and form the basis of sound poultry science in the context of public awareness.

Poultry farmers in rural communities

A last but not unimportant social aspect is the role of farmers in today's rural communities. With affluent citizens moving to the countryside, for example to renovated farm houses, even in the countryside farmers may become a minority. Aspects of living in the countryside, like farm odours or the appearance of modern buildings, are increasingly criticized by non-food-producing countrymen. This often leads to complicated building procedures, contradictory legal requirements and difficulties in adapting the farm to implement new findings or to extend the farm for the next generation. Non-farmers are also increasingly vocal on issues such as animal welfare and sustainability. Generally citizens will more and more affect how farmers do their job and even whether they continue to farm.

ECONOMIC ASPECTS

An intrinsic part of the sustainability of the poultry sector in Europe is its viability in economic terms and its ability to continue to provide farms and farming for the next generation. Key to this is the resilience of farmers and of the rest of the poultry chain in the short and longer term.

Despite negligible change in overall EU meat consumption, EU poultry meat consumption is expected to grow by 4.3% per year over the next few years, reaching 24.1 kg/person by 2022, mostly driven by increased consumption in the new EU member states. Relevant attributes are the affordability and health aspects of poultry products. Consumption differs between member states: for example, in 2012 averages (kg/person/year) were Spain 30, Hungary 29, the UK 29, Czech Republic 23, the Netherlands 22, Slovakia 20, Germany 19, Italy 19 and Sweden 12. Average EU turkey consumption in 2012 was 3.4 kg/person, including Austria 6.1, Germany 5.7, France 5.3, Italy 4.8 and the UK 4.2. For broiler meat, the EU is 105% and the UK is 89% self-sufficient (avec, 2014). The UK could improve its self-sufficiency if consumers ate more of the brown meat (Farmers Weekly, 2013). Egg consumption is currently stable in the UK, with a figure of 11.5 billion in 2013 (British Egg Information Service, 2014), and in Europe as a whole (Global Poultry Trends, 2013).

In 2012, global poultry meat and egg production were 103.7 and 66.4 million t, respectively (1250 billion eggs). The top ten egg-producing countries (2012) were China, the USA, India, Japan, Russia, Mexico, Brazil, Ukraine, Indonesia and Turkey. In chicken meat the 2012 top ten were the USA, China, Brazil, Russia, Mexico, India, Iran, Indonesia, Turkey and Argentina (FAOSTAT, undated). Through the period 2013–2022 poultry is predicted to remain the fastest growing meat sector in terms of production (+2.2% each year) and to have the highest volume of production by the end of the period, exceeding pig meat. Two-thirds of the rise will occur in Asia (avec, 2014).

Appleby (2012, p. 53) has commented that:

> A common tendency in developed countries in the second half of the 20th century was the drive for efficiency in agriculture, for cutting the cost of producing each egg or kilogram of meat. This was initiated by public policies - before, during and after World War II – in favour of more abundant, cheaper food. It subsequently became market driven, with competition between producers and between retailers to sell food as cheaply as possible, and thereby acquired its own momentum.

The poultry sector has developed accordingly. Whereas a poultry farmer would only have to maintain 1000 breeders in order to make a living around 1940, nowadays a farmer needs to farm more. Farm/hatchery employment data are available from some countries, for example France, the Netherlands and Germany employed 1.2, 3.9 and 4.6 persons/farm or hatchery in 2011 (Van Horne and Bondt, 2013). From 1993 to 2013 in the UK, food prices in general increased by 68% but chicken only by 31%, less than half. If compared to house-price increases since 1971, a chicken would now be expected to cost £51, whereas the supermarket price is between £4 and £5. Poultry price increases have been mitigated by the level of consolidation within the industry and production levels have gone up by 25% (Bradnock, 2013).

In general, the opportunity to produce feed at places other than where poultry production takes place, using energy-efficient transport including by sea, has enabled many poor agricultural regions to develop into professional regions, first in Europe (e.g. Lower Saxony in Germany, Peel/Brabant and Veluwe in the Netherlands, Brittany in France) and the USA (e.g. Alabama, Tennessee,

Arkansas) and increasingly in the emerging economies (Livestock Environmental Assessment and Performance Partnership, 2014b).

Feed represents a major component of poultry production costs: around 70%. Feed price differs between EU member states, for example in 2011, €32.0/100 kg in Hungary, 32.8 in Denmark, 32.8 in France, 33.8 in the Netherlands, 34.5 in Germany, 34.5 in Poland, 34.6 in Spain, 35.4 in the UK and 40.1 in Italy. The EU average (€34.5/100 kg) is higher than for major competitors Argentina (23.1), the USA (25.7), Brazil (26.5), Ukraine (28.0), Thailand (32.9) and Russia (33.9) (Van Horne and Bondt, 2013). Over the last decade, commodity prices have been at historically high levels, while profitability for poultry businesses has been under pressure. It is expected that feed price will become more volatile and will increase over the coming decades (avec, 2014). 'It is the volatility that is the problem and the inability of the industry to recover each cost increase quickly enough from the marketplace, resulting in losses' (Farmers Weekly, 2013).

Financial management must, of course, operate within the political context discussed above. It is therefore relevant to note (Appleby, 2012, p. 55) that:

> In no other country has legislation advanced as far as in Europe. That is partly because of different attitudes to animal welfare, and partly because of different legal systems. For example, in the USA there are only three federal laws that apply to animal welfare; two (on slaughter and general welfare) specifically exclude poultry and the other (on transport) has never been applied to poultry. In the country as a whole the industry and the retail sector have achieved more in improving how poultry are kept than has any legislation to date.

Van Horne (2012) and Van Horne and Bondt (2013) have investigated how EU legislation and its implementation impact on the competitiveness of the EU egg and poultry meat sectors, and to what extent poultry production costs differ between countries within and outside the EU. Legislation adds costs related to environmental protection, food safety and animal welfare in the following areas: (i) environment: manure disposal costs, reduction of ammonia emission in manure application, during manure storage and in the poultry house; (ii) food safety: *Salmonella* control, hygiene measures, sample collection, testing and vaccination, meat-and-bone meal ban, growth-promoter ban, genetic modification ban for crops; and (iii) animal welfare: stocking density housing cost, beak treatment (mainly layers) and housing (layers). On average EU law added 5% to total production costs (€cents 4.79/kg live weight) of broilers in 2011, and 15% to total production costs of eggs in 2012. Figure 1.2 shows that without import levies the poultry businesses of EU member states would not be able to stay in business. This is most threatening when it concerns eggs other than table eggs, egg powder and (de)frozen meat. The authors stress that the import levies are an indispensable barrier to protect the EU poultry farming sector.

As part of this, 'costs are generally higher in systems perceived to have higher welfare: greater space allowances in cages, as well as production costs in different systems, increase costs: housing, labour, feed, hygiene, mortality and predictability of performance' (Appleby *et al.*, 2004, p. 53). Extra costs, both for EU (and sometimes additional national) legislation and for perceived welfare,

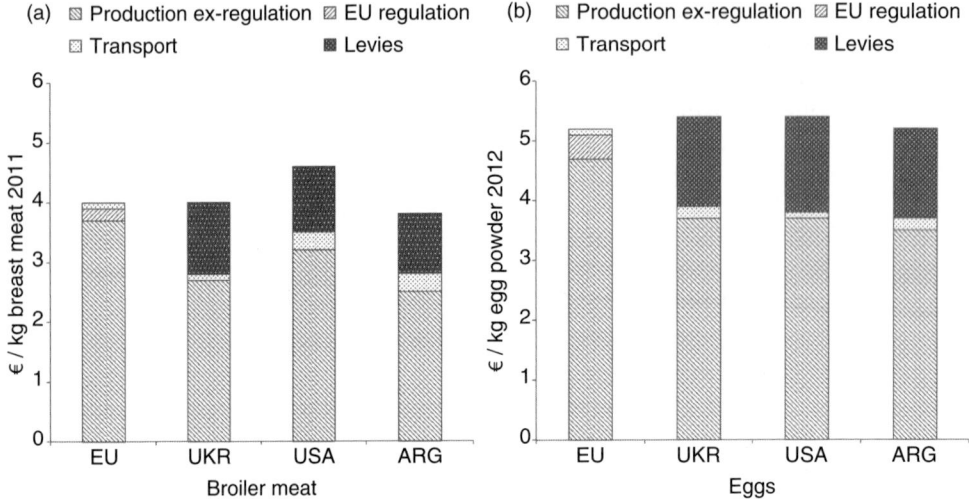

Fig. 1.2. Comparison of lean production cost (without EU Directive impacts), EU legal impact cost, transport cost and import levies of the EU, Ukraine (UKR), the USA and Argentina (ARG) for (a) broiler meat and (b) eggs (after Van Horne, 2012, p. 9; Van Horne and Bondt, 2013, pp. 51–54).

are mostly incurred up to and including the farm level. However, at the primary level little extra value is generated: that is mostly generated at the processing level (Plakké *et al.*, 2008) and later.

Initiatives to compensate farmers for extra perceived welfare costs are currently being undertaken; however, in 2014, these were only at the planning and agreement phase and have not yet been implemented. In the Netherlands, retail and poultry sectors have signed a so-called 'convention' that should ensure that fresh Dutch chicken in the supermarket comes from chickens with more space allowance, but which also grow a maximum of 50 g/day, as from 2015 to 2020 (Groenkennisnet, 2013). However, not all supermarkets comply, and details on dates and rates of commitment have not been settled. In Germany business stakeholders are currently discussing a retail commitment to compensate poultry farmers for perceived higher welfare production.

DISCUSSION

The resilience of the European poultry industry to thrive, i.e. to survive the challenges ahead and to realize the opportunities, is not a 'given'. The higher cost of poultry production in Europe compared to elsewhere, due to a combination of legal, labour, feed and building costs, as well as the need to respond to social demands and perceptions to ensure consumer purchases, make the future a challenging one.

Van Horne and Bondt (2013) suggest that strengths of the EU poultry sector are:

- A large internal market with demand for both white and dark meat. New markets for dark meat could increase UK self-sufficiency.
- A strong focus on fresh poultry, as developing countries cannot compete in this segment.
- The increased competitive position of a few European countries where the industry has grown in recent years and is working with modern poultry farms and processing plants. This then puts countries, regions and farmers with old buildings or plants at a disadvantage.

Clear weaknesses in the EU (after Van Horne and Bondt, 2013) are:

- The high cost to European poultry farms and in industry, partly as a result of EU and national legislation.
- A small and fragmented industry, in many countries, operating with only a regional or national focus.
- A suboptimal supply chain with medium or low efficiency in part of the industry.

In addition to this there are clear threats with regard to the future from:

- Free-trade agreements with third countries including reduction of import levies and larger quotas. Figure 1.2 suggests that if such agreements happen, the European poultry industry could be marginalized. Trade agreements are often negotiated as 'packages' where agriculture is at a disadvantageous position compared to 'industrial' assets such as the car industry or banking sector. Within their borders, the EU and EU member states can make and implement laws, and if they impact a declining part of the voting public it hurts only a few; by contrast, in international negotiations 'matters of principle' will directly impact opportunities to make money in areas where many people are employed.
- Currency developments, as a stronger Euro would make other countries more competitive on the EU market. The UK, Denmark and Sweden have their own currency development.
- Import of fresh meat or eggs from neighbouring countries in Eastern Europe, e.g. Ukraine.
- Animal disease pressure, especially the risk of poultry diseases in high density poultry areas with a high impact on animal health, animal welfare and public health, and by the increased use of outdoor poultry production.
- The high and volatile price of feed ingredients due to increased global population and wealth, competition between feed, food, fuel and fibre, climate change and, for Europe, the increasing difficulty of purchasing non-genetically-modified (GM) feed. The UK and German poultry sectors have indicated that they can no longer guarantee fully GM-free feed, partly because of the stringent 'contamination' definitions.
- Increasing regulations/legislation at the national and EU level, including country-of-origin labelling and animal welfare.
- Attracting sufficient dedicated, skilled people.
- The growing concern in society about sustainability and welfare issues.

Opportunities can also be identified (Van Horne and Bondt, 2013):

- The increase of poultry meat consumption in the EU (good health perception, low price, convenient product).
- Growing demand for speciality poultry meat and eggs, especially in northwest Europe (e.g. Beter Leven, Label Rouge, Freedom Food, free range, Poule de Brest, Kip van Morgen, dual purpose).
- Further product development at the processing, retail and food service level.
- Optimization of supply chain management.
- Opportunities to increase the use of by-products of human food production.
- Further consolidation and internationalization of the EU poultry industry.
- Further improvement of the image of poultry meat and eggs as healthy, convenient products with a low carbon footprint compared to other animal protein sources.

One way to improve the resilience of the poultry industry is further advances in technology and organization. New and improved technologies, techniques and innovations will without doubt be developed, as European science is of a high level, and the poultry sector is innovative and used to surviving without subsidies. Europe has a good reputation for achieving balanced genetic change, developing tailor-made management information, and implementing improvements in processing, transport and product-addition, as its tight cost–benefit margins under heavy regulatory load have also been stimulants for improvements. There are ample opportunities to improve further and simultaneously the efficiency, environmental impact, welfare, health and safety of poultry and poultry products. In addition the processing, retail and food service sector can profit from new or improved product applications. However, some of these improvements may not be perceived as improvements by society and could lead to further increasing legal requirements, and a long and insecure path for implementation of innovations.

The success of the poultry sector, in building up food production in often less thriving regions and making poultry meat and eggs available for all people in society, demonstrates intrinsic strength, but the sector also has intrinsic weaknesses. Other agricultural sectors, in originally richer regions, have easier and stronger links to the influential circles of society. While agriculture in general has not been very successful in having a dialogue with society, this is particularly true for poultry.

On the one hand there is the buying power of the citizen in affluent European societies with increasing poultry produce consumption, food safety expectations and the positive aspects of poultry regarding environment, health and affordability. On the other hand, there is the voting power of the modern citizen, who is becoming more critical of farming in general and poultry production in particular.

Groups such as animal welfare organizations have played a critical role in putting animal welfare and animal production on the societal and political agenda. Where at first the poultry sector may have reacted defensively, they have also recognized that animal welfare is an intrinsic and important part of poultry production. Multiple improvements have materialized over the last decades, including poultry breeding becoming a broader programme with up to one-third emphasis on animal welfare, and management adapted to the changing needs of the birds. Bradnock (Farmers Weekly, 2013) comments: 'The industry has a very good relationship with the mainstream animal welfare groups. The more radical groups don't get the same media attention that they used to, while the British public are better informed'. This is not the case in all north-west European countries, for example in Germany where the situation is similar to that in the UK two decades ago, before dialogue started.

Knowledge about welfare has improved over time. At first, welfare often got defined as the Five Freedoms, the ideal states of animal welfare (FAWC, 2011), and welfare research concentrated on behavioural research and improvement of housing conditions. Later the Three Essentials of Stockmanship (FAWC, 2007) were also outlined. Recently, apart from the strengthening of the peer reviewed literature, professional knowledge is also gaining momentum. Science has developed further, and welfare now includes the balance and whole spectrum ranging from animal handling, physiology, tailor-made feeding, management, health care, behaviour and transport to optimal hatchery management. Science also developed further from fact-finding research (does the bird have good welfare?) to solution-based research (what management, treatments and feeding suit these birds?). Welfare is not a high-level option anymore, as farmers realize that welfare impacts directly and positively on the bird's ability to produce, thrive and maintain value.

For the bird's welfare it is best if there is a 100% match between the welfare it experiences and the perception of welfare by society. That should be the major drive for all parties to close possible gaps between welfare perception and the welfare a bird can experience.

The players in the poultry sector, including the poultry associations, have been increasing their efforts towards transparency and dialogue with society during recent years. Although biosecurity is important, at local and national levels initiatives are frequently set up to 'open the doors', such as the Open Days organized by the Science and Information Centre for Sustainable Poultry Production of Vechta University (Germany), production of 'learning kits' for schools, showing video clips on farming, and explaining poultry production via websites. Poultry industries also increasingly publish popular and scientific articles, becoming more transparent about the way they work. These activities supporting transparency are relatively new and need further development.

A further important development, in which representative poultry societies are playing an important role in organizing the sector and enabling dialogue between stakeholders, is the Codes of Good Practice and Quality Schemes that have been developed and that are underway. These show how partners in the poultry chain operate, and how they improve the quality, safety, health and welfare of poultry and poultry products.

Another way in which the poultry sector aims to improve its image is with the development and marketing of speciality products with additional welfare, sustainability, regional or typical-product value. These products are well perceived in the press and politics and in articles about added-value livestock production. It is positive that these products sell for a higher price, so that additional production and further chain costs can be covered. Unfortunately, meat chicken speciality products have not yet achieved high market shares, in contrast to free-range eggs or welfare-label pork produce. This may be due to the extra requirement that meat chickens in welfare label schemes should have a slower growth rate.

It is important that speciality products are not marketed in a way that is detrimental to the image of conventional European poultry meat and eggs, as the safety and environmental impact and the health and welfare of the birds producing the latter have improved over time.

There are a number of conflicts in the various aspects the public would appreciate. Slower growing chickens (perceived to have better welfare) have a higher environmental impact. Outside production systems have a higher disease risk, which is in contradiction with food safety and human health. People prefer birds not to be beak trimmed for welfare reasons, but if layers injure each other by pecking that seriously impacts their welfare, even fatally.

Finally, there are conflicting items in legal requirements for farms, and difficulties in implementing innovations. On the one hand, it is the responsibility of the poultry sector to show its improvements and commitment. The extent to which the poultry sector will achieve sufficient acceptance from society will depend on its ability to understand and work with the concerns, desires and powers in today's society, to organize itself, and to (re)gain trust. On the other hand, other parties in society also have a responsibility to ensure that their image of the poultry sector is in line with reality.

The message could be: poultry meat and eggs are of high quality in Europe (and in individual countries, emphasized when appropriate), safe, and produced by professional people working with birds with care, skills and knowledge. Poultry welfare is taken seriously, and there have been improvements in recent years, although there are important issues still to address. There is also a variety of speciality poultry products available, with specific qualities that make them worth an additional price.

REFERENCES

Appleby, M.C. (2012) Politics and economics. In: Sandilands, V. and Hocking, P.M. (eds) *Alternative Systems for Poultry.* Poultry Science Symposium Series, vol. 30. CAB International, Wallingford, UK, pp. 53–61.

Appleby, M.C., Mench, J.A. and Hughes, B.O. (2004) *Poultry Behaviour and Welfare.* CAB International, Wallingford, UK.

avec (2014) *Annual Report 2014.* Brussels, Belgium.

Bradnock, P. (2013) *Industry Efficiency Sees Poultry Prices Rise by an Average of Just 1.5% a Year – Half the Average of Other Foods. Retirement release.* British Poultry Council, London.

British Egg Information Service (2014) Available at: http://www.egginfo.co.uk (accessed 16 April 2014).

British Poultry Council (2014) *Economic Impact Assessment: The British Poultry Industry 2013.* Oxford Economics, Oxford, UK.

De Jong, I. and Van Harn, J. (2012) *Management Tools to Reduce Footpad Dermatitis in Broiler,* TechNote0912-AVN-34. Aviagen, Huntsville, Alabama.

Delgado, D., Rosegrant, M., Steinfeld, H., Ehui, S. and Courbois, C. (1999) *Livestock to 2020. The Next Food Revolution*. International Food Policy Research Institute, Washington, DC, Food and Agriculture Organization, Rome, International Livestock Research Institute, Nairobi.

Department for Environment, Food and Rural Affairs (2010) *Poultry in the United Kingdom*. The Genetic Resources of the National Flocks, London.

Department for Environment, Food and Rural Affairs (2013) Award details 'A Mathematical Modelling approach to defining factors which cause keel fractures in free range laying hens'. BB/K001906/1. Available at: http://www.bbsrc.ac.uk/pa/grants/AwardDetails. aspx?FundingReference=BB/K001906/1 (accessed 22 May 2014).

European Commission (2013) *Renewable Energy Progress Report*. COM(2013)175. Brussels, Belgium.

European Commission Health and Consumers (2014) Available at: http://ec.europa.eu/food/food/ sustainability/index_en.htm (accessed 16 April 2014).

European Union Committee (2014) *Counting the Cost of Food Waste: EU Food Waste Prevention*, 10th report. UK Parliament House of Lords, London.

FAOSTAT (undated) FAOSTAT: Inputs – Land. Available at: http://faostat3.fao.org/faostat-gateway/go/to/download/R/RL/E (accessed 14 May 2014); Population – Annual Population. Available at: http://faostat3.fao.org/faostat-gateway/go/to/download/O/OA/E (accessed 14 May 2014).

Farmers Weekly (2013) *Profile of Peter Bradnock*. British Poultry Council, Sutton, UK.

FAWC (Farm Animal Welfare Council) (2007) *FAWC Report on Stockmanship and Farm Animal Welfare*. London.

FAWC (Farm Animal Welfare Council) (2010) *Opinion on Osteoporosis and Bone Fractures in Laying Hens*. London.

FAWC (Farm Animal Welfare Council) (2011) Five Freedoms. Available at: http://www.fawc.org.uk/ freedoms.htm (accessed 30 March 2014).

FAWC (Farm Animal Welfare Committee) (2012) *Opinion on the Welfare Implications of Breeding and Breeding Technologies in Commercial Livestock Agriculture*. Department for Environment, Food and Rural Affairs, London.

Foresight (2011) *The Future of Food and Farming*. Final project report. The Government Office for Science, London.

Fraser, D., Weary, D.M., Pajor, E.A. and Milligan, B.N. (1997) A scientific conception of animal welfare that reflects ethical concerns. *Animal Welfare* 6, 187–205.

Ghirotti, M. (1999) Making better use of animal resources in a rapidly urbanizing world: a professional challenge. *World Animal Review* 92, 1–14.

Global Poultry Trends (2013) Static egg consumption in Europe. Available at: http://www. thepoultrysite.com/articles/2777/global-poultry-trends-static-egg-consumption-in-europe (accessed 20 May 2014).

Groenkennisnet (2013) Available at: http://www.groenkennisnet.nl/dierenwelzijnsweb/Pages/ NewsLoader.aspx?npid=4429 (accessed 23 May 2014).

Hiemstra, S.J. and Ten Napel, J. (2013) *Study of the Impact of Genetic Selection on the Welfare of Chickens Bred and Kept for Meat Production*, DG SANCO 2011/12254. IBF International Consulting, Brussels.

Intergovernmental Panel on Climate Change (2014) Summary for policy makers. In: Agard, J. and Schipper, L. (eds) *Climate Change 2014: Impacts, Innovation and Vulnerability*. Assessment Report 5. IPCC, Geneva.

Kapell, D.N.R.G., Hill, W.G., Neeteson, A.M., McAdam, J., Koerhuis, A.N.M., *et al.* (2012) Genetic parameters of foot-pad dermatitis and body weight in purebred broiler lines in 2 contrasting environments. *Poultry Science* 91, 565–574.

Laughlin, K.F. (2007) *The Evolution of Genetics, Breeding and Production*, Temperton Fellowship, vol. 15. Harper Adams University, Newport, UK.

Leinonen, I., Williams, A.G., Wiseman, J., Guy, J. and Kyriazakis, I. (2012) Predicting the environmental impacts of chicken systems in the United Kingdom through a life cycle assessment: broiler production systems. *British Poultry Science* 91, 8–25.

Livestock Environmental Assessment and Performance Partnership (2014a) *Environmental Performance of Animal Feeds Supply Chains. Version 1*. Guidelines for quantification. Draft for public review. Food and Agriculture Organization, Rome.

Livestock Environmental Assessment and Performance Partnership (2014b) *Greenhouse Gas Emissions and Fossil Fuel Depletion from Poultry Supply Chains*. Guidelines for quantification. Draft for public review. Food and Agriculture Organization, Rome.

Miao, Y.W., Peng, M.S., Wu, G.S., Ouyang, Y.N., Yang, Z.Y., *et al.* (2013) Chicken domestication: an updated perspective based on mitochondrial genomes. *Heredity* 110, 277–282.

Neeteson, A.M., Swalander, M. and Ralph, J. (2013) A European perspective of Turkey welfare. In: Bentley, J.S. (ed.) Proceedings of the 7th Turkey Science and Production Conference, Chester, UK, pp. 19–21.

Neeteson-van Nieuwenhoven, A.M., Knap, P. and Avendaño, S. (2013) The role of commercial pig and poultry breeding for food safety. *Animal Frontiers* 3(1), 52–57.

Plakké, T., Duijghuijsen, R. and Leenstra, F. (2008) *Toekomstvisie pluimveehouderij 2015–2020*. Stichting Fonds voor Pluimveebelangen, Zoetermeer, the Netherlands.

Rollin, B.E. (1993) Animal production and the new social ethic for animals. In: Baumgardt, B. and Gray, H.G. (eds) *Food Animal Well-Being*. Purdue University, West Lafayette, Indiana, pp. 3–13.

Silversides, F.G., Singh, R., Cheng, K.M. and Korver, D.R. (2012) Comparison of bones of 4 strains of laying hens kept in conventional cages and floor pens. *Poultry Science* 91, 1–17.

United Nations Environment Programme (2012) *Global Environmental Outlook GEO Environment for the Future We Want*. UN, Nairobi.

United Nations General Assembly (2005) *Resolution Adopted by the General Assembly 60/1 10. World Summit Outcome 1. Values and Principles*. United Nations, New York.

Vandenburg, S.W.K., Taal, C., de Boer, I.J.M., Bakker, T. and Viets, T.C. (2012) *Environmental Performance of Wild-Caught North Sea Whitefish*. EAFE Workshop Green and Blue Growth, Bilbao, Spain, 20 pp.

Van Horne, P.L.M. (2012) *Competitiveness of the EU Egg Industry*, Report 2012-065. LEI-Wageningen UR, The Hague, the Netherlands.

Van Horne, P.L.M. and Bondt, N. (2013) *Competitiveness of the EU Poultry Meat Sector*. Report 2013-068. LEI-Wageningen UR, The Hague, the Netherlands.

Welfare Quality Advisory Committee (2009) *A Report on Welfare Quality® by Its Advisory Committee*. Copenhagen.

Williams, A.G., Audsley, E. and Sandars, D.L. (2006) *Determining the Environmental Burdens and Resource Use in the Production of Agricultural and Horticultural Commodities, Main Report*. DEFRA Research Project IS0205. Cranfield University, Bedford, UK.

World Commission on Environment and Development (1987) *Our Common Future*. Oxford University Press, Oxford, UK, p. 27.

Chapter 2
Consumer Perceptions of Poultry Meat and Eggs: Bridging the Gap Between Public Perceptions and Reality

Patricia Parrott,[1]* Keith Walley[1] and Philip Clarke[2]

[1]Harper Adams University, Newport, Shropshire, UK; [2]Poultry World, Sutton, Surrey, UK

INTRODUCTION

Since 2000, the EU poultry industry has been impacted by many issues including changing legislation, competition from low-cost producing countries, increasing animal feed prices, tight margins, media attention and lingering recession. These factors have together served to alter both the market and the marketing landscape for poultry meat and eggs.

Producers and retailers have responded to these changes in many different ways. As companies have sought economies of scale there has been significant consolidation within the industry, many companies now offering a tiered product range to suit consumer demands, while others have sought to exploit market niches through the development of non-intensive production systems. What is clear, however, is that the sector is now more market focused than ever before and as a consequence producers and retailers have to be much better informed regarding the needs and wants of consumers.

As a consequence, Harper Adams University has for the last 18 years been measuring, monitoring and interpreting consumer behaviour relating to poultry meat and eggs. The research now constitutes a valuable longitudinal database that has produced some useful insights into UK consumers and the behaviours that they exhibit toward poultry meat and eggs and it is this body of knowledge, in the form of a case study, that will be used to provide the underpinning for the majority of this chapter. The authors acknowledge a gap between public perceptions of the sector and reality, consider the extent and nature of the disconnect and make a number of suggestions for bridging the gap. To begin, however, it is useful to consider the broader European market for poultry meat and eggs in a little more detail.

*Corresponding author: tparrott@harper-adams.ac.uk

THE EUROPEAN MARKET FOR POULTRY MEAT AND EGGS

In 2012 the poultry industry in the EU-27 produced 12.9 million t of poultry meat (Van Horne and Bondt, 2013). Some 76% of production was broiler meat and the biggest producers were France and the UK (AHDB, 2013; BPC, 2014). The EU egg supply in 2012 was 7.3 million t of which 6.1 million t were for human consumption (AHDB, 2013).

The EU poultry meat and egg producers have to comply with a raft of legislation. Much of the legislation originates in Brussels and embraces all poultry meat and egg producers over a given size based in EU member states. Some producers, however, are subject to additional legislation enacted by sovereign member states and applicable by virtue of companies being based in those states.

In recent years legislation has been developed and applied to various practices associated with poultry meat and egg production in the EU:

- testing for *Salmonella*;
- on-shell stamping;
- disposal of end of lay hens;
- prohibition of low-grade cooking oils in animal feeds and the use of therapeutic antibiotics;
- animal welfare through cage-space requirement and enrichment;
- environmental regulations;
- 'tallow' in egg-layer diets;
- the withdrawal of some egg yolk colourants;
- prohibition of moulting;
- the requirement for non-genetically modified (GM) feeds; and
- broiler regulations regarding stocking densities.

One of the key areas of poultry production that is covered by EU legislation is animal welfare. In the EU, Directive 2007/43/EC sets the minimum welfare requirements for broiler production in terms of stocking density. The directive sets a maximum stocking density of $33 \, kg/m^2$ under normal conditions or >33–39 to >39–42 kg/m^2 if other requirements are met. However, some countries and retailers choose to go beyond these standards by implementing more stringent legislation and additional requirements. For instance, the Red Tractor Farm Assurance Scheme (AFS, 2011) has standards covering every element of production from hatch to slaughter and operates a maximum stocking density of $38 \, kg/m^2$. Other assurance schemes may operate at lower stocking densities and the Freedom Food scheme which is administered by the RSPCA (2012) stipulates a maximum stocking density of $30 \, kg/m^2$. Organic standards often stipulate a longer growing period and an even lower stocking density. Obviously, poultry produced under the lower stocking density schemes is presented to the consumer as premium quality and often has an associated higher price.

In many instances the legislation that impacts EU poultry producers is the politicians' and bureaucrats' responses to increased public awareness and concern for animal welfare in commercial poultry production. Yet, how much the public are aware of the production systems is questionable and is evidenced by research work undertaken by Hall and Sandilands (2007), which showed consumers had very little prior knowledge about production methods under which broiler chickens were reared. It has been suggested that this increased interest in and concern for animal welfare has been born out of anthropomorphism (Kiesler, 2006; Karlsson, 2012), the tendency to ascribe human values and emotions to animals, a concept that some (e.g. DoRazario, 2006; Rose, 2013) have referred to as the 'Disneyfication' of animals. While the impact of this phenomenon is open to some debate, what is clear is that EU consumers have become very interested in animal welfare and the ethics of the production process are now an important factor impacting their consumer behaviour.

The effect of animal welfare in the consumer purchase of chicken is not consistent across the EU but impacted by culture, religion and particularly income. As consumer income rises and people become more affluent food can fulfil both a functional and emotional role as well as fulfilling a complex mix of social needs. Indeed, when making purchase decisions consumers will switch between 'value' and 'values' (Lister, 2012; James, 2013) according to the economic situation and promotional offers. Consumer income levels will influence consumers' decisions to purchase higher welfare products from the different schemes available (Van Horne and Achterbosch, 2008) particularly for fresh chicken meat and so welfare may well be traded off against price.

While the politicians and bureaucrats have responded to consumer concern regarding animal welfare in the poultry sector with legislation, many retailers have introduced their own assurance schemes to ensure welfare standards and assuage consumer concern. Indeed, the retailers' efforts to secure a competitive advantage over their rivals have led to the retailers actually driving innovation in terms of animal welfare. So the introduction of 'enrichment' into broiler growing programmes in the form of natural lighting, pecking objects and perches, and the introduction of 'enriched' eggs into the market place has been the result of retailers' efforts to meet consumer demand for higher welfare chicken (Morrisons, 2013). Other examples of the perceived 'super welfare' level can be seen in the egg sector, e.g. the Rondeel eggs system and the Happy Egg brand which have been developed specifically to meet and respond to consumer demands regarding welfare.

Legislation is widely seen as an additional cost of production not just for producers but also for slaughterhouses and other companies in the supply chain. Indeed, Van Horne and Bondt (2013) estimate that the legislation relating to producing poultry meat in the EU raises the cost of production by 4.8 eurocents per kg of live weight. Poultry meat and egg producers located outside the EU are often not subject to so much legislation and, as a consequence, do not incur the additional production costs.

When the costs associated with legislation are combined with the costs associated with quality assurance schemes and lowered import levies it is apparent that there is a trade distortion and that, were it not for the existing levels of import tariff on poultry meat and egg products, EU producers would be left at a significant cost and competitive disadvantage. It is clear, therefore, that in order to redress the imbalance it is necessary for producers to have a better understanding of the markets that they serve and this is one reason why work such as that reported here has become much more important in recent years.

The UK accounts for 12.8% of EU poultry production (BPC, 2014; European Commission, 2014), which makes it the second largest producer in the EU, as well as being the main consumer market for chicken meat in the EU. Consumption of eggs in the EU was 215 per capita in 2012, whilst in the UK it was only 182 eggs per capita. The UK poultry sector is an important component of the EU economy in terms of production and consumption and because of this it has been the focus for a longitudinal research project which, in the form of a case study, will now be used as the basis for the remainder of this chapter.

A CASE STUDY OF THE UK POULTRY INDUSTRY

UK production

Poultry meat production in the UK has increased significantly since 1990. Although production dipped during the 2000s (due partly to the decline in turkey output) the decline stabilized in 2009. Since then, and despite a backdrop of high feed prices, there has been steady growth in broiler, turkey and boiling fowl meat with annual production levels above 1.6 million t (DEFRA, 2012, 2014; European Commission, 2014). Currently, chicken accounts for 83% of the UK poultry meat output and turkey accounts for 11%.

The broiler industry in the UK is vertically integrated, with a short supply chain that is concentrated into a few large companies (from around 20 companies in 1993, to five companies in 2013) with a high level of control throughout the chain. The UK broiler industry is not supported by financial aid from the UK or EU governments compared to other agricultural sectors and is customer driven. The chicken value chain offers full traceability with several assurance schemes of differing levels, food safety, quality control, efficient planning and logistics, and high levels of research and development. In the UK there are around 2567 broiler farms (Crane *et al.*, 2011) and on average each farm holds around 69,500 birds per crop (Crane *et al.*, 2010) with 81% of units operating barn rearing systems, 6% free-range, whilst the remaining units comprise unconventional housing and 1.7% organic units.

Egg production from laying hens in the UK has changed over the past decade from producing over three-quarters (78%) of shell eggs in intensive caged systems to approximately half of all eggs (50%) in 2013, with the remaining eggs from free-range (46% including an estimated 2% organic) and barn (3.5%) systems (DEFRA, 2013). In general, demand for free-range eggs has been increasing but more recently there was an unexpected fall due to the recession and it was only in 2014 that growth in the segment resumed. Prices have narrowed between intensive and free-range and the market has attempted to differentiate free-range offerings further recognizing that it is moving into a mature market. This, coupled with rising feed prices, has also placed financial pressure on organic and free-range egg producers.

UK consumption

UK consumption of poultry meat in 2012 was approximately 2 million t and was based on strong demand for home-produced chicken and turkey. Although some 0.3 million t of production went for export markets, around 0.7 million t was supplied by imports, mainly through the Netherlands, which makes the UK a net importer of poultry meat.

Annual consumption levels of poultry meat have mirrored the trend in the decline in production levels falling from 29.3 kg per capita in 2001, to a low of 25.9 kg in 2008 (when there was intense media attention to methods of poultry production), and a subsequent increase to 28.5 kg per capita in 2011. According to EBLEX (2012), the demand for poultry meat has been increasing, with poultry being the most popular protein consumed due to it being low in fat, healthy, and cheaper than red meats in the current economic climate. Broiler meat is the major component of poultry consumption per capita making up more than 80% of the volume (DEFRA, 2012) and 49% of all meat consumed in the UK (BPC, 2014). In addition, in the UK, poultry offers a cost-effective source of protein and poultry prices have risen more slowly than food prices in general over the last decade. Low-income households and large households are tending to increase their expenditure on turkey meat. The UK chicken restaurant sector is also expected to display strong growth in the future (Keynote, 2012).

The UK egg market breakdown shows 51% of eggs being sold through the retail sector, 25% through food service and 21% through the processing sector. The majority of retail sales of eggs (85%) are sold through the major supermarkets, with some retailers offering a wide range of egg products from the different egg production systems, to others only offering eggs from non-caged systems. Egg consumption in the UK has increased by 12.3% over the last decade to 185 eggs per capita at present (BEIC, 2014; FarmingUK, 2014).

After addressing a series of challenges including cholesterol and *Salmonella* scares the UK poultry industry has positioned itself as an efficient and competitive industry without any reliance on subsidies. Lower margins and the changes in regulation regarding housing, management and increased costs associated with welfare were countered with consolidation in the sector in the search for economies of scale in order to maintain or increase market share.

UK CONSUMER PERCEPTIONS

Background to the study

A longitudinal programme of research relating to consumer views of the poultry sector for both poultry meat and eggs in the UK has been underway since 1996 with the latest survey round being completed in 2012. The results of the work have been documented in various papers including Jones and Parrott (1997), Parrott (2001a, b), Hingley and Parrott (2008), Parrott *et al.* (2013a, b) and

Walley *et al.* (2014). Over this period of time, the research has served to monitor UK consumer reaction to various challenges to the UK poultry sector including issues of food safety, changing legislation, higher welfare regulations, sustainability, economic pressure during recession, and changes in the supply chain.

Methodology

The design of the fieldwork has varied over the years that the project has been running but essentially the methodology used is quantitative in nature and survey-based. The survey makes use of a questionnaire that was originally developed in consultation with key poultry industry personnel and is designed to gather data concerning consumer attitudes and perceptions towards the UK poultry industry and the factors that influence consumer choice and purchase of poultry meat and eggs. Specifically, the questionnaire gathers data relating to a wide range of consumer views relating to quality dimensions and measures, country of origin, production systems, welfare, marketing mix variables, branding, assurance schemes and food safety, and their influence on purchase behaviour.

Great care has been exercised to ensure that the survey gathers consistent or comparative data across the 18-year history of the project, however, on occasions the questionnaire has been updated to maintain currency and take account of contemporary issues. For instance, in recent times when it became apparent that total UK food and drink waste is around 15 million t per year, with UK households generating 4.4 million t of avoidable food waste (i.e. food fit to eat) (DEFRA, 2014) and that consumers are still misinterpreting food date labelling, then questions relating to food wastage were included in later versions of the survey.

The research instrument utilizes a variety of question formats although in all cases a 'don't know' option is provided so that consumers may provide realistic responses. Questions relating to the importance of factors in the consumer decision process use a five-point scale where $1 =$ not very important and $5 =$ very important. Analysis of these data is achieved via the calculation of a mean score for each factor. Questions that gather data relating to consumers' perceptions and attitudes make use of a standard five-point Likert scale where $1 =$ strongly agree and $5 =$ strongly disagree. Analysis of this type of data is achieved by the use of a 'balance of opinions' approach, which seeks to measure the strength of opinion towards the various statements. This method has the advantage of trading off 'agree' and 'disagree' data, and also removes the impact of 'don't know' responses in order to determine a net 'strength of opinion' value. A more detailed explanation of the approach can be found in Walley *et al.* (2009).

In the most recent survey round the questionnaire was delivered by hand to a variety of housing types across the UK. The sample size for each survey wave varies (1997, n = 462; 2000, n = 362; 2008, n = 327; 2012, n = 348), but even the smallest sample size gives statistical significance of 90% confidence with +/−5% accuracy (West, 1999).

UK consumer perceptions of poultry meat and eggs

Data relating to the factors impacting the purchase of poultry meat and eggs in 2012 are presented in Tables 2.1 and 2.2, respectively. The factors have been ordered according to their importance showing the calculated mean score with the most important at the top of the list and least at the bottom.

It is apparent from Table 2.1 that factors such as use-by date, taste, odour and texture are the factors that are most important in the UK consumer's purchase decision for poultry meat while packaging and brand name are the least important. It is also apparent from Table 2.2 that production system and best before date are the most important factors in the UK consumer's purchase decision for eggs, while packaging and brand name are again the least important.

Although data were gathered on a range of factors it should be noted that a number of factors are actually related. An example of this might be animal welfare being related to production system and rearing method or taste, odour and texture being a function of product quality. It is also apparent that a number of the factors are important in the purchase decision for both poultry meat and eggs. As a consequence, the discussion that follows will now turn to consider some of these issues in greater detail. It should also be noted that while the discussion is based primarily on the most recent survey wave and the data presented above, some interpretation is derived from our overall data set, which it is not possible to present in a short chapter such as this. Readers are, therefore, advised to see the other publications arising from the programme of research

Table 2.1. Factors impacting the purchase of poultry meat in the UK.

Factor	Mean importance[a]
Use-by date	4.30
Taste	4.18
Odour	3.99
Texture	3.91
Country of origin	3.84
Tenderness	3.83
Leanness	3.80
Welfare assurance	3.76
Rearing method	3.72
Colour	3.72
Price	3.68
British Farm Standard	3.65
Fat content	3.57
RSPCA monitored assurances	3.54
Diet the poultry were fed	3.32
Weight	3.04
Retailing store	2.86
Producer name	2.67
Packaging	2.56
Brand name	2.45

[a]Mean score shown where 1=not very important and 5=very important

Table 2.2. Factors impacting the purchase of eggs in the UK.

Factor	Mean importance[a]
Production system	4.19
Best before date	4.12
Animal welfare	3.98
Taste	3.95
Country of origin	3.77
Price	3.37
Date laid	3.34
Place of origin	3.33
Size of egg	3.29
Lion mark	3.19
Diet the poultry were fed	3.11
RSPCA logo	3.05
Class of egg	3.01
Colour of egg	2.79
Retailing store	2.35
Producer name	2.28
Packaging	2.22
Brand name of egg	2.18

[a]Mean score shown where 1 = not very important and 5 = very important

(Jones and Parrott, 1997; Parrott, 2001a, b; Hingley and Parrott, 2008; Parrott *et al.*, 2013a, b; Walley *et al.*, 2014).

ANIMAL WELFARE

In the UK the importance of animal welfare in the consumer's purchase decision is widely acknowledged. For example, Mintel (2013) found that 45% of adults reported that animal welfare certification raised their trust in a food product while a survey undertaken by Keynote (2012) showed that respondents aged 55–64 and social grade A were more concerned about welfare standards, production conditions and the provenance of their meat than those who were younger and from lower social income groups. Further, the National Chicken Council (2012) acknowledges consumer demand to ensure that animals being raised for food are respected, treated fairly and that producers have an ethical obligation to ensure that their animals are properly cared for.

The data gathered during the latest survey round of our longitudinal research programme does not contradict the view that animal welfare is important to UK consumers. Indeed, as animal welfare was overtly assessed in respect of eggs and attained a score of 3.98 it was confirmed by the respondents as the third most important factor impacting the purchase decision for eggs. What the survey data do permit, however, is consideration of various aspects of the animal welfare issue.

The first key aspect of animal welfare in poultry production is the system used to produce the meat and eggs. In the case of eggs the respondents confirmed that production system was the most important factor impacting the purchase decision with a score of 4.19 while in the case of poultry-meat rearing method was scored at 3.72, which was ninth in a list of 20. Whilst the mean score for this factor is seen to be fairly important, it highlights the difference between hens and broilers. This could be related to the negative publicity that hen housing has received in the past.

The data gathered using the Likert scales provide some interesting insight into the consumer's views regarding animal welfare. In the first instance, while there is evidence to suggest that the level of knowledge and understanding of methods of production is poor, with particular confusion about the barn system, it is apparent that many UK consumers strongly relate the welfare of poultry to the use or non-use of cages. Indeed, consumers often view the use of cages as 'intensive farming' and consider the use of battery cages as 'sheer cruelty'. Some UK consumers also have concerns about hormone and antibiotic use, but concern regarding *Salmonella* and cholesterol in the national flock appears to be receding.

Consumer views about hens being kept in cages has shown an interesting trend in strength of opinions over the duration of the research programme and this is despite 'don't know' answers being disregarded. In 1997, only 16% of the respondents strongly disapproved of keeping hens in cages. However, 41% chose to neither agree nor disagree, which, in other words, meant that they were not prepared to commit to an opinion either way, which thereby culminated in a balance of opinion of 3.4% of respondents disapproving of hens being kept in cages. In the later surveys conducted in the years 2008 and 2012, more than two-thirds of the respondents strongly disapproved of keeping hens in cages with a very small percentage (7%) of respondents neither agreeing nor disagreeing, culminating in a balance of opinion in 2012 of 71.4% disapproving of keeping hens in cages. It would seem logical to assume that it is the emerging awareness of ethical issues associated with animal welfare in general that has led to disapproval of keeping hens in cages and the growth in the market for free-range eggs (IGD, 2008; Cooperative Bank, 2012; DEFRA, 2013).

A second key aspect of animal welfare in poultry production is welfare assurance; however, as assurance also relates to food safety and quality then this will be covered in detail later in this chapter, in a separate section. In the current context it is sufficient to note that several assurance schemes have been developed over the period of this longitudinal research programme that focus on welfare and that to some extent they are promoted via labelling on the packaging of poultry meat and eggs. Unfortunately, only one-third of respondents to the research programme consistently look for welfare guarantees on packaging and so this would appear not to be a particularly effective means of promoting the schemes. Respondent data also showed a declining trend to being 'prepared to pay more' for meat with welfare assurance since the survey started. This could be due to consumers expecting that poultry meat assurances should be built in to the supply chain and having an expectation and trust in the retailer to provide this (since the majority of fresh poultry meat is purchased through this route) and therefore being less inclined to be prepared to pay more.

A final point worth considering regarding animal welfare in the poultry sector is that there is evidence to suggest that consumer perceptions of animal welfare in the poultry sector and producer perceptions are not always the same. For example, Appleby *et al.* (2004) reported that farmers rearing animals under intensive systems perceive their animals to be well treated and healthy whereas Velde *et al.* (2002) suggest that consumers perceive animals raised in this way to be the product of an intensive production system that gives the birds a lack of space, air and light. As a consequence of these conflicting perceptions it is useful to think of welfare as being either real welfare or perceived welfare. Real welfare is scientifically based and harder to establish but is the basis of the producer's perceptions. Perceived welfare is not necessarily scientifically based, is often subjective and ill-founded, but is the basis for consumer purchase decisions.

FOOD SAFETY

The survey confirms food safety in the poultry sector as being a moderately important factor in the purchase decision via a number of proxy variables. Thus with reference to poultry meat use-by date, fat content and diet the poultry were fed score 4.30, 3.57 and 3.32, respectively. Similarly, with reference to eggs' best-before date, date laid and diet the poultry were fed score 4.12, 3.34 and 3.11, respectively.

The survey also provides insight into the changing importance of food safety in the consumer purchase decision over time. When the longitudinal research programme was initiated, food safety was seen by consumers as more important than it is now. The agri-food sector had been beset by issues relating to *Salmonella*, *Campylobacter*, BSE, *E. coli* 0157, antimicrobial resistance and growing public unease with the use of antibiotics in animals. As a consequence, statements such as 'I think about health scares' and 'I am concerned about antibiotic residues' were developed and added to the survey instrument.

In recent years, however, much has been done in the poultry sector to alleviate some of these concerns. For instance, the Lion Quality Assurance scheme was successfully reintroduced in the UK in 1998 (FarmingUK, 2014; World Poultry, 2014) in an attempt to combat consumer concerns and regain trust. As a consequence, there has been a decline in the strength of opinion towards the proxy variables relating to the importance of food safety in the consumer purchase decision, which is positive for the sector.

It is important to note, however, that the data from the survey also suggest that there is a continuing lack of knowledge regarding farming practices amongst consumers, which leads to various misconceptions about the poultry industry. It would appear, therefore, that there is potential for the industry to further improve consumers' knowledge and understanding regarding food safety in the poultry sector.

PRODUCT QUALITY

In many purchases product quality is a fundamental component of the purchase decision yet, just like animal welfare and food safety, it too is multivariate in nature. In the case of poultry meat the research programme has measured product quality using proxy variables including taste (4.18), odour (3.99), texture (3.91), tenderness (3.83), leanness (3.80), colour (3.72) and fat content (3.57), while with eggs it is measured by taste (3.95), size of egg (3.29), class of egg (3.01) and colour of egg (2.79). While it is often useful for managers and others to speak of product quality in a general sense, the variation in the scores suggests that consumers can and do distinguish between these variables and consider them independently within the context of the purchase process and so it is useful to consider some of them on an individual basis in this discussion.

The questionnaire sought to find out whether consumers believed that the taste of poultry meat was influenced by the production system. Using the balance of opinion data, a majority of respondents across the longitudinal research programme consistently believed that there were differences in the taste of poultry meat according to the production system, with poultry reared in a free-range production system thought to taste better than poultry reared in an intensive production system. It is interesting to note, however, that while the respondents do also believe that poultry reared using organic systems taste better than birds reared using intensive production systems, the strength of agreement is much less than with free-range production systems. The diet fed to the hens seemed to have little effect on the purchasing behaviour.

Similarly, the questionnaire sought to find out whether consumers believed that the taste of eggs was influenced by the different methods of production systems. On balance of opinion, almost two-thirds of the respondents believed that there were differences in the taste of eggs and that those eggs from free-range systems were perceived to taste better than eggs from caged hens.

Opinions regarding the nutritional content of eggs produced by the different production systems have changed over time. In the more recent surveys respondents were less inclined to agree with the statement 'Free range eggs are more nutritious than eggs from caged hens' than in the earlier surveys.

While colour of egg is one of the least important factors influencing the purchase of eggs the survey does provide further insight into how this impacts consumers. Basically, the survey incorporated a number of Likert statements regarding egg and egg yolk colour and from this it is possible to conclude that the preferred yolk colour would be a dark yellow, which was perceived to be an indication of freshness, of having the most flavour, and of being likely to have come from free-range hens. Eggs with pale yellow yolks were believed to come mainly from caged hens and dark yellow yolks from non-caged hens.

It would appear that many of the factors influencing buying behaviour through product quality may be perceived to be related to a particular method of production system (i.e. free range or organic etc.) or, due to the impact of packaging on the consumer, the prices being charged, or indeed, there may be very limited decision making and 'any egg' type will satisfy their needs. Evidence from earlier research (Parrott, 2001a, b) suggests that with increasing affluence, consumer buying behaviour is often based on emotional factors, involving intangible benefits over and above tangible factors such as nutritional content and may not always take place on an entirely rational and conscious level. It is suggested that the consumer is fickle (Lister, 2012) and those who profess an interest and awareness of some of the more 'ethical' influences may not follow through on their ideals when actually making their purchase decisions.

ASSURANCE SCHEMES

Assurance schemes have been in existence for many years and the survey data confirm that they do have a role to play in influencing consumer decision making for poultry meat and eggs. In respect of poultry meat welfare assurance (3.76), the British Farm Standard (3.65) and RSPCA monitored assurances (3.54) all have a moderately important impact on the consumer's purchase process. Similarly, the Lion mark (3.19) and the RSPCA logo (3.05) have a moderately important impact on the consumer's purchase process for eggs. The nature of the schemes varies with some focusing on animal welfare while others are somewhat broader in scope.

Several welfare frameworks have been developed, which essentially encompass the five freedoms (FAWC, 2012). These have been endorsed by the RSPCA (2012) to develop the Freedom Food assurance scheme, which is utilized by many retailers to demonstrate a commitment to compassion in animal welfare. This has also been incorporated into the Red Tractor Farm Assurance Poultry Scheme (AFS, 2011), which has been well publicized in the UK.

The Lion assurance scheme was reintroduced in the UK in 1998 with a strict code of practice, which incorporates both farm assurance and quality assurance in order to build brand value and regain trust and loyalty in the egg sector (FarmingUK, 2014; World Poultry, 2014). The scheme led to widespread improvements in egg-laying flocks and following early promotional campaigns it was apparent within a short space of time that consumers felt comfortable enough with the safety data to launch a subsequent promotional campaign featuring a child eating a boiled egg with a runny yolk asking 'Why has my egg got a Lion on it, mum?'. This, coupled with research on cholesterol and eggs, led to a further promotional campaign, 'Fast Food and Good for You'. The development and continuation of the scheme has subsequently enhanced the ability of the sector to promote and develop different methods of production and a wide range of innovative products and campaigns in an ever increasingly competitive environment.

The Lion scheme along with its respective promotional campaigns and greater innovation among suppliers seems to have arrested the decline in egg consumption in the UK and seeks to benefit all parties in the supply chain. For consumers it provides assurance of less risk of *Salmonella*, fresher eggs and eggs from higher production standards. For retailers, the Lion scheme provides them with an insurance allowing them to offer their customers due diligence in the food chain through improved food safety and traceability.

PRICE

Price plays some part in influencing the consumer decision for virtually all products and services and poultry is no exception. Indeed, according to DEFRA

(2014) price is the most important factor influencing consumer product choice and while price is not the most important factor influencing the purchase of poultry meat and eggs (those being use-by date and production system, respectively), the survey data do confirm price as being a moderately important factor scoring 3.68 and 3.37 for poultry meat and eggs, respectively.

Virtually every country in the EU has undergone economic recession in recent years. Averaged across the EU, food prices have risen 17% in real terms since 2007 (DEFRA, 2014). Food prices in the UK for the same period have risen 12% (DEFRA, 2014), which places them on a par with prices last seen in the late 1990s in terms of cost of food relative to other goods and puts pressure on household budgets (DEFRA, 2014).

Against this backdrop, and contrary to data gathered in earlier surveys, the latest survey shows that respondents were less concerned with price than many other factors and were even prepared to trade up and pay more for products that offered them higher value. In this context different consumers perceived value to be offered by things such as different production systems (e.g. free-range or organic versus intensive cage-rearing systems), assurance schemes, local or regional origin, packaging and promotions.

An interesting final observation regarding price relates to consumer attitudes and their behaviour. In recent years data from the research programme have shown that over 50% of respondents agree with the statement that they would pay more for eggs from a welfare-assured scheme whilst retail sales data have not shown a concomitant rise in sales of eggs produced in welfare-assured schemes. This phenomenon is not new but producers should be aware that what consumers say is not always matched by what they do.

COUNTRY OF ORIGIN

Originally, country of origin was assumed to be 'the national origin' of products (Schooler, 1965), but since its initial inception the concept has fragmented and become more complex (Chu et al., 2010). Today, authors make use of a variety of terms including 'country of brand' (Uddin et al., 2013), 'country of headquarters' (Showers and Showers, 1993) and 'country of assembly' (Biswas et al., 2011).

At first glance it might appear difficult to relate country of origin to the purchase of poultry meat and eggs but the two are linked fairly strongly in the minds of consumers. In the latest survey round 'country of origin' as a factor impacting the purchase of poultry meat scored 3.84 while 'country of origin' and 'place of origin' as factors impacting the purchase of eggs scored 3.77 and 3.33, respectively. In one sense the link between country of origin and poultry meat and eggs is not surprising as an ethnocentric tendency to favour products and services originating from a person's home country are well documented (Chattalas and Takada, 2013; Xu et al., 2013). Indeed, for the duration of the longitudinal research programme two-thirds of respondents have held a strong desire (92.7%) to buy British poultry meat in order to support the home market.

Our research does indicate, however, that ethnocentricity is not the only reason that UK consumers prefer to buy poultry meat and eggs originating from the UK. For many consumers the 'Made in Britain' label is also a proxy for good standards of animal welfare, food safety and product quality. Indeed, the well documented (e.g. Veterinary Record, 2013) contamination of meat products with horsemeat, widely referred to as 'the horsemeat scandal' (Lawrence, 2013), which occurred in January 2013, may well have caused many UK consumers to become even more interested in local and regional provenance.

Producers should be aware that while many respondents have indicated a strong balance of opinion towards the statement 'British farmers have a high standard of animal welfare' throughout the research programme, in the most recent survey waves there has been a decline in the strength of agreement. At the same time there has been an increase in respondents choosing to answer 'neither agree nor disagree', which would appear to suggest that consumers are beginning to recognize that they do not know as much about poultry welfare as they had thought in the past.

Producers might also wish to bear in mind that while country of origin is an important factor in the purchase decision for poultry meat and eggs, most of the interest in sourcing information is from consumers in the 65+ years age group. Among younger age groups, there is virtual apathy regarding country of origin as far fewer actually check country of origin information than report wanting origin details. Further, only 14% of adults identify traceability as a choice factor, which Mintel (2013) suggests is because consumers expect brands and retailers to take responsibility for product quality rather than relying on food origin.

A DISCONNECT WITH THE CONSUMER

While the poultry industry in the UK has responded very well to the challenges of the last 25 years it is important it does not rest on its laurels. Indeed, the longitudinal research programme discussed here provides an insight into an issue that has begun to develop and which may constitute an important challenge for the industry over the next 25 years.

Basically, while it is readily apparent from the existing data set that the contemporary consumer is increasingly concerned about issues associated with animal welfare, food safety and product quality, there is also a developing suspicion that they are less connected with the land than ever before (Wilson, 2014). Indeed, it is in an attempt to educate consumers about the reality of farming that LEAF (Linking Environment And Farming) has introduced 'Open Farm Sunday' where members of the public are invited on to farms to see how they operate and how food is produced.

In the poultry sector, as with many other agricultural sectors, various assurance schemes have been introduced in order to reassure consumers. To date this has proved a very successful strategy, however, it is likely that this situation will be subject to change over time. It is highly likely that consumer expectations will continue to rise but also that they will begin to acknowledge their limited understanding of the poultry sector. Assurance schemes may well begin to lose their effectiveness and, as a consequence, the poultry sector must look at new ways to bridge the gap between public perceptions and farming reality.

BRIDGING THE GAP BETWEEN PUBLIC PERCEPTIONS AND REALITY

Extent of the disconnect

There is a significant disconnect between what consumers say they want, what they believe is involved in modern food production and what goes on in practice. This is evidenced by the numerous surveys that are put into the public domain by organizations such as FACE (Farming and Countryside Education) and the RASE (Royal Agricultural Society of England) that show schoolchildren believe potatoes grow on trees or that ice cream comes from a van, rather than a cow. These stories are lapped up by the media – both the trade press and the mainstream press.

One likes to think that the presenters of these reports have cherry-picked the most extreme examples of ignorance, just to capture a headline while trying to make a more serious point about the findings of their studies. However, evidence from those few poultry farmers who are brave enough to open their gates to the general public suggests such misconceptions are actually quite widespread. On the meat side in particular, there is genuine surprise when consumers find out that broilers are not kept in cages, whereas most consumers' mental picture of a 'nest box' would involve lots of straw and a mother hen brooding her young.

Equally, there is clearly a huge lack of knowledge about the legislation that governs these things. This is hardly surprising, and certainly not unique to agriculture. Every sector of the UK economy – be it car making, medicine or banking – is governed by a raft of regulation of which only those 'at the coalface' can be expected to understand or have detailed knowledge.

The fact that consumers may not be *au fait* with things like the Integrated Pollution Prevention and Control directive or the Welfare of Laying Hens directive in itself is not the problem. It is the more broad-brush messages about how these directives impact at farm level, how they benefit the environment, or improve the life of caged hens, that need to be explained.

What is the nature of the disconnect problem?

In a nation of animal lovers, it is little surprise to find animal welfare ranked so highly in consumers' buying decisions. We know that they have genuine concerns about things like food safety and product quality and these considerations are often reflected in national legislation governing production methods.

It is a concern, however, that the government still has a tendency to gold-plate EU legislation, for example in requiring stricter stocking rate limits for broilers than most other member states, and unilaterally seeking a ban on beak trimming.

It is also a concern that much of this regulatory behaviour is a reaction to what can only be described as a vocal minority. Organizations that purport to represent consumer views may only represent the views of the handful of ardent activists that care enough to become members of those groups. But if nobody else is talking, their views will resonate with legislators.

The same goes for the retailers. In an era of intense competition on the high street, the main supermarkets are all eager to differentiate themselves. It is not just about price, but also appealing to consumer values. But are they necessarily getting the right messages?

There are plenty of examples in the poultry sector where supermarkets have insisted on elevated standards of production, even though the perceived welfare benefits are at best negligible and, at worst, counterproductive. One example is requiring windows to be put in broiler houses, at considerable expense to the farmer or integrator. The industry view is that these improve the environment for farm workers but have a negative impact on the carbon footprint of the industry while making bird management more difficult for stockmen.

Another example is perches for broilers – a nice idea to the uninitiated, but recently described by one respected veterinarian as 'the work of the devil'. The modern broiler has little perching instinct, finds it physically difficult to get on to a perch, and is inclined to fall off resulting in broken keel bones and other damage.

Often these changes have been driven by consumer and welfare pressure groups who may have strong ideals but little idea about what conditions are actually like on farms. It is therefore important to bridge that gap so that, when these groups are demanding changes in the regulations or specifications, they are doing so from a position of knowledge.

How to address the disconnect?

Clearly there is no simple solution to this issue – there are too many complexities and vested interests at play. But there is little doubt that the whole industry does need to take a more open approach.

The fact that only two broiler growers in the country have ever embraced Open Farm Sunday, inviting the public on to their farms to see how chicken meat is produced and to answer their questions, speaks volumes for the closed nature of the poultry sector. The intensive egg sector has an even worse record in this respect. Further, the fact that many poultry farms do not even have a sign or farm name at the end of the drive suggests, at best, that the industry is secretive and, at worst, that it has something to hide.

There are, of course, genuine concerns about biosecurity and about 'letting the wrong sort of people' on to a unit. But the first can be overcome with some fairly basic procedures, while the second has to be offset against the downside of not explaining to the public what the sector does and why. Those producers that have taken the plunge invariably describe the experience as wholly positive, providing better learning and understanding on both sides.

It is not just about bringing people on to farms. It is also about going into the community, and involving lobbyists and decision makers. One recent example was an egg producer who arranged a visit to a hatchery to see infra-red beak trimming in action, and took her local Compassion in World Farming representative with her, to see what a low impact process it is.

Producers should not forget that reaching out may also involve making use of social media such as Facebook and Twitter. While not everyone engages with social media a disproportionate number of the younger generation do use them and if we are to communicate successfully with future generations of consumers then so must the industry.

It is also important to involve the media. There is a real desire to learn about food, so it is always going to attract media attention. It might also help ensure there is more balanced coverage when the next food scare comes along. The media certainly still loves a good food scare – even though there is evidence that the public is now a little more sceptical, and the market is certainly more robust. It is potentially positive however, if a journalist already has a relationship with a producer or processor who they are able to approach for a different point of view. It needs to be considered whether this relationship is worth initiating by the industry rather than the journalist.

The industry's representatives also have a major part to play in this. As well as encouraging their members to make themselves more accessible to the public, they have a role to play in explaining to retailers, food service companies and legislators what the industry is already doing in terms of welfare, food safety and environmental protection. To some extent, they do this already. But there is always more that can be done, especially in terms of getting some of these people out of their offices and on to production and processing sites, to see the industry in action.

Having a more open approach is an essential part of reconnecting consumers with where their food comes from. Farm assurance schemes, such as the Red Tractor for meat and the Lion scheme for eggs, have certainly provided a means of reassuring consumers. And, to a certain extent, consumers are happy to let the retailers take some of the burden off them to ensure responsible sourcing. But the industry must play its part too, to ensure future changes to legislation or retailer specification is more in tune with reality.

CONCLUSION

The EU poultry industry has been subject to many issues since 2010 and this has led to fundamental changes in the market and the marketing landscape for poultry meat and eggs. A market focus is essential and this requires companies to be well informed regarding the needs and wants of consumers.

Using the UK as a case study, this chapter, therefore, makes use of a database that has been compiled from data collected over an 18-year period, to demonstrate how UK consumer attitudes and opinions about poultry meat, eggs and the poultry sector have evolved over time and to identify issues that are currently important to UK consumers.

The main findings include:

- Animal welfare is an important consideration in the consumer decision-making process and in the context of poultry meat and eggs it is closely related to method of production.
- Consumer knowledge and understanding of methods of production is poor, with particular confusion about the barn system.

- In the consumer's mind the use of cages is often associated with 'intensive farming', which has negative connotations.
- Consumers are beginning to expect welfare assurance to be part of the product that they are buying and are not willing to pay a premium for it.
- Consumer buying behaviour is often based on emotional factors over and above objective factors.
- Contrary to earlier findings, the latest survey showed that respondents were less concerned with price than with many other factors and were even prepared to trade up and pay more for products that offered them better value.
- There is, however, evidence to suggest that what consumers say is not always matched by their actions and so they may not actually pay more for greater value; consumers are fickle and those who profess an interest and awareness of some of the more 'ethical' influences may not follow through on their ideals when actually making their purchasing decisions.
- UK consumers prefer poultry meat and eggs produced in the UK, but this is not just down to ethnocentricity as the UK origin is also a proxy for good standards of animal welfare, food safety and product quality.
- Consumer knowledge regarding farming practices in the poultry sector is limited and leads to various misconceptions about the poultry industry.
- Consumer perceptions of animal welfare in the poultry sector and producer perceptions are not always the same.

The final two points appear to suggest a disconnect between the consumer and the poultry industry that may be an issue of increasing concern in the future. The chapter, therefore, considers the nature of this disconnect and makes a number of suggestions regarding bridging the disconnect:

- Poultry meat and egg producers to adopt a more 'open' approach with the public and share more information with the consumer.
- More poultry meat and egg producers to engage with events like Open Farm Sunday.
- Poultry meat and egg producers need to be more proactive in going into the community, and involving lobbyists and decision makers.
- Poultry meat and egg producers should involve the media more.
- Industry representatives should take the lead in making the poultry industry more accessible to the public.

In conclusion, the UK poultry sector has been very successful in addressing a number of issues that have impacted the industry over the last 20 years. However, there is an apparent disconnect between the poultry industry and the consumer that may well constitute the next major issue facing the sector.

REFERENCES

AFS (Assured Food Standards) (2011) Poultry Standards – Broilers and Poussin. Available at: http://www.assuredfood.co.uk/resources/000/618/003/Poultry Standards – Broilers and Poussin.pdf (accessed 12 December 2012).

AHDB (Agriculture and Horticulture Development Board) (2013) Poultry Pocketbook 2013. Available at: http://www.smartstore.bpex.org.uk.10.5.2013.13.17.48.pdf%26i%3D302937&ei =ki6DUrSOYGX1AWBv4GgAg&usg=AFQjCNFhciqzEujypiicUM8rXeGlcqbUfw&bvm=bv. 67720277,d.ZGU (accessed 13 December 2013).

Appleby, M.C., Mench, J.A. and Hughes, B.O. (2004) *Poultry Behaviour and Welfare.* CAB International, Wallingford, UK.

BEIC (British Egg Industry Council) (2014) Egg Information. Available at: http://www.egginfo.co. uk/industry-data (accessed 1 August 2014).

Biswas, K., Chowdhury, M.K.H. and Kabir, H. (2011) Effects of price and country of origin on consumer product quality perceptions: an empirical study in Bangladesh. *International Journal of Management* 28(3), 659–674.

BPC (British Poultry Council) (2014) Economic Impact Assessment: The British Poultry Industry 2013. Oxford Economics. Available at: http://www.britishpoultry.org.uk/wp-content/uploads/ 2014/03/Economic-Impact-Assessment-2013.pdf (accessed 12 May 2014).

Chattalas, M. and Takada, H. (2013) Warm versus competent countries: national stereotyping effects on expectations of hedonic versus utilitarian product properties. *Place Branding & Public Diplomacy* 9(2), 88–97.

Chu, P.-Y., Chang, C.-C., Chen, C.-Y. and Wang, T.-Y. (2010) Countering negative country-of-origin effects: the role of evaluation mode. *European Journal of Marketing* 44(7/8), 1055–1076.

Cooperative Bank (2012) The Cooperative's Ethical Consumer Market Report. Available at: http:// www.co-operative.coop/corporate/Investors/Publications/Ethical-Consumerism-Report (accessed 20 June 2013).

Crane, R., Davenport, R. and Vaughan, R. (2010) Farm Business Survey 2009/2010 Poultry Production in England. University of Reading, UK. Available at: http://www.fbspartnership. co.uk/documents/2009_10/PoultryProduction_2009_10.pdf (accessed 11 April 2013).

Crane, R., Davenport, R. and Vaughan, R. (2011) Farm Business Survey 2010/2011 Poultry Production in England. University of Reading, UK. Available at: http://www.fbspartnership. co.uk/documents/2010_11/Poultry_Production_2010-11.pdf (accessed 11 April 2013).

DEFRA (Department for Environment, Food and Rural Affairs) (2012) Poultry and Poultry Meat Statistics. DEFRA, London. Available at: http://www.defra.gov.uk/statistics/files/defra-stats-foodfarm-food-poultry-statsnotice-121025.pdf (accessed 11 April 2013).

DEFRA (Department for Environment, Food and Rural Affairs) (2013) UK Egg Production Statistics. Available at: https://www.gov.uk/government/organisations/department-for-environment-food-rural-affairs/series/egg-production-and-prices (accessed 25 June 2013).

DEFRA (Department for Environment, Food and Rural Affairs) (2014) Food Statistics Pocketbook 2013 – In Year Update. DEFRA. Available at: https://www.gov.uk/government/uploads/system/ uploads/attachment_data/file/307106/foodpocketbook-2013update-29apr14.pdf (accessed 12 May 2014).

DoRazario, R.C. (2006) The consequences of Disney anthropomorphism: animated, hyper-environmental stakes in Disney entertainment. *Femspec* 7(1), 15.

EBLEX (Agricultural Horticultural Development Board) (2012) UK Yearbook 2012 – Cattle. Available at: http://www.eblex.org.uk/documents/conte/markets/m_uk_yearbook12_ cattle240812.pdf (accessed 11 April 2013).

European Commission (2014) Agricultural Production – Animals. Eurostat. Available at: http://epp. eurostat.ec.europa.eu/statistics_explained/index.php/Agricultural_production_-_animals (accessed 13 June 2014).

FarmingUK (2014) Egg Consumption Expected to Rise in 2014. Available at: http://www.farminguk. com/News/Egg-consumption-expected-to-rise-in-2014_27005.html (accessed 13 June 2014).

FAWC (Farm Animal Welfare Committee) (2012) Five Freedoms. Available at: http://www.defra. gov.uk/fawc/about/five-freedoms (accessed 11 April 2013).

Hall, C. and Sandilands, V. (2007) Public attitudes to the welfare of broiler chickens. *Animal Welfare* 16, 499–512.

Hingley, M.K. and Parrott, P. (2008) *Consumer Attitudes to Poultry*. Harper Adams University College, Shropshire, UK.

IGD (2008) Ethical Shopping – Are Shoppers Turning Green? March. Available at: http://shoppervista.igd.com/Hub.aspx?id=32&tid=4&rptid=54 (accessed 26 June 2014).

James, D. (2013) Consumers place more value on poultry price than welfare. *Poultry World* 166(3), 12.

Jones, W. and Parrott, P. (1997) *Consumer Perceptions of the Poultry Industry*. Report No. 6, Temperton Fellowship. Harper Adams University, Newport, UK.

Karlsson, F. (2012) Critical anthropomorphism and animal ethics. *Journal of Agricultural & Environmental Ethics* 25(5), 707–720.

Keynote (2012) Poultry Market Report 2012. Available at: http://www.keynote.co.uk/market-intelligence/view/product/10618/poultry (accessed May 2014).

Kiesler, T. (2006) Anthropomorphism and consumer behaviour. *Advances in Consumer Research* 33, 149.

Lawrence, F. (2013) Horsemeat Scandal: Where Did the 29% Horse in Your Tesco Burger Come From? *The Guardian*, 22 October. Available at: http://www.theguardian.com/uk-news/2013/oct/22/horsemeat-scandal-guardian-investigation-public-secrecy (accessed 12 June 2014).

Lister, S. (2012) *Poultry in the Public Eye – The Consumer: Industry Interface*. Report No. 20, Temperton Fellowship. Harper Adams University, Newport, UK.

Mintel (2013) Consumer Trust in Food – UK – June 2013. Available at: http://academic.mintel.com/display/659917 (accessed 12 May 2014).

Morrisons (2013) Research Uncovers What Increases Chicken Wellbeing. Available at: http://www.morrisons.co.uk/corporate/Media-centre/Corporate-news/Research-uncovers-what-increases-chicken-wellbeing (accessed 26 February 2013).

National Chicken Council (2012) Animal Welfare for Broiler Chickens. Available at: http://www.nationalchickencouncil.org/industry-issues/animal-welfare-for-broiler-chickens (accessed 28 February 2013).

Parrott, P. (2001a) *Eggs – Consumer Buying Behaviour*. Harper Adams University, Newport, UK, Discussion Paper.

Parrott, P. (2001b) *Poultry Meat – Consumer Buying Behaviour*. Harper Adams University, Newport, UK, Discussion Paper.

Parrott, P., Walley, K. and Custance, P. (2013a) *Consumer Defined Dimensions of Egg Quality*. Paper presented to XV European Symposium on the Quality of Eggs and Egg Products, Bergamo, Italy, 15–19 September.

Parrott, P., Walley, K. and Custance, P. (2013b) *Consumer Defined Dimensions of Poultry Meat Quality*. Paper presented to XXI European Symposium on the Quality of Poultry Meat, Bergamo, Italy, 15–19 September.

Rose, S. (2013) Are animals in Hollywood films too human? *The Guardian* Thursday 25 April.

RSPCA (Royal Society for the Prevention of Cruelty to Animals) (2012) RSPCA Welfare Standards for Chickens. Available at: http://www.rspca.org.uk/sciencegroup/farmanimals/standards/chickens (accessed 11 April 2013).

Schooler, R.D. (1965) Product bias in the Central American Common Market. *Journal of Marketing Research* 2(4), 394–397.

Showers, V.E. and Showers, L.S. (1993) The effects of alternative measures of country of origin on objective product quality. *International Marketing Review* 10(4), 53–67.

Uddin, J., Parvin, S. and Rahman, M.D. (2013) Factors influencing importance of country of brand and country of manufacturing in consumer product evaluation. *International Journal of Business & Management* 8(4), 65–74.

Van Horne, P. and Achterbosch, T. (2008) Animal welfare in poultry production systems: impact of EU standards on world trade. *World Poultry Science Journal* 64, 40–52.

Van Horne, P. and Bondt, N. (2013) Competitiveness of the EU Poultry Meat Sector LEI. Wageningen, the Netherlands. Available at: http://www.britishpoultry.org.uk/eu-competitiveness-report-of-the-poultry-meat-sector-2 013 (accessed May 2014).

Velde, H.T., Noelle, A. and Woerkum, C.V. (2002) Dealing with ambivalence: farmers' and consumers' perceptions of animal welfare in livestock breeding. *Journal of Agriculture and Environmental Ethics* 15, 203–219.

Veterinary Record (2013) Horsemeat Review Highlights Confusion about FSA's Role. News & Reports, 20 July, 57–58. Available at: http://eds.b.ebscohost.com/eds/pdfviewer/pdfviewer?sid=761aaa45-45aa-4858-83f8-e017f0030139%40sessionmgr111&vid=3&hid=109 (accessed 12 June 2014).

Walley, K.E., Custance, P.R. and Parsons, S.T. (2009) Controversies in food and agricultural marketing: the consumer's view. In: Lindgreen, A., Hingley, M.K. and Vanhamme, J. (eds) *The Crisis of Food Brands – Sustaining Safe, Innovative, and Competitive Food Supply*. Gower Applied Research, Surrey, British Columbia, Canada, pp. 197–220.

Walley, K., Parrott, P., Custance, P., Meledo-Abraham, P. and Bourdin, A. (2014) A review of UK consumers' purchasing patterns, perceptions and decision making factors for poultry meat. *World's Poultry Science Journal* 70, 493–502.

West, C. (1999) *Marketing Research*. Macmillan Business Masters, Macmillan Press, Basingstoke, UK.

Wilson, H. (2014) Do You Know Where Your Food Comes From? Available at: http://food.uk.msn.com/food/do-you-know-where-your-food-comes-from (accessed 12 June 2014).

World Poultry (2014) UK Egg Consumption Beats Chocolate This Easter. *World Poultry*, 22 April. Available at: http://www.worldpoultry.net/Layers/Eggs/2014/4/UK-egg-consumption-beats-chocolate-this-Easter-1507065W (accessed 12 June 2014).

Xu, H., Leung, A. and Yan, R.-N. (2013) It is nice to be important, but it is more important to be nice: country-of-origin's perceived warmth in product failures. *Journal of Consumer Behaviour* 12(4), 285–292.

PART II
The Economics of Sustainable Production

CHAPTER 3
Global Context on Price Volatility and Supply Chains – Is Europe Competitive?

Nan-Dirk Mulder*
Rabobank International, Utrecht, the Netherlands

INTRODUCTION

The global poultry industry is being driven towards change by challenging global fundamentals in food and fuel demand and supply. One of the major challenges is the upwards variation in input costs, brought about through higher and more volatile grain and oilseed prices. In the future, business models applied within the poultry industry will require adjusting to reflect this change, particularly as grain and oilseed prices represent from 50 to 70% of production costs (Mitchell, 2008). The significance of the country of poultry production, due to variations in both global supply and demand for poultry meat, will reflect these changing input costs and therefore will differ greatly between countries; for example, it is likely that production in the Americas and in the Black Sea regions will increase whereas Asia will face increasing difficulties in meeting demand and therefore have a greater need for imports (Rabobank, 2014). These changes are likely to lead to stronger linkages between Asian countries and the Americas with investments in both directions.

An important example of the changing global direction of poultry production is in the EU, which became a net importer of poultry meat for the first time in 2007 as a result of increased domestic consumption, currency movements and high domestic production costs. The EU is therefore continuously losing international market share to Brazil (Guerrero-Legarreta and Hui, 2010), one of the leading poultry producers in the world and the world's leading exporter of poultry meat. Brazil has a unique set of natural resources, including high availability of arable land and the appropriate climate and water availability to grow maize and soybeans. It also has low labour costs and well managed vertically integrated companies (Oliveira *et al.*, 2012). The cost of broiler production in EU countries is higher than in many other key poultry meat producing and exporting

*Corresponding author: Nan-Dirk.Mulder@rabobank.com

countries. In the EU average production costs vary greatly between member states, with production in the Netherlands alone varying by up to 8% around the average (Horne, 2009). This suggests that cost differences between farms within a country can potentially be greater than the cost differences between member states, and therefore no single member state has competitive advantage over another. These differences between winners and losers in the industry will continue to increase, with the winners being those who adequately implement the new realities in their business models. This is an issue with which the industry has been struggling for years and will remain an issue in the future.

Major issues affecting the industry include supply chains, feed quality issues, disease and consumer perceptions, but grain volatility has the most obvious impact on the observed increase in input process due to its large effect on cost. It is therefore important to question how will Europe react and deal with this volatility, what is the global perspective on feed price volatility and how competitive is the industry globally. This chapter will focus on where the industry needs to be heading in the future to remain competitive.

CHANGING FUNDAMENTALS AS THE DRIVING FORCE FOR CHANGE

The world food industry will face big challenges in the years ahead. The FAO expects world food demand to grow by 70% by 2050, with world population growing from the current 7 billion to 8 billion in 2030 and 9 billion by 2050 (Alexandratos and Bruinsma, 2012). Income levels are expected to increase worldwide, which will stimulate meat consumption, especially in lower income countries, which in turn will have a huge impact on the supply chain (Magdelaine et al., 2008). Competition for land, water and energy will intensify due to limited availability of these resources. Land availability per world capita has declined in the last few decades and this is a particular problem in emerging countries where expansion of big cities competes with agriculture. The areas in Asia which face some of the largest growth in food demand already have a very high cultivation level of the available arable land. As high water availability becomes increasingly a competitive advantage, Asia may be at a loss compared to more water-rich production areas in the Americas and Black Sea region.

It is apparent that in such a changing global food scenario, the focus of the industry must change and adapt and sustainability will become an important topic in the new market environment of the next decade. The growth in demand for poultry coupled with limited resources will require more emphasis on existing resources throughout the value chain. The industry must deal with a situation in which land resources are limited and although there are available land reservoirs for agricultural supply – especially in Brazil, Russia and Africa, but also in the EU and the USA because of set-aside programmes – the expansion into these land areas will be difficult and slow. The emphasis must therefore change to focus on output, with a drive towards either higher yields or yields that are better focused on changed consumer demand, in order to help reduce resource use (Global Food Security, 2013). For example, increased agricultural commodity prices in the EU were caused by a combination of issues including low global inventory levels, weather-induced reduction of supply, outside investor influence, oil prices and structural changes in demand for grains and oilseeds due to changes in population dynamics and the development of the biofuels sector (Banse *et al.*, 2008).

With regards to demand, urbanization, economic growth, changing diets and expanding populations have all driven increases in food and feed demand, particularly in developing countries. Food and feed remain the largest sources of demand growth in agriculture, although demand for agricultural commodities for use as feedstock by the biofuel sector, for example wheat for bioethanol production and soybean for biodiesel production, represent the largest source of new demand (Baier *et al.*, 2009). Agricultural commodity prices increased due to the weather-induced lower supply and low global inventory prices, which meant that the shortfall in demand could not be met by driving up prices. A result of this was that some countries took protective policy measures to keep supply within domestic markets and discourage exports. In addition, the higher oil prices were spread along the supply chain leading to higher grain prices as the production costs of fertilizers and transport increased (European Commission, 2008). World agriculture has a great challenge as better efficiency and yields can only be achieved by better farm management – both for arable and animal production – by using better genetics, better feed (fertilizers or animal nutrition), equipment (including housing) and by using better disease protection. Depending on regional circumstances, optimal agronomics and farm management will play a significant role in optimizing inputs for best performance. The whole process of improving inputs, in an integrated way, will help the industry to reach the challenging target for production by 2050.

It is true that experiences from the past have shown that the road to a new balance will not be an easy one. The increase in feed prices since 2006 can be attributed to the challenge of a growing demand for food, feed and fuel, while resources were limited. The latter is illustrated by historic low stock-to-use ratios in cereals and a shift in world trade, with a growing importance of the volatile Black Sea supply together with a move in Asian markets towards net imports, especially in China for maize (Gale, 2012). Any change in supply and demand in these markets will always have a huge impact on market prices while also attracting speculators. This has resulted in a trend of increasing and more volatile prices in the international grain and oilseed markets. These types of shocks to the supply chain have a huge impact on the meat industry. This volatility results from a mixture of structural and temporary factors, ranging from general global population growth to adverse weather conditions and exchange-rate movements. It is important to also remember the different mind-sets in other parts of the world, for example in China issues such as antibiotic use are of lesser importance to the consumer than they are in the EU.

GROWING MEAT DEMAND AS A MAIN DRIVER FOR CHANGE

A heightened demand for meat, together with increased interest in biofuels and other food products, is one of the main driving forces behind a growing demand for grains and oilseeds. Global meat demand is expected to grow by 45% with poultry demand growing by 60% as world population expands and average incomes increase (Alexandratos and Bruinsma, 2012). Poultry meat has no cultural consumption limitations (unlike beef and pork) and has a strong health and convenience image. The poultry industry benefits from very competitive cost of production and this production efficiency gives it a relatively good sustainability footprint compared to other proteins. The cost of production of chicken meat is approximately €0.75/kg compared to approximately €1/kg for pork, €1.75/kg for beef and €2.25/kg for salmon. The production cycle length is much lower for chicken meat, with both breeding and finishing taking less than 1 year compared to pork in which breeding takes approximately 15 months and finishing takes approximately 4 months, beef in which breeding takes approximately 23 months and finishing takes an additional 15 months and salmon in which breeding takes approximately 1 year and finishing takes approximately 18 months (Rabobank Analysis, 2009).

The EU is self-sufficient in poultry meat; after an increase in production between 1996 and 2002, EU-15 poultry meat production fell slightly in 2003 but has since gradually increased. This decrease observed in 2003 was in part due to the avian influenza crisis; between 1999 and 2003, following the spread of the H5N1 strain to EU borders, an outbreak was detected on EU territory, which resulted in falls in consumption and the imposition of trade bans on some of the EU member states. The European Commission devised market support measures to alleviate any fall in consumer confidence resulting from the avian influenza outbreaks, as well as compensating for losses and providing assistance with vaccination costs (FAO, 2006).

Following the enlargement of the EU, poultry production and consumption dramatically increased due to the increased capacity brought about by the member states. The poultry meat trade is predominantly dictated by the demands for different cuts of meat; for instance in the EU there is a high demand for breast fillets and lower demand for lower value cuts, leading to the import of breast fillets and export of lower value cuts. However, Brazil exports breast meat to the EU, whole birds to the Middle East, leg meat to Russia and deboned meat to Japan (Baracho *et al.*, 2006).

There are two principal methods of broiler meat production in the EU: (i) integrated (common in France, Germany and Spain), which has the advantage of higher capacity utilization, lower risk and income volatility and quicker technology transfer to farmers; and (ii) non-integrated production (common in the Netherlands, Poland and Belgium), which can be beneficial because it provides performance incentives for farmers who can therefore benefit from competition among partners in the supply chain, such as hatcheries and slaughterhouses (FAO, 2010). The negative financial impact of integrated production is that broiler farmers may sacrifice opportunities to receive high revenue when market conditions are favourable. In non-integrated production, fluctuations in input and output prices have a direct consequence for the income for broiler farmers. These non-integrated businesses tend to be dependent on loans and are therefore vulnerable to changes in interest from banks, resulting in little incentive for maximizing efficiency within the production chains (Bamiro *et al.*, 2006).

In the slipstream of a growing meat trade, companies from developed markets and exporting countries will face strong pressure to benefit from global growth. US companies are facing a more challenging local production and trade environment whilst also being encouraged by shareholders to internationalize their business models (Nelson, 2005). The success of the Brazilian industry may well drive further internationalization of the industry with companies moving to multinational structures (with support of national investment funds), in which the three directions of internationalization – access to low-cost production, local market growth and synergy in distribution – will be all exploited (Sluis, 2006).

GLOBAL CHALLENGES, REGIONAL DIFFERENCES – HOW DOES IT AFFECT EUROPE AND THE REST OF THE WORLD?

Not all companies will make the move to internationalize, as in many regions there are still many opportunities in local markets as demand grows. In some cases, modern distribution may still be in an early stage of development and the level of fragmentation is low, therefore export potential is still not being utilized. For these reasons, it is advantageous for companies in the EU to first utilize internal opportunities via both better integration with local industry and a move to a regional business model, before potentially moving towards a pan-European business model. Companies in Eastern Europe still have many growth opportunities in their domestic market, although the growth potential in Russia will potentially slow down somewhat after 2015, when Russia becomes fully self-sufficient (European Commission, 2014). Ukrainian companies still have significant local growth potential, and additional growth could be realized as they may become an exporter of poultry products in the medium to long term (FAO, 2014). An opening of the EU market in the medium term might present local industries with a great growth opportunity.

It will be a big challenge for the poultry industry to keep up with the expected global demand in growth in the next decade, estimated at around 30%, particularly as this growth is not evenly spread around the world. It is clear that most of the production growth will continue to be in local areas as fresh-product demand remains strong. However, a projected scarcity of product in global markets will increase the importance of international trade. As a result, countries in Asia and large parts of Africa with no natural competitiveness in poultry production will need to reconsider their supply strategy in the future (Winkler *et al.*, 2006). The position of Asian companies will change in the medium term as local market growth coupled with limited resources will lead to growing awareness among Asian importers about the importance of food supply security. It can be expected that more Asian countries, like China, will follow the Japanese model regarding import security via local joint ventures. China's import position may be forced to change due to the limited internal availability of resources for grains and oilseeds (Zhang, 2011). Although most of the supply of chicken will continue to be produced in China, import companies will start to acquire companies or set up joint ventures with exporting countries in Asia and Latin America to secure supply

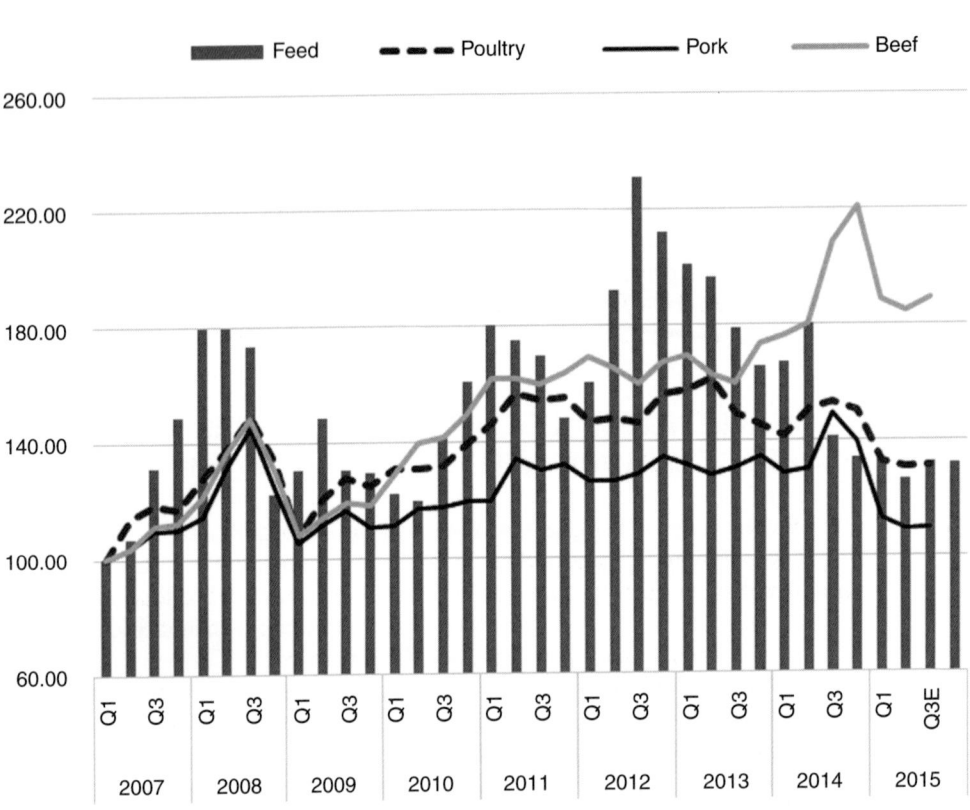

Fig. 3.1. Rabobank global protein to feed price monitor (OECD/Food and Agriculture Organization of the United Nations, 2014; Rabobank, 2014).

(Ghose, 2014). Such a strategy might also force other importers like Middle Eastern countries and the EU to react and follow a similar path.

VOLATILITY – WHERE ARE WE NOW AS AN INDUSTRY?

Rabobank is the largest agricultural bank with a global presence and is unique in that it is the only bank focused on food and agriculture. The grain and oilseed markets have fluctuated greatly over the last few years. The price of wheat and maize has remained consistent at approximately US$100–150/t between 2000 and 2006. Prices then rose to approximately US$200/t and continued to rise, reaching a maximum of approximately US$450 and US$300/t for wheat and maize, respectively, between 2008 and 2009 (Rabobank Analysis, 2013). Maize and wheat prices then dropped by approximately half between 2009 and 2011, before peaking again at approximately US$250–350/t between 2011 and 2013 and then declining to around US$150–200/t in 2014. The price of soybeans has been relatively consistent between 2000 and 2003, at approximately US$200/t, but doubled between 2003 and 2004 (reaching a maximum of US$400/t) before dropping again to US$200–250/t between 2004 and 2007. From 2007 the price of soybeans has increased, reaching a peak of over US$600/t in 2008. The price reduced slightly between 2009 and 2011 before then increasing again and has fluctuated greatly between 2011 and 2014, with the highest prices in 2013 (US$650/t), and dropping as low as US$400/t in 2012. The years between 2007 and 2014 can be viewed as the commodity boom years. As from 60 to 70% of the cost of poultry production is based on cereal price, this variability is highly important (Rabobank, 2014). The protein markets have been bullish, with higher prices and lower feed costs, where the cost of protein is not passed on to the consumer (Fig. 3.1). This increases the pressure on poultry producers further.

Another large source of volatility for international companies is the exchange rate. Between 2007 and 2014 the exchange rate index for the Argentine Peso dropped dramatically, reaching an index of approximately 40 whereas the Chinese Yuan and Thai Baht increased gradually to an index of approximately 130 and 120, respectively, by 2013–2014. The exchange rate of the Russian Rouble and Euro both increased, reaching a peak index of approximately 110 and 120, respectively, by mid-2008, before decreasing to around 80 and 90, respectively, and increasing again slightly to an index of 90 and 110 before plateauing. The Japanese Yen gradually increased between 2007 and 2012, reaching a peak index of over 150, and then decreased to an index of approximately 120 in 2013 where it then plateaued. The Brazilian Real increased to an index of approximately 130 in 2008, then decreased dramatically to an index of 90 at the beginning of 2009 before increasing gradually to a peak index of approximately 140 by mid-2011 and then gradually decreasing back to an index of approximately 90 between mid-2013 and 2014 (Rabobank Analysis, 2013). These volatilities are so large and variable between currencies that it is difficult to predict with any accuracy where the markets will be in the future.

A GLOBAL PERSPECTIVE ON FEED PRICE VOLATILITY

There is predicted to be a 20% increase in food demand in the next 10 years (Alexandratos and Bruinsma, 2012), which poses a significant global food supply challenge. Most of this increase in demand will be linked to the declining land availability (Godfray et al., 2010). This is a major challenge for the industry: how to increase production with fewer resources. Increasing efficiency brings other challenges; increasing pressure on resources while raising questions about welfare and sustainability. The range of animal welfare legislation in force in the EU contrasts strongly with the few provisions for animal welfare in developing countries, with most of the legislation only applying to EU producers (with the exception of those surrounding slaughter). These animal welfare provisions fulfil an EU consumer preference for higher welfare (Eurobarometer, 2007); however, they may arguably disadvantage EU production compared with third country production (Bagnara, 2009).

Egg production is well poised to take advantage of the new reality, as efficiency is very high and production flexible compared to meat proteins. Egg demand is also increasing year on year. Shell eggs produced under different systems in the EU must be labelled according to the system used, but for developing countries this labelling is not required (European Commission, 2009a). The majority of the competition in the egg industry is for egg products used in food processing as opposed to shell eggs, the focus for which is more on price than on welfare concerns (Sumner *et al.*, 2011). The rapid modernization of the animal protein industry is therefore being driven by higher incomes and feed prices coupled with the need to deal with food safety issues, animal disease threats and modern distribution. These factors drive increased efficiency and yield, the formation of larger companies, and well-managed, modern value chains. In combination these areas can drive up compound feed demand. Land use, dictated by the efficiency of cultivation yield, must be divided into livestock and poultry production to meet meat demands, and direct grain and oilseed production to meet food and fuel demands. Combined, these demands are predicted to increase by over 20% in the coming 10 years (Alexandratos and Bruinsma, 2012), so yield must be increased accordingly, driving volatility in the markets.

Grain and oilseed demand has increased globally and supply has not expanded to fit this demand. Grain stocks are dwindling; the grain stocks to use ratio decreased from approximately 0.35 in 1999 to approximately 0.18 in 2007 where they have then plateaued (Rabobank Analysis, 2013; USDA, 2013). The low stocks of grain and oilseeds bring with them increased volatility as concerns about yield, political issues and weather become more important. Additionally, biofuels are a relatively new source of demand for feedstuffs and production has increased rapidly, particularly from 2006 to 2010 where there was a 24% annual growth rate. For example, 12% of global maize production and 16% of sugarcane supply is used for ethanol production, which has a direct market impact on animal protein production (Shikida *et al.*, 2014). Additionally, approximately 17% of soya oil, 10% of palm oil, 25% of rapeseed oil and 1% of sunflower oil global supplies are used for biodiesel production. Increasing demand from the ethanol and biodiesel industries in the USA, Europe and China for maize and other cereal grains has increased input prices from 2007 onwards. Although prices reduced from April 2008, the underlying structural pressures still remained. Between 2006 and 2010 in North America, biofuel production increased from approximately 5000 t (1000 t oil equivalent) to 25,000 t, whereas in South and Central America it increased from 10,000 t to 45,0000 t. This trend is similar in other areas; in Europe production increased from 11,000 to 55,000 t and in Asia Pacific it increased from approximately 11,000 to 60,000 t (FAO, 2014; Rabobank, 2014; USDA, 2014). Now that the market has matured and the USA has reached its ethanol mandate for renewable fuel use in cars, production is declining, but in other parts of the world the impact of biofuels is still substantial, although ambitions have been lowered (Ray *et al.*, 2013). Biofuel production has an indirect impact on the animal protein market as it results in meals being produced as a by-product; for example, a by-product of ethanol production is distillers dried grains with solubles (DDGS), which can subsequently be used as high-protein livestock feed (Salim *et al.*, 2010).

Maize use for biofuels has a direct effect on feed costs, as exactly the same material is used for feed and food. Maize demand from Brazil, Russia, India and China increased dramatically between 2003 and 2013, from approximately 475 to 675 million t (Mt). In the USA and EU this demand was less pronounced, from 475 to 550 Mt, and in the USA (excluding ethanol) and the EU there was an increase between 2004 and 2006 from approximately 450 to 475 Mt followed by a decline to approximately 425 Mt by 2013 (Rabobank Analysis, 2013; USDA, 2013). The demand for maize in the USA and the EU is expected to decrease as feed conversion ratio (FCR) continues to improve and as biofuel production declines, whereas in China demand is expanding very rapidly, related to the demand for animal protein. These changes in demand for feed will alter trade streams with an increasing move towards the east for both pigs and poultry (Ciampitti and Vyn, 2014).

Animal protein demand is expanding in the developing world, with the move from a vegetable-based diet to an animal protein-based diet. Growth in global demand for animal protein has consistently increased year on year, although the extent of this increase has reduced each decade (Topliff *et al.*, 2009). For example, global demand increased by 39% between 1980 and 1990, 35% between 1990 and 2000 and 20% between 2000 and 2010. It is predicted that global demand for meat protein will increase by a further 45% by 2030. Growth is expected to reach around 60% over the next 20 years for poultry and 48% for eggs (FAPRI-MU, 2014; OECD/Food and Agriculture Organization of the United Nations, 2014; Rabobank, 2014). The EU is currently self-sufficient in egg production, but although production has increased, egg consumption has increased at a higher rate, which means the egg surplus has reduced. There is little trade in eggs and imports are very small and mainly just in powder form due to issues with logistics. The absence of cultural and religious limitations will create a higher demand for poultry meat and eggs due to high consumer preferences (Mahiuddin *et al.*, 2008; Higenyi *et al.*, 2014). Egg production systems are categorized by layer housing method; although most of the EU uses conventional cage systems, there is also significant production in barn and free-range systems, enriched cages and organic production. The proportions of hens raised under each system varies between countries and member states (Windhorst, 2005). In the commercial egg industry in the USA, numerous independent producers market on a local basis and apply price competition as a component of their marketing strategy. It is estimated that the top ten egg producers (each with 5 million layers) represent 44% of the industry, therefore they have high efficiency of production, marketing and distribution and are hence a large exporter of eggs and egg products (Shane, 2003). In India, the growing population will increase the local market for poultry products meaning export efforts are not required for Indian producers. None the less, some companies are exporting egg powder to the EU and Japan due to their lower production costs, meaning India could potentially have a competitive advantage when it comes to the world market for egg powder (Horne and Achterbosch, 2008).

Poultry feed for broilers consists of 60% grains (primarily maize and wheat) and 25% protein-rich ingredients such as soybean meal, however the inclusion of soybean meal in feed rations of layers is lower at around 15-20%, resulting in feed costs for broilers being higher than that for layers. There are large differences in the cost of broiler production between countries. Broiler costs in Europe are highest in Norway at over €1.10/kg live weight and lowest in Ukraine at less than €0.90/kg live weight. In the USA, broiler production in 2011 cost approximately €0.56/kg live weight, made up of approximately 70% feed costs, 10%

day-old chick costs and 20% operating costs. In Italy and the UK however, 2011 costs for broiler production stood at almost double this, at €1.07 and €0.97/kg live weight, respectively, made up of approximately 70% feed costs, 15% day-old chick costs and 15% operating costs (Rabobank, 2014).

Figure 3.2 illustrates that the predicted 65% increase in demand for meat and eggs arises largely from Asia and only 6% from the USA/EU over the period 2012–2022. The key challenge is how to produce enough to feed this growing demand with limited resources. Most new agricultural land (45%) is located in sub-Saharan Africa and not Asia (3%), where there is poor soil quality and poor water availability. More grain is already being supplied from comparatively new production regions such as Brazil and Ukraine while typical producers such as the USA have stagnating or declining production.

DEALING WITH VOLATILITY

The big global challenge is to manage the relationships and interactions between many factors: weak economic circumstances, stronger retail power, consumer

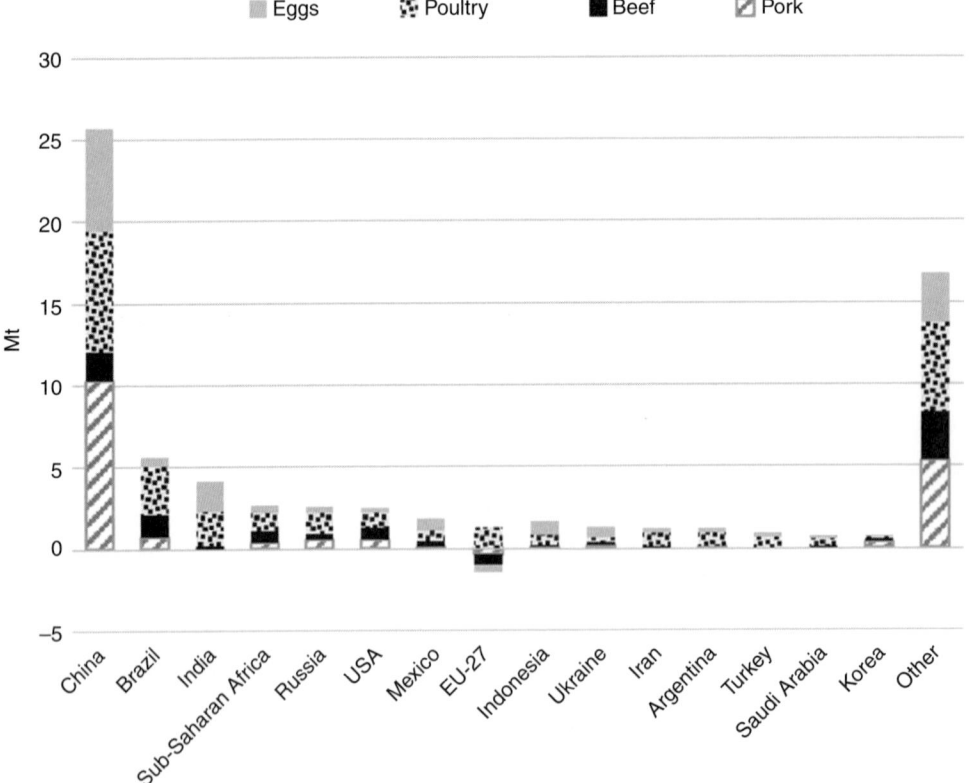

Fig. 3.2. Projected market growth for meat and eggs 2012–2022 (FAPRI-MU, 2014; OECD/Food and Agriculture Organization of the United Nations, 2014; Rabobank, 2014).

concerns about welfare and safety, animal disease outbreaks and growth in emerging global markets. The impact of these factors is heightened by volatility arising from exchange rates, supply of agricultural products and higher input prices. New business models are required to deal with this global challenge, and these models need to be adjusted as the challenge changes. The main challenges for the global grain market are the slowdown in ethanol growth leading to a need for new outlets for grain production in the USA, a higher dependency on volatile supplies from new markets and the rapid growth and modernization of the Asian protein industry. These new markets will alter the value chains via added market intelligence, added value and value chain management, cost price leadership and improved risk mitigation and flexibility. The environment will remain volatile for grains and oilseeds and industry will need to deal with this. The greatest volatility is likely to be in the production of grain and in the livestock farming sectors, but the consumer demand for stable prices means that forward planning, particularly with regard to pricing, is vital to deal with this new market reality. There is a need to ensure that we as an industry, buy at the best price and communicate across the value chain, and develop bargaining power to decrease risk. Efficiency is of the utmost importance and this makes differences between farming systems of vital importance.

In the USA/EU between 2007 and 2013 consumption of meat dropped by 0.6%, but meat consumption is increasing in emerging markets where the dynamics are very different. Between 2007 and 2013 meat consumption increased by 3.4% in China, 3.3% in the rest of Asia and 2.3% in the rest of the world. The compound annual growth rate (CAGR) of poultry meat has consistently increased between 2000 and 2013: from 2000 to 2004 it increased by 0.8%, from 2004 to 2007 by 0.1% and from 2007 to 2013 by 1.5%. However, the same cannot be said for other meats, for example between 2007 and 2013 the CAGR decreased by 1.5% and 1.3% for beef and pork, respectively. The countries with the fastest growth in demand for poultry meat are Turkey, Sweden, Romania, Finland, Russia, Poland and Spain with the UK and France having the lowest growth in demand (Rabobank, 2014). The volatility with regards to prices of grains and oilseeds coupled with that of the eggs, livestock and poultry markets, results in a need for stable sale prices. The end consumer requires a stable price, but the power of the feed companies means the producer is taking the brunt of feed price volatility and therefore there is a need to improve branding and cut costs in order to remain competitive.

There is a direct relationship between income and food demand in a region, as when salaries increase for those on low incomes, people move from grain to meat. There are hence different attitudes towards meat consumption (Drewnowski and Specter, 2004). For example, in countries such as Nigeria and India where the average annual income is less than US$5000, food is viewed as a necessity, but in in the EU and USA where average income is more than US$40,000, food is seen as a stimulation and the public has become increasingly concerned about animal welfare and health issues. This has led to new initiatives in the USA and EU such as welfare labelling (e.g. free range) (Harper and Henson, 2001; European Commission, 2009b). In other countries it is cultural practice as opposed to income that dictates meat consumption, for example in Brazil and China there is significant emphasis on the social aspect of food, and in Japan and Singapore the nutritional value of food is of high importance (Dindyal and Dindyal, 2003). Thus, the emerging meat consumption market is heavily governed by location and time. Furthermore, consumer concerns are growing and food quality and safety has become an increasingly more important issue. This concern will continue to increase, largely due to negative publicity about food safety, incidences of disease, poultry welfare and practices in factories, perceptions about buying local produce and impact on the environment (Rabobank Analysis, 2013).

Markets are changing, and the ability to deal with volatility is a key success factor. As a consequence of these changes there will be larger companies with vertical integration, which are more efficient, have better yield and improved chain management. For example, China is starting to put feed mills on to farms to improve control over production. Globalization has increased international investment, with the most obvious example of this being the Chinese investment in Smithfield Foods (Mattioli et al., 2013), but it is expected that there will be more of this type of development. Companies from emerging markets are taking the lead in industry globalization. The largest companies used to be in the USA but now these are more evenly distributed. China will become even more important as demand continues to rise; at the moment China is renting 5% of the total available farm land in Ukraine (Spillius, 2013) and this is expected to increase.

IS EUROPE COMPETITIVE?

In the EU, chicks are 30% more expensive than in more competitive parts of the world and the cost of production is high, resulting in a big import industry in Europe, particularly in Eastern Europe. There is a consumer shift to poultry meat in Europe despite the reduction in total meat intake; poultry consumption is increasing at around 1–1.5% per annum, which should continue in years to come (Rabobank, 2014). The fastest growth areas for poultry meat consumption are Russia and Turkey. This has led to a common perception that we are missing opportunities in Europe and that the north-western countries are leading the way. A possible explanation for this is that European companies tend to be national whereas elsewhere companies tend to be international (FAO, 2012). As a result, in 5 years it is believed that there will be a change to a regional EU industry rather than a pan-European industry.

EU animal welfare legislation in the poultry sector covers production, transport and slaughter and states the maximum stocking density of broiler chickens and allows only enriched battery cages for laying hens (Council Directive, 2007). Yet consumer concerns are still high in the EU and issues such as food safety, welfare, environmental sustainability and local production need to be taken into account. As a result, new concepts have been put in place to change the north-west European market. In the Netherlands the 'Chicken of Tomorrow' initiative has been put in place to aid in improving public perception of poultry production (Bock et al., 2014). The cost of the fillet has been increased to meet increased production costs of approximately 5–10% (European Commission, 2011), and a new dedicated breeding line has been put in place. The Netherlands has also altered rearing systems towards producing slow-growing birds (max 50 g gain/day) with a minimum finishing age of 45 days. In 2015 there will also be lower bird density (from 42 kg/m² to 38 kg/m²), straw bedding in the houses and 6 h dark/day, as well as reduced target ammonia emissions (from 35 to 22 g/animal placed/year) and heightened focus on sustainable energy (Rabobank, 2014). In Germany the animal welfare initiative system (Initiative Tierwohl) will begin in 2015 and 80% of supermarkets have accepted the standards put in place by this system and will guarantee to compensate for additional costs (Averós *et al.*, 2013). This system includes reducing bird stocking density to 35 kg/m² and improving welfare standards at farming and breeding stages (Rabobank, 2014). Similar animal welfare initiatives have also been implemented in the UK as Freedom Foods and the Red Tractor mark and in France as Label Rouge. These new initiatives in the market are rising in profile, but there is concern that these alone are not enough to make a significant improvement to bird welfare and public perception of the poultry industry. If the consumer is asking for a direction this should be seen as an opportunity for the industry. Although there are some productivity and meat quality benefits from the implementation of the welfare legislations, these benefits are difficult to quantify and arguably do not fully offset the cost disadvantages that arise, for example in the move to enriched cages for layers. Other legislation such as the IPPC Directive (integrated pollution and prevention control) and *Salmonella* requirements also have negative effects on competitiveness.

CONCLUSIONS

There are considerable opportunities for the global poultry industry due to increased international trade, demand growth in all regions of the world and growth in modern distribution. On the whole, the poultry industry is in a good competitive position compared to other proteins. Industry players are well positioned to benefit from these challenges but they will need to take the right strategic direction. They must shape their business models to be ready to deal with the challenges and fluctuations in the global market environment in order to be winners in the next decade.

The price of global grains is expected to decrease, but volatility will still remain very high. This volatility is largely dictated by the growing and rapidly modernizing Asian animal protein industry, the limited global land availability (especially in Asia) and the higher dependence on new exporters, especially Brazil and Baltic Sea Region (BSR) countries. The poultry industry needs to adjust to this new volatile reality by prioritizing efficiency, increasing market power through consolidation and internationalization, reconsidering supply chain management and restricting supply in times of increasing feed prices.

The EU will move from national to regional production within the next 5 years and towards a pan-European model within 10 years. Consumer concerns will drive industries globally to change to more modern systems in Asia and more welfare-driven systems in Western markets. Demand growth in Asia will lead to the development of global food supply chains. As a result, Asia is a game-changer for the industry and will be the major change globally going forward to 2050. The egg industry is well positioned to benefit from new market circumstances due to low price and flexibility.

REFERENCES

Alexandratos, N. and Bruinsma, J. (2012) World agriculture towards 2030/2050: the 2012 revision. ESA Working Paper No. 12-03. *Agricultural Development Economics Division, Food and Agriculture Organization of the United Nations*, 1–154.

Averós, X., Aparicio, M.A., Ferrari, P., Guy, J.H., Hubbard, C., Schmid, O., Ilieski, V. and Spoolder, H.A.M. (2013) The effect of steps to promote higher levels of farm animal welfare across the EU. Societal versus animal scientists' perceptions of animal welfare. *Animals* 3, 786–807.

Bagnara, G.L. (2009) The impact of welfare on the European poultry production: political remarks. *Poultry Welfare Symposium*, Cervia, Italy, 18–22 May 2009.

Baier, S., Clements, M., Griffiths, C. and Ihrig, J. (2009) Biofuels impact on crop and food prices: using an interactive spreadsheet. *Board of Governors of the Federal Reserve System International Finance Discussion Papers, Number 967*, March 2009.

Bamiro, O.M., Phillip, D.O.A. and Momoh, S. (2006) Vertical integration and technical efficiency in poultry (egg) industry in Ogun and Oyo States, Nigeria. *International Journal of Poultry Science* 5(12), 1164–1171.

Banse, M., Nowicki, P. and van Meijl, H. (2008) *Why are Current World Food Prices so High?* LEI report 2008-040. LEI Wageningen UR. The Hague, the Netherlands, June 2008.

Baracho, M.S., Camargo, G.A., Lima, A.M.C., Mentem, J.F., Moura, D.J., Moreira, J.V. and Nääs, A. (2006) Variables impacting poultry meat quality from production to pre-slaughter: a review. *Revista Brasileira de Ciência Avícola* 8, 201–212.

Bock, B.B., Hacking, N. and Miele, M. (2014) Coordinated European Animal Welfare Network (EuWelNet): report on the main problem areas and their sensitivity to be addressed by knowledge transfer for each of the specific aspects of the legislation chosen for this project. 3 Feb, 2014. Grant agreement SANCO 2012/G3/EUWELNET/SI2.635078.

Ciampitti, I.A. and Vyn, T.J. (2014) Understanding global and historical nutrient use efficiencies for closing maize yield gaps. *Agronomy Journal* 106(6), 2107–2117.

Council Directive (2007) Council Directive 2007/43/EC of 28 June 2007 laying down minimum rules for the protection of chickens kept for meat production. *Official Journal of the European Union*, L 182/19.

Dindyal, S. and Dindyal, S. (2003) How personal factors, including culture and ethnicity, affect the choices and selection of food we make. *The Internet Journal of Third World Medicine* 1(2), 4.

Drewnowski, A. and Specter, S.E. (2004) Poverty and obesity: the role of energy density and energy costs. *The American Journal of Clinical Nutrition* 79(1), 6–16.

Eurobarometer (2007) Attitudes of consumers towards the welfare of farmed animals, wave 2. *Special Eurobarometer* 229(2), March 2007.

European Commission (2008) *High Prices on Agricultural Commodity Markets: Situation and Prospects*. Brussels, July 2008.

European Commission (2009a) *Feasibility Study on Animal Welfare Labelling and Establishing a Community Reference Centre for Animal Protection and Welfare: Part 1: Animal Welfare Labelling*. Brussels, January 2009.

European Commission (2009b) *Options for Animal Welfare Labelling and the Establishment of a European Network of Reference Centres for the Protection and Welfare of Animals*. Brussels, October 2009.

European Commission (2011) *Study on the Competitiveness of the European Meat Processing Industry*. Rotterdam, October 2010.

FAO (2006) Symposium summary: market and trade dimensions of avian influenza. *21st Session of the Intergovernment Group on Meat and Dairy Products*, Rome.

FAO (2010) *Agribusiness Handbook: Poultry Meat and Eggs*, pp. 1–76. Available at: https://www.responsibleagroinvestment.org/sites/responsibleagroinvestment.org/files/FAO_Agbiz%20 handbook_Poultry_Meat.pdf (accessed October 2015).

FAO (2012) *Food Outlook Global Market Analysis Highlights. Trade and Markets Division*. November 2012. Available at: http://www.fao.org/docrep/019/i3473e/i3473e.pdf (accessed October 2015).

FAO (2014) *Ukraine Meat Sector Review*. FAO Investment Centre Country Highlights. Rome, 2014.

FAPRI-MU (2014) Food and Agricultural Policy Research Institute (FAPRI) at the University of Missouri (MU) FAPRI-MU 2014 Outlook Update. Available at: http://www.cmegroup.com/education/files/fapri-mu-2014-ag-outlook.pdf (accessed October 2015).

Gale, F. (2012) A tale of two commodities: China's trade in corn and soybeans. *China's Agricultural Trade: Issues and Prospects Symposium*, July 2007, Beijing, China. International Agricultural Trade Research Consortium, No. 55021.

Ghose, B. (2014) Food security and food self-sufficiency in China: from past to 2050. *Food and Energy Security*, October 2014.

Godfray, H.C.J., Crute, I.R., Haddad, L., Lawrence, D., Muir, J.F., Nisbett, N., Pretty, J., Robinson, S., Toulmoin, C. and Whiteley, R. (2010) The future of the global food system. *Philosophical Transactions of the Royal Society B* 365, 2769–2777.

Guerrero-Legarreta, I. and Hui, Y.H. (2010) Processed poultry products: a primer. *Handbook of Poultry Science and Technology*, Vol. 2: *Secondary Processing*. John Wiley & Sons, Hoboken, New Jersey, pp. 3–10.

Harper, G. and Henson, S. (2001) Consumer concerns about animal welfare and the impact on food choice. EU FAIR CT98-3678. *Centre for Food Economics Research, Department of Agricultural and Food Economics* 1, 38.

Higenyi, J., Kabasa, J.D. and Muyanja, C. (2014) Social and quality attributes influencing consumption of native poultry in eastern Uganda. *Animal and Veterinary Science* 2(2), 42–48.

Horne, P.L.M., van (2009) *Production Cost of Poultry Meat: An International Comparison*. LEI Report 2009-004. The Hague, March 2009 (in Dutch).

Horne, P.L.M., van and Achterbosch, T.J. (2008) Animal welfare in poultry production systems: impact of EU standards on world trade. *World's Poultry Science Journal* 64, 40–52.

Magdelaine, P., Spiess, M.P. and Valceschini, E. (2008) Poultry meat consumption trends in Europe. *World Poultry Science* 64(1), 53–64.

Mahiuddin, M., Khanum, H., Wadud, M.A., Howlider, M.A.R. and Hai, M.A. (2008) Consumer attitude towards poultry meat and eggs in Muktagacha powroshava of Mymensingh district. *Journal of Agroforestry and Environment* 2(2), 159–164.

Mattioli, D., Cimilluca, D. and Kesmodel, D. (2013) China makes biggest U.S. play: Asian meat giant strikes $4.7 billion deal for Virginia's Smithfield Foods. *The Wall Street Journal* 30 May 2013.

Mitchell, D. (2008) A note on rising food prices. *The World Bank Development Prospects Group: Policy Research Working Paper* 4682, 1–20.

Nelson, R. (2005) *Analysis of US Poultry Meat Trade with the EU: Past, Present, Future*. Report E35166 USDA Foreign Agricultural Service.

OECD/Food and Agriculture Organization of the United Nations (2014) OECD-FAO Agricultural Outlook 2014, OECD Publishing. Available at: http://dx.doi.org/10.1787/agr_outlook-2014-en (accessed October 2015).

Oliveira, C.A., Corte, V.F.D., Finger, M.I.F. and Waquil, P.D. (2012) Developments of the Brazilian chicken meat industry in international trade. *International Association of Agricultural Economists 2012 Conference*, Foz do Iguaçu, Brazil, 17349.

Rabobank (2014) Annual Report 2014: Rabobank Group. Available at: https://www.rabobank. com/en/images/annual-report-2014-rabobank-group.pdf (accessed October 2015).

Rabobank Analysis (2009) Report 2009: Rabobank Group. Available at: https://www.rabobank. com/en/images/UK_Report_2009_RG_tcm43-136963.pdf (accessed October 2015).

Rabobank Analysis (2013) Sustainability Report 2013: Rabobank Group. Available at: https:// www.rabobank.com/en/images/Sustainability_Report-2013-Rabobank-Group-_.pdf (accessed October 2015).

Ray, D.K., Mueller, N.D., West, P.C. and Foley, J.A. (2013) Yield trends are insufficient to double global cop production by 2050. *PLoS One* 8(6), e66428.

Salim, H.M., Kruk, Z.A. and Lee, B.D. (2010) Nutritive value of corn distillers dried grains with solubles as an ingredient of poultry diets: a review. *World's Poultry Science Journal* 66, 411–431.

Shane, S.M. (2003) The US Poultry Industry 2003 in review. *World Poultry* 19, 12.

Shikida, P.F.A., Finco, A., Cardoso, B.F., Galante, V.A., Rahmeier, D., Bentivoglio, D. and Rasetti, M. (2014) A comparison between ethanol and biodiesel production: The Brazilian and European experiences. *Liquid Biofuels: Emergence, Development and Prospects, Lecture Notes in Energy* 27, DOI: 10.1007/978-1-4471-6482-1_2.

Sluis, W., van der (2006) Brazil will continue to be a major poultry producer and exporter. *World Poultry* 22, 5.

Spillius, A. (2013) China 'to rent five per cent of Ukraine'. *The Telegraph* 24 September 2013.

Sumner, D.A., Gow, H., Hayes, D., Matthews, W., Norwood, B., Rosen-Molina, J.T. and Thurman, W. (2011) Economic and market issues on the sustainability of egg production in the United States: analysis of alternative production systems. *Poultry Science* 90(1), 241–250.

Topliff, M., de Roest, K., Roguet, C., Chotteau, P., Mottet, A., Sarzeaud, P., Beaumond, N., Magdelaine, P., Hoste, R., van Horne, P. and Deblitz, C. (2009) The impact of increased operating costs on meat livestock in the EU. *European Parliament Directorate-General for Internal Policies, Policy Department Structural and Cohesion Policies B* 1–260.

USDA (2013) United States Department of Agriculture Economic Research Service: Recent Convergence Performance of Futures and Cash Prices for Corn, Soybeans, and Wheat. Available at: http://www.ers.usda.gov/publications/fds-feed-outlook/fds13l-01.aspx (accessed October 2015).

USDA (2014) United States Department of Agriculture Economic Research Service: Agricultural Act of 2014: Highlights and Implications. Available at: http://www.ers.usda.gov/agricultural-act-of-2014-highlights-and-implications.aspx (accessed October 2015).

Windhorst, H.-W. (2005) Development of organic egg production and marketing in the EU. *World's Poultry Science Journal* 61, 451–462.

Winkler, H., Davidson, O., Kenny, A., Prasad, G., Nkomo, J., Sparks, D., Howells, M. and Alfstad, T. (2006) *Energy Policies for Sustainable Development in South Africa*. Energy Research Centre, University of Cape Town, South Africa, April 2006.

Zhang, J. (2011) China's success in increasing per capita food production. *Journal of Experimental Botany* 62(11), 3707–3711.

Chapter 4
Industry Challenges Surrounding Sustainability

Steve Ellis[1]* and Richard Kempsey[2]

[1]2 Sisters Food Group, Birmingham, UK; [2]Stonegate Ltd, Wolverhampton, UK

INTRODUCTION

Advancing science in any field requires accurate and detailed knowledge of the setting in which the science will be applied. Therefore two leaders representing the egg and meat sectors of the poultry industry were invited by the Symposium Scientific Committee to present their views on how sustainability impacts on their businesses, and also to highlight their key challenges and priorities.

Steve Ellis is Managing Director of 2 Sisters' UK Poultry Division. He joined 2 Sisters in 2011 to lead the integration of the Northern Foods acquisition. Steve was previously at Molson Coors where he was a member of the UK Board as Strategy Director and Sales Director. Following the acquisition in 2013 of Vion he has led and integrated the combined UK poultry business, responsible for 13 factories and 8000 colleagues and processing approximately 6 million birds per week. 2 Sisters Food Group is one of British businesses' most compelling success stories of the past 20 years. 2 Sisters is also one of UK agriculture's largest customers through wheat consumption, primarily for animal feed but also for products such as biscuits and pizzas.

Richard Kempsey is Technical Director of Stonegate, which packs and delivers eggs from over 250 producers to supermarkets. In 1926 a small group of farmers from the village of Stonegate in East Sussex collaborated to form an egg cooperative. The farmers pooled all the eggs they produced together and packed them, they then took the finished product to markets to sell. This was the beginning of Stonegate. Following the merger between Horizon Farms and Stonegate in 2000 and major acquisitions, including Thames Valley Foods in 2001, the modern day Stonegate evolved, now working closely and supportively with at least 250 egg producers to collect, grade and package the most fragile of food

*Corresponding author: stephen.p.ellis@btinternet.com

products. Richard also farms 22,000 free-range hens (Columbian Blacktail and Clarence Court) on his own farm in Shropshire.

SUSTAINABILITY IN THE POULTRY BROILER SECTOR

This section focuses on how the 2 Sisters Food Group UK approaches sustainability in the UK poultry industry and highlights their key challenges and priorities.

What do we mean by sustainability?

It is important to consider sustainability from three different and sometimes competing perspectives.

1. Economic sustainability: being able to supply a product demanded by customers and consumers at a price that covers cost and delivers a margin to sustain investment.
2. Ethical sustainability: ensuring adherence to animal welfare standards, providing a safe and engaging workplace for our colleagues and ensuring we consider impacts on the wider population.
3. Environmental sustainability: considering how we feed a growing population with scarce resources and minimize the environmental impact of production.

It is a critical time for the UK poultry industry at the moment and choices made in the next 12–18 months will have a bearing on the long-term sustainability of the industry. Poultry has some really positive fundamentals when it comes to being sustainable but there are some big challenges ahead; for example, the move to British sourcing for almost all the major retailers has brought with it some short-term challenges for the UK processors. It is imperative to focus on a balance of all three of these key areas of sustainability in order to provide long-term success.

Economic sustainability

Poultry consumption continues to grow in the UK driven by positive demand fundamentals. Poultry has many advantages over other meats, such as being low fat, low cost, tasty and versatile and lacks any religious obstructions (Magdelaine *et al.*, 2008). Poultry production has as a result improved; in 2014 the UK produced over 900 million broilers a year, producing approximately 1,400,000t of chicken meat (DEFRA, 2014). Poultry demand in the UK in terms of grams per person per week overtook red meat for the first time since 1974 in 2012 (DEFRA, 2012).

    One of the key drivers of demand for poultry is the value it can offer versus other meats. For example, based on the price of chicken and beef in a Tesco store, a family could have three roast chicken dinners for the price of one beef dinner. Based on Tesco figures from September 2014, a whole medium chicken costs approximately £3.50/kg and chicken breast portions cost approximately £7.01/kg compared to a beef medium joint and beef medallion steak, which cost approximately £11/kg and £12.50/kg, respectively, a pork shoulder joint, which costs approximately £6.99/kg and salmon fillets, which cost approximately £10.67/kg.

The average UK household now spends approximately £360 on meat, fish and poultry each year (DEFRA, 2012). As illustrated in Fig. 4.1, today's shoppers are very price-conscious and are consistently searching for the best value ingredients to feed the family. Figure 4.1 looks at percentage change in volume versus percentage change in average price and shows that there is almost a direct correlation between the price of the product and the volume consumed. As the price of poultry rises the volume declines, and as the price drops the volume increases; approximately 1% increase in price directly results in 1% decrease in demand as consumers look for either cheaper protein or a cheaper meal solution. Growth in the UK poultry sector is set to slow over the next decade, but the product will still dominate the meat industry, largely due to the value it can offer consumers versus other meats.

These data also illustrate the relationship between volume consumed and wheat and soybean prices; when wheat and soybean prices increased in 2013 driving an increase in retail price consumers actively sought out cheaper meals (Garnett *et al.*, 2014) by purchasing more frozen chicken imported from Brazil, chilled pizzas, frozen ready meals and sausages. This highlights that there is significant cross-protein switching based on price. Price and impact on consumers must therefore be considered when choices are made in the poultry industry around the environmental and economic areas.

Another key economic driver in the UK poultry industry is the huge demand for British produce. The 'horsegate' scandal led to demand for transparency right through the food supply chain and heightened public interest into understanding where their food has come from. As a result, following the horsemeat scandal,

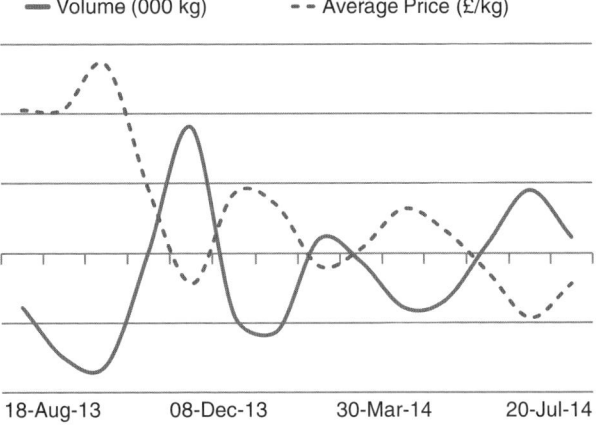

Fig. 4.1. Relationship between poultry meat consumption and price between August 2013 and July 2014 (Source KANTAR Primal Poultry).

UK supermarkets decided to move their fresh and primal poultry (raw chicken, either a whole carcass or portions of a whole carcass, i.e. the first cutting after slaughter) into British origin (Ryan, 2013). Some supermarkets went even further than this and also moved their ingredients, ready meals and added-value chicken to British origin. The result of this was a significant rise in the number of birds being slaughtered in the UK: from 856 million in 2011 to 900 million birds slaughtered per year currently (DEFRA, 2014). Whilst the switch to British poultry sourcing from the majority of UK retailers is positive for the long term, the industry is dealing with some short-term challenges that are a consequence of the speed of this switch. Britain predominantly consumes the white-meat or breast of the chicken (Baracho *et al.*, 2006), which means that 30–40% of the bird is left after satisfying the retail demand. Prior to the move to British-only sourcing, the processors could balance the carcass through importing white meat from the EU. The industry now has to grow the 'whole bird equivalents' of the demand for white meat and trade the imbalance to other channels. The majority of the imbalance is exported to world commodity markets. The revenue per bird achieved is a summation of all of the sales to retail plus the revenue achieved through these other channels. There have been several factors that have reduced the revenue per bird sold. First, the increased carcass imbalance has led to increased volumes traded. When supply increases to commodity markets pricing decreases. Second, the price commanded on world commodity markets is already lower than supply to UK retail. The cost of production in the UK is higher due to the cost of feed, more EU regulation and higher welfare standards than in other countries (van Horne and Bondt, 2013). Therefore, selling more to export markets dilutes the average price. Last, this has been exacerbated due to trade restrictions from South Africa and the ban on EU poultry imports to Russia imposed in reaction to EU sanctions, both of which are key markets of European dark meat exports. Therefore the immediate impact of an increase in demand for British chicken has been to reduce the price that the producer receives for the whole chicken. They may receive a higher price from the retailer for British versus EU sourced meat, however, the price received for the whole bird equivalent has reduced.

Environmental sustainability

Against a backdrop of population increase and scarce resources, there are some fundamental positive drivers around environmental sustainability for poultry versus other sources of protein.

The increase in world population, predicted to reach 9 billion in 2050, brings a huge demand for increased food production with scarce resources in terms of land and water usage. Food production is the main driver of biodiversity loss; the Nature Climate Change Report 2014 states that if current trends continue, food production alone will reach, if not exceed, the global targets for total greenhouse gas (GHG) emissions in 2050. GHG from food production will increase by 80% if meat and dairy consumption continues to rise at its current rate (Harrabin, 2014).

To combat this it is suggested that people should limit their meat consumption and have a balanced diet across different protein sources. It is advised that people should have just two 85 g portions of red meat and five eggs per week as well as a portion of poultry a day (Mason, 2014).

Chicken production continues to improve efficiency with year on year improvements in feed conversion rates through advances in breeder selection and feeding efficiency. This has delivered approximately 1 day's extra growth per year. Chicken is also very efficient in terms of land-use and poultry farms can be run using biomass generators, solar and wind power.

Therefore, versus other proteins, chicken is efficient to produce. More work is required to consider other sources of protein in feed to limit demand for soya production.

Ethical sustainability

The three key areas driving ethical sustainability are animal welfare, the impact on human health and the welfare of the people who work in the supply chain.

The basis of animal welfare are the 'five freedoms': freedom from hunger and thirst; freedom from discomfort; freedom from pain, injury or disease; freedom to express normal behaviour; and freedom from fear and distress. The science of measuring animal welfare is constantly evolving and focuses on delivery of key outcome measures: mortality, antibiotic usage, podermatitis, hock marking, birds dead on arrival and birds rejected by factory meat inspectors. A measure of natural behaviour is still required to complete this set of outcomes and 2 Sisters is working with the Food Animal Initiative to pioneer research in this area.

Higher welfare practices can include decreasing stocking density, providing free range, and use of slow-growing breeds. For example, in the Netherlands chicken breeds with slower growth are now being selected over faster growing breeds (Bokkers and de Boer, 2009).

The poultry sector is making every effort to continually improve standards of animal welfare and it has become increasingly important to convey this message to the public following the horsemeat scandal and other health scares. As an example, 2 Sisters has fitted windows to all the poultry sheds and provides an enriched environment for all their birds even if the customer requirements are lower. Not only is this good for the chickens but it also provides a better working environment for our people.

This is just one example of how 2 Sisters is constantly reviewing ways to improve working environments. Reducing accidents and improving employee engagement are two of the key measures for the group.

In terms of human interaction the key focus is on the UK Food Standards Agency (FSA) *Campylobacter* targets for 2015, but also includes reducing antibiotic usage and managing risk of an avian influenza outbreak.

The FSA have focused on *Campylobacter* reduction as a key target, by conducting retail audits of *Campylobacter* levels by retailer, which they intend to publish. This is at the forefront of the challenges to the UK poultry industry, and there is a huge amount of work being done to deliver a solution to *Campylobacter*.

Campylobacter is considered to be responsible for more than 280,000 cases of food poisoning each year in the UK with more than 72,000 of these confirmed to be *Campylobacter* poisoning. Four in five of these cases come from contaminated poultry. *Campylobacter* is thought to cause more than 100 deaths a year in the UK and costs the UK economy about £900 million (FSA, 2014).

Whilst *Campylobacter* can be easily killed by proper handling and cooking, the industry has worked together to find a solution to eradicating *Campylobacter* before it reaches the home. The different interventions focus on work at farm, in factory and at home. 2 Sisters is investing £10m from farm to fork to find interventions that could eradicate *Campylobacter* in chicken.

The farm initiatives include farmer training, increased biosecurity practices and 'no thinning'. Thinning can be practised within the limits of EU welfare laws, which dictate a maximum of 30 kg meat production/m^2 in buildings with natural ventilation systems and 38 kg of meat produced/m^2 in buildings with artificial ventilation systems (Sani and Gobbi, 2006). The 2 Sisters Food Group takes 25% of the birds out at about 32 days and then grow the rest of the birds on to 38 kg/m^2. At 2 Sisters Food Group, £6m is being spent on a 12-month no-thinning trial across one integration, starting in October 2014. Background base data were collected for the 12 months prior to the start of the trial to provide an objective comparison. All the broiler chickens in one factory will be part of this trial. We believe it is important to run the trial at this scale to truly understand the impact on *Campylobacter* levels.

The key interventions in the factory that are being trialled are focused on either heat (e.g. Faccenda – Sonos Steam) or cold (2 Sisters Food Group – Blast Surface Chilling). By providing a secondary treatment of heat or cold, it is hoped that any *Campylobacter* will be killed off.

At home, 'Roast in the Bag' packaging has been rolled out and labelling has been improved to promote correct handling and 'no need to wash' messaging.

The use of antibiotics critical to human health is another key factor in human interaction and a key focus for the World Health Organization (WHO). At 2 Sisters we are committed to ensure that we use all medicines, including antibiotics, in a responsible manner across the farms that supply us, and lead the industry in delivering a strategy that protects both animal and human interests, now and in the future. Foremost in our agenda is removing antibiotics, which the WHO identified as the highest priority critically important to human medicine.

It is important to note that many of the actions taken under ethical sustainability can come into conflict with both economic and environmental sustainability drivers as they reduce the amount of chicken produced per square metre, requiring a significantly higher retail price and reducing demand as shown in Fig. 4.2. In Fig. 4.2(a), reducing stocking density reduces the weight off the farm for the same cycle length and same fixed costs, equal to the dark blue triangle, increasing average price; in Fig 4.2(b), removing the practice of thinning means few chicks are placed at the start of the cycle, reducing weight off the farm for the same cycle length and same fixed costs, equal to the light blue triangle, increasing average price; and in Fig 4.2(c), choosing slower growing breeds increases the cycle time and therefore higher fixed costs to achieve the same weight off the farm, therefore increasing average price.

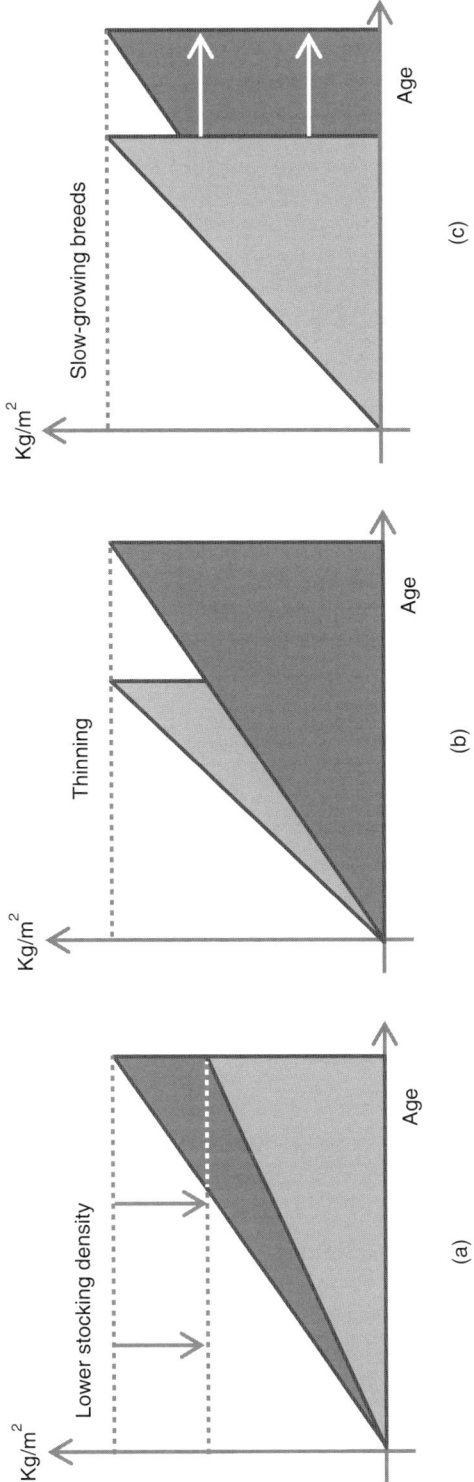

Fig. 4.2. Impact of ethical options for chicken rearing by (a) reducing stocking density, (b) removing the practice of thinning and (c) choosing slower growing breeds.

The ethical choices we make have an impact on both environmental and economic sustainability. If we are to feed a growing population with high quality and great value food we need to balance the choices we make. It is vital that we generate improved understanding around the science behind welfare of chickens. 2 Sisters is pioneering behavioural measurement by trialling cameras on its farms to monitor chicken behaviour throughout the day to truly understand impacts on natural behaviour. The focus must be on delivering what will make the biggest beneficial impact to the chicken, rather than perceived benefit. For example, putting perches in broiler farms has very limited benefit. We have found that bales for perching are much more likely to be used by the chickens. Our objective is to improve the welfare of chickens, reduce *Campylobacter*, remove critical antibiotics and do so in an economically sustainable way.

IMPORTANT LESSONS TO LEARN FROM THE PORK INDUSTRY

The pork industry provides some important lessons for the poultry industry in how the ethical choices impact economic sustainability. In 1999 the UK banned sows' stalls and imposed higher welfare standards on British pork farmers than the rest of Europe. Changing British welfare standards independent of import requirements reduced the competitiveness of UK agriculture and caused a 40% decrease in the UK pig herd over the next few years (Houston, 2012). Pork consumption, ham consumption and bacon consumption remained at the same levels. Therefore, the impact of introducing higher welfare measures has technically exported the welfare issue. A group of cross-party MPs who reviewed the legislation found that approximately two-thirds of imported pig meat could have been reared in conditions banned in the UK.

The key issue is that a sustainable chicken industry requires to be competitive in all areas of the market: the fresh retail market; ingredients; supply to foodservice and supply to manufacturing processors (e.g. ready meals). A key element of economic sustainability is being able to compete effectively across all the sectors to provide good flock utilization and carcass balance. Increasing the welfare required in one sector independent of import restrictions will significantly challenge the economic sustainability of the industry.

DELIVERING SUSTAINABILITY

The challenge facing the poultry industry is find the 'sweet spot' of the right sustainability actions across ethical, environmental and economic sustainability perspectives.

2 Sisters has taken decisive steps to reduce the level of *Campylobacter* in chicken. The £10m investment in a range of initiatives from farm to in-home is industry leading. The objective is to eradicate *Campylobacter* at levels where colony forming units of *Campylobacter* per gram exceed 1000 cfu/g (Koolman et al., 2014). We believe that, with the substantial investment we are making, we will find interventions or a combination of interventions to deliver this objective.

2 Sisters is also committed to the responsible usage of antibiotics. We will look to replace antibiotics with other interventions wherever possible. This includes use of vaccinations and changes to husbandry and biosecurity techniques. Our strategy is continually refined through the use of trial locations to see what works best. Our first commitment is to preserve the effectiveness of antibiotics critical to human health by removing all antibiotics defined by the WHO as being of highest priority.

There are several different welfare standards of chicken grown in the UK from organic, free range, higher welfare indoor (lower stocking density and/or slower growing breed), British Red Tractor and British Red Tractor with added enrichment. 2 Sisters has moved all production of standard chicken to the higher British Red Tractor with enrichment standard. The vast majority of all the 17.8 million chickens grown every week in the UK are grown at a British Red Tractor standard (with or without added enrichment). The British consumer is very price conscious and the extra cost driven by the higher welfare standards does prove to be a barrier to purchase, as consumers are very willing to switch between protein categories to whichever one offers best value. Our focus at 2 Sisters is to improve the welfare of the majority of chickens we grow. We will measure this improvement through the standard outcome measures but are also pioneering a measure of natural bird behaviour. We can then focus on farming interventions to ensure that the actions we take actually make a difference to the chickens we grow.

One of the key ways we can improve the economic sustainability of all the welfare standards we grow is to improve flock utilization and carcass balance. Selling 100% of the chickens we grow in the UK to UK customers will improve the revenue per bird received for the processors. Conversely, this will mean that the cost price to UK customers can actually be reduced – and will stimulate more demand. This will be good for the UK poultry industry, good for UK customers and consumers and good for the chickens as the UK welfare standards are higher than elsewhere in the world. Whilst 100% carcass balance and flock utilization may be unachievable, there is huge scope to find ways to improve the current position dramatically. Without an improved carcass balance, increasing welfare standards in the UK independent of EU or world standards and import controls could have huge implications for the industry, as the demise of the UK pork industry has shown.

Chicken is an environmentally efficient protein to grow. This can be further improved through increased use of renewable energy. At 2 Sisters Food Group the target is to reduce overall energy use by 25% and become more efficient by replacing conventional energy usage with renewable energy. Reduction in water usage and waste are also key targets to continue to reduce environmental impact across the poultry supply chain from breeders to processing factory.

CONCLUSIONS

There are three key areas of sustainability: ethical, economic and environmental. To deliver long-term sustainability it is important to find the right balance between the three different perspectives on sustainability and find the 'sweet spot' between the three. The key ethical priorities for the industry are to reduce the incidences of *Campylobacter*, replace antibiotics deemed critical to human health and develop a behavioural outcome measure so we can better understand how to improve the welfare of the majority of chickens grown in the UK. For environmental sustainability there needs to be increased use of renewable energy and improved delivery of an efficient total supply chain from feed mill to processing plant. To achieve economic sustainability the UK should aim to sell 100% of the chicken it grows in the UK to the UK. Applying the same standards expected on UK chicken to imports will ensure we do not follow the experience of the pork industry.

A PERSONAL VIEW OF SUSTAINABILITY IN THE POULTRY EGG INDUSTRY

As a young scientist starting at university I was enthralled by Rachel Carson's book *Silent Spring* (Carson, 2002). Acclaimed as the catalyst for the modern environmental movements, *Silent Spring* roundly condemned the overuse of pesticides claiming that between 1950 and 1962 the amount of DDT found in human tissue had tripled. Rachel's book was hugely contentious toward the chemical industry and Dr Robert White-Stevens, a former biochemist and spokesman for that industry during the 1960s, told the public, 'If man were to follow the teachings of Miss Carson, we would return to the Dark Ages, and the insects and diseases and vermin would once again inherit the earth' (Stein, 2013). But John F. Kennedy initiated a review of pesticide use at the US Science Advisory Committee and DDT was subsequently banned and the formation of new bodies such as the Environmental Protection Agency followed (EPA Order 1110.2, 1970). The debate about sustainability and not just productivity in agricultural products developed its own roots during this same period.

My own career in egg production now spans 30 years and during that period I have been part of just as dramatic a change in egg production systems as was witnessed in pesticide use in the 1970s. Driven by the strong European welfare and ethical bodies, which themselves had grown enormously in funding and support since the 1970s, there was a rapidly growing demand from the UK consumer for free-range eggs. From being just 0.5% of the UK market in 1987 free-range production has grown at a very rapid rate reaching 46% of the UK market by 2013 (Egginfo, 2014). This was mirrored almost exactly by similar trends in northern EU states, but the southern EU states lagged behind and still do.

The same welfare organization's lobbying led to the Laying Hens Directive (EU, 1999), which outlawed the use in the EU of battery-cage production. New colony-cage production, with lower stocking density and enrichment of the hen's environment, fully replaced cage production by 2013 at a capital cost for the UK layer sector alone of over £320 million (Driver, 2012).

So consumers have voted with their feet with their purchasing decisions on eggs but this is only a partial assistance in helping to clear up any confusion over what constitutes a sustainable egg, for what is best for the environment may not be best for the bird or for people.

Let us set the scene with the global challenges first.

The global challenge of feeding that 9 billion 2050 world population with all the egg and poultry product is required without using a single extra hectare of land for feed raw materials and without using a single extra gallon of water is huge.

Food insecurity is the first of the global challenges: 925 million people – almost one in seven of the world's population – currently go to bed hungry (FAO, 2012). The challenge of feeding a growing population while lifting millions out of poverty is daunting. Yet if sustainable farming practices are adopted, agriculture can continue to provide critical ecosystem services, such as water regulation and carbon controls, while still producing higher yields of food. This has shown to be possible in developing nations. Niger, for example, has witnessed a farmer-led 're-greening' movement that has reversed desertification and brought increased crop production, income, food security and self-reliance to impoverished, rural food producers. This transformation has been driven by the practice of agroforestry, which integrates trees into food-crop landscapes to maintain a green cover year-round, improving soil quality, erosion control and carbon sequestration.

The slower world population growth rate compared with the last two decades and the rising economic growth in less developed states will also help to address the food insecurity challenge as we go forward to 2050.

Food price volatility, particularly of raw material input prices, has been a major challenge on EU and global markets over the last decade as limited resources, increased trading on commodities markets and weather effects have greatly increased price volatility. Whilst all food price volatility will remain a major challenge to all poultry producers going forward, some global initiatives have had some success in mitigating some of the excesses of the commodities speculators.

Improved global governance on commodities markets has played an important role in warding off additional food price spikes since July 2012 and the Agricultural Market Information System (AMIS, 2012) created by the G20 in 2011 has proved an effective new weapon in the arsenal to fight against excessive price volatility, providing up to date, reliable information and increasing transparency in the international food markets.

Collaborative private approaches on a global scale are also having an impact. The Consumer Goods Forum, a voluntary grouping of 400 retailers from 70 countries, has agreed to take deforestation out of their supply chains by 2020 for four commodities (CGF, 2014).

However, we are still left, as an industry, with major challenges in this area. Further action is needed at all levels to hedge risk, secure adequate supply chains and to research and secure alternative protein raw materials. Crucially this will require cross industry and government working if global markets in plant proteins for example become challenged.

Having presented the global challenges we can now start with outlining what our objectives for sustainability in poultry production are.

EU poultry production systems need to fully meet the needs of the present consumer whilst improving the ability of future generations to meet their own needs by:

- continually increasing productivity to help to meet future food demands;
- decreasing all impacts on the environment;
- helping to improve human health;
- improving poultry health and welfare;
- measuring outcomes; and
- improving the social and economic well-being of consumers and those involved in the supply chain.

DRIVE CONTINUAL IMPROVEMENT IN PRODUCTION EFFICIENCY

The first of our poultry objectives of increased productivity has been easily met by our well integrated industry over the last 50 years and we have no reason to believe that the same level of productivity increase illustrated by the US Egg Industry Center data below is not achievable for the forthcoming period to 2050.

The US Egg Industry Center quantified historic productivity increases in US egg production by reviewing data from 97.2 million hens for the period 1960–2010 (Xin *et al*., 2013). The results were impressive:

1. Today's hens use a little over half the amount of feed to produce a dozen eggs and have 42% better feed conversion efficiency.

2. At the same time, today's hens produce 58% more eggs over the laying period and are living longer.

3. The egg production process now releases significantly less polluting emissions, including 71% lower GHG emissions.

4. Today's hens use 32% less water per dozen eggs produced.

IMPROVING POULTRY HEALTH AND WELFARE

The EU poultry industry has an ongoing challenge in the area of poultry health and welfare. The industry is engaged in deep ethical, social and welfare debates on poultry husbandry systems.

For the egg industry the debate has been structured largely round 'cage versus free-range production' and 'organic production versus conventional production'.

In many of the debates over agriculture and food, one might be led to believe that there are only two sides – those who support organic farming, and those who support conventional farming – with no common ground between them. When you dig down a little deeper the reality is more complicated than might be expected, for a lot of 'organic' produce is grown using conventional farming techniques and a lot of 'conventional' produce benefits from practices developed by organic farmers.

In the consumer's eye the perceived top of the poultry production husbandry pyramid is probably pastured poultry production. This is an extensive organic system where the poultry follow grazing livestock round the farm. This system has been promoted by Joel Salatin, an evangelical promoter in the USA of simple pasture-based systems of poultry production. The title of his book *A Farmer's Advice for Happier Hens, Healthier People and a Better World* would lead one to believe that he has answers to many of the sustainability challenges (Salatin, 2011).

Examination of the US pasture-based poultry systems based on Joel Salatin's ideas are indeed interesting as they help to shape the debate. A land-based poultry rearing system undoubtedly ticks a lot of the boxes as it can be run sustainably in conjunction with other livestock production and contributes fertility directly back to the grazing land. But is it welfare friendly? As the birds are reared in open shelters they are exposed to the stresses and strains of inclement weather and invariably in the USA they remain a seasonal enterprise when grazing conditions are right. Undoubtedly these systems in the USA command support from the local markets that they supply. The USDA (US Department of Agriculture) Sustainable Agriculture Network report *Profitable Poultry: Raising birds on pasture* states that small scale, typically with 2000–3000 broilers raised per season, pasture-based poultry enterprises have net returns of 50% of the sale price of each bird (SARE, 2012). At the time of this report (2014) these pastured birds were sold dressed to the public at eight times the price for conventionally raised chickens and Joel Salatin states his market for his eggs and chickens is 'families, restaurants and a smattering of retail boutiques'.

Looking now at the egg sector, poultry scientist Professor Christine Nicol sparked widespread controversy in November 2013 when an interview with her at a symposium resulted in UK national newspaper reports claiming that cage hens enjoyed better welfare than free-range birds (McDermott, 2013; Silverman, 2013).

The School of Veterinary Science at Bristol in a 2010 report stated that stress and mortality levels are lower in hens raised in 'enriched cages'. Plus, they are less likely to suffer from bone fractures or pecking than free-range chickens (Sherwin *et al.*, 2010).

The British Free Range Egg Producers Association *Ranger* magazine (Farming UK, 2013) reported Professor Nicol as saying:

> I am not convinced that the ability to do natural behaviour offsets the health and injury risks that free-range hens sustain. I don't think consumers have the slightest idea about the scale of these problems or the actual levels of mortality, fractures, pecking and disease that we find. I think consumers expect free-range birds to be in good physical condition as well as having options to do a range of behaviours. [. . .] My own view is that free-range systems still offer the best potential for good welfare if these problems can be solved. I'm not saying it's great, there's still a lot of room for improvement but the birds have space, they've got a little perch, they've got things they can scratch on.

In the same *Ranger* article Compassion in World Farming (CIWF) said:

> The welfare potential of these cages will never be as high as free-range systems. Compassion believes that laying hens should be able to perform all of their natural behaviours, including stretching and flapping their wings, perching up high, foraging, scratching, dust bathing and laying their eggs in a comfortable nest. Only free-range and organic systems can provide fully for all of these behaviours. The challenge for the industry is realizing the potential of the free-range system . . . so that they actually do what consumers think they do, which is provide all hens with good welfare.

There is a wide range of views expressed in these debates and a relative dearth of reviewed metrics and science on health and welfare for the wide range of systems used by EU poultry producers. Given these limitations it would be wise for all parties to enter into an informed debate on the issues raised and to continue to undertake independently verified research to verify metrics and claims about systems.

One of the key strengths of the EU poultry supply chain is its integrated structure with quick and effective communication between all parts of the chain. In this environment the chain can react swiftly and successfully to changing consumer purchasing patterns.

However, these same chains are not easily visible to the modern consumer and the vast majority of EU consumers have no first-hand experience of the wide-ranging husbandry systems employed by the poultry industry nor have they been necessarily taught about poultry production at school level or even been given a chance to debate the issues.

ENVIRONMENTAL FOOTPRINT

A carbon footprint and life cycle assessment (LCA) is used to systematically record and analyse the impact on the environment throughout the entire life cycle of a product or service. This involves an end-to-end analysis of the product or service. The analysis considers all raw materials, transport, production processes, usage and disposal of the product. Standards for LCA that are followed in Europe include PAS 2050 and ISO 14067 for GHG.

LCAs should help us to understand and manage the supply chain for GHG emissions but the complexity of the food chain makes this a very difficult process in practice and there is huge debate over appropriate cut-off points. For example, how should LCA be allocated to different cuts of chicken sold to the consumer? The legs and breast could be assessed separately and this could be done by weight or value, which would greatly influence the result.

LCAs are also subject to political manoeuvring. Take for example the importation of South American soybean meal for use in poultry rations in the EU. Straight examination of LCAs for Brazilian soybean and EU-produced protein legumes (peas and beans) to PAS 2050 standards would give comparable results in carbon emissions. However, according to FAO data, 45.6% of GHG emissions in Brazil were estimated to come from direct land-use change and forestry in 2010 (Lindquist *et al.*, 2012). These direct land use changes (dLUCs) are presently not considered under PAS 2050 and the whole area of land use change apportionment is political.

The effects of biodiesel production in Brazil have been demonstrated to have potentially very large indirect land use changes. Lapola *et al*. (2010) demonstrated that sugarcane ethanol and soybean biodiesel each contribute to nearly half of the projected indirect deforestation of $121,970 \, km^2$ by 2020, creating a carbon debt that would take about 250 years to be repaid using these biofuels instead of fossil fuels.

Guidance from legislators and politicians is urgently needed in this area as incorporation of dLUC figures, or more politically sensitive indirect land use figures (iLUC) does materially affect comparison of LCA figures for different raw materials, particularly for egg and poultry producers who rely on high protein soymeal for cost-effective rations.

Whilst we wait for political guidance on LCA standards our best available LCA measurement tool, particularly for free-range egg laying and outdoor poultry production, is provided by the relatively simple online Cool Farm Institute's LCA measurement tool that can be completed directly by the grower.

The Cool Farm Institute's mission is to enable millions of growers globally to make more informed on-farm decisions that reduce their environmental impact. The Institute provides the online Cool Farm Tool (CFT) as a quantified decision support tool that is credible and standardized (Coolfarm, 2014).

The Institute is supported by global food businesses including PepsiCo, Unilever, Heineken, Marks & Spencer, Tesco, Yara and Fertilizers Europe. It was originally developed by Unilever and researchers at the University of Aberdeen with the aim of increasing growers' understanding of the carbon footprint of their production systems and helping them to adapt management to improve their performance. The tool identifies hotspots and makes it easy for farmers to test alternative management scenarios and identifies those that will have a positive impact on the total net GHG emissions. Unlike many other agricultural GHG calculators, the CFT includes calculations of soil carbon sequestration, which is a key feature of agriculture that has both mitigation and adaptation benefits.

The Cool Farm Institute's vision is to be a highly credible and capable partner for agricultural GHG management – 'credible' through using best available science and multi-stakeholder processes for methodology development and quality assurance, and 'capable' through providing leading agricultural GHG management products and services.

GENETIC ADVANCES

Laying hens have for many years been selected for feed efficiency and egg production and substantial response has been achieved as is demonstrated by the production figures above. However, newly developed tools, such as high-density SNP chips, which allow animals to be genotyped for tens of thousands of genetic markers across the genome, can provide additional information on the genetic basis of efficiency and thus enhance the selection progress. For laying hens a

SNP chip which can measure 64,000 markers is currently available from the USA for around US$70.

The new technology has allowed global breeding companies to analyse feed efficiency and egg productivity data from multiple generations of their layer lines with pedigree and marker-based methods. Marker information improves the accuracy of estimated breeding values (EBV), especially for long-term prediction. Confidence in the new technology has given one global breeding company the confidence to state that within 5 years their hybrid layers will be able to produce 500 eggs by the time the hens reach 100 weeks of age. These advances come from both genetic productivity gains and increases in the length of time that the hen can be kept in lay.

However, the rapid and sustained genetic improvement in laying hen hybrids of over 1% per annum for the last 50 years brings with it many challenges for the industry.

Before the recent marker technology introduction, laying breeding was limited to normally five or six selection criteria, which were almost always based on productivity factors such as egg weight and feed conversion. Up until now breeder companies have not had the resources to continually examine in detail the environmental and welfare impacts of their new crosses. With most of the additional genetic improvement now coming from lengthening the period that the hen is in lay, breeder companies need to use their new-found marker technologies to successfully address for instance bone strength and calcium metabolism to mitigate any welfare effects of the extra production gained by the hen.

The layer breeder industry has consolidated rapidly over the last 15 years into a few global companies that are able to afford the very large investments required to incorporate genetic marker-based methods into their breeding programmes.

The USDA Field to Market collaborative group has developed a supply chain system for agricultural sustainability (USDA, 2014a). They have developed outcomes-based metrics to measure environmental, health and socio-economic impacts of agricultural production. Starting with crop production they have a 'spider's web' created using one axis for each of five efficiency indicators: land use, energy use, soil loss, irrigation water use and climate impact. This gives a good, visual, outcome-based metric to compare in this case different crops, i.e. maize, soybean, wheat and cotton.

The main challenges to sustainable supply chains come from safety, security and stability. Measuring sustainability metrics can be seen in this environment to be good management practice as it directly relates to risk management. One evolving work product from the Field to Market scheme is the Fieldprint Calculator that farmers can use to assess the relative farming practices on their farms (USDA, 2014b).

WAYS FORWARD

A good start to this is to target the objectives of the EU's 'A resource-efficient Europe' (Flagship Initiative of the Europe 2020 Strategy) (EU, 2011).

1. Better technical knowledge on the environmental impacts of food.
2. Stimulating sustainable food production.
3. Promoting sustainable food consumption.
4. Reducing food waste and losses.
5. Improving food policy coherence.

Putting an industry viewpoint to these objectives, we would recommend the following.

1. The focus should remain on *defining and measuring* the sustainability metrics of egg and poultry production. Whilst EU poultry producers have increasingly used more and more sophisticated key performance indicators (KPIs) in their production systems, the emphasis initially to allow simple comparison across poultry sectors and member states needs to be on specific *critical efficiency metrics* such as feed conversion.
2. These critical impact efficiency metrics then need to be *benchmarked.*
3. Additional tools are still needed to *help growers to analyse husbandry systems* and for food companies to *explain how natural resources* are being managed. This is best substantiated by measuring each metric *using the best available scientific methods at prescribed frequency* so that the results can be subsequently reported with a confidence that will engage the consumer.
4. There is a need to *develop outcomes-based metrics* to measure system sustainability. These need to be supported by the *adoption of goals for each outcome-based metric* so that improvement can be easily measured and communicated.
5. Producers need to adopt and to *implement improvement strategies*. These can be communicated in overall company corporate social responsibility (CSR) policies or in individual production system policies communicated on the company's website, for example.
6. There is a need for producers to measure the *environmental and socio-economic impacts* of egg and poultry production. To support fully these sustainability metrics there must be then a willingness to adjust and adapt system practices as necessary.

If farmers, scientists, agricultural companies and policy makers can work together to make this agenda happen at scale, we will strengthen our ability to preserve the resources that underpin global agriculture and human well-being, whilst meeting that challenge of feeding 9 billion people.

REFERENCES

AMIS (2012) Agricultural Market Information System: About AMIS. Available at: http://www.amis-outlook.org/amis-about/en (accessed 10 October 2014).

Baracho, M.S., Camargo, G.A., Lima, A.M.C., Mentem, J.F., Moura, D.J., Moreira, J.V. and Nääs, A. (2006) Variables impacting poultry meat quality from production to pre-slaughter: a review. *Revista Brasileira de Ciência Avícola* 8, 201–212.

Bokkers, E.A.M. and de Boer, I.J.M. (2009) Economic, ecological, and social performance of conventional and organic broiler production in the Netherlands. *British Poultry Science* 50, 456–557.

Carson, R. (2002) *Silent Spring*. Mariner Books, Boston/New York, ISBN: 0618249060 [1st publ. Houghton Mifflin, 1962].

CGF (2014) The Consumer Goods Forum: Strategic Focus. Available at: http://www.theconsumergoodsforum.com/strategic-focus/sustainability/our-sustainability-pillar (accessed 10 October 2014).

Coolfarm (2014) Cool Farm Tool. Available at: http://www.coolfarmtool.org/CoolFarmTool (accessed 10 October 2014).

DEFRA (2012) Food Statistics Pocketbook 2012 – In Year Update. Available at: https://www.gov.uk/government/uploads/system/uploads/attachment_data/file/183302/foodpocketbook-2012edition-09apr2013.pdf (accessed February 2015).

DEFRA (2014) Poultry and Poultry Meat Statistics. Available at: https://www.gov.uk/government/uploads/system/uploads/attachment_data/file/417253/poultry-statsnotice-26mar15.pdf (accessed February 2015).

Driver, A. (2012) Government faces legal challenge over battery egg imports. *Farmers Guardian*, 3 January 2012.

Egginfo (2014) Industry Data. Available at: http://www.egginfo.co.uk/industry-data (accessed 10 October 2014).

EPA Order 1110.2 (1970) Environmental Protection Agency Order 1110.2. December 4, 1970. Available at: http://www2.epa.gov/aboutepa/epa-order-11102 (accessed 10 October 2014).

EU (1999) Council Directive 1999/74/EC of 19 July 1999 laying down minimum standards for the protection of laying hens.

EU (2011) Communication from the Commission to the European Parliament, the Council, the European Economic and Social Committee and the Committee of the Regions, 26.1.2011. A Resource-Efficient Europe – Flagship Initiative under the Europe 2020 Strategy. Available at: http://ec.europa.eu/resource-efficient-europe/pdf/resource_efficient_europe_en.pdf (accessed 10 October 2014).

FAO (2012) Food and Agriculture Organization: The State of Food Insecurity in the World 2012. Available at: http://www.fao.org/docrep/016/i3027e/i3027e00.htm (accessed 10 October 2014).

Farming UK (2013) *UK Free Range Egg Production Welfare Standards amongst Some of the Best in the World. Farming UK*, 18 November 2013. Available at: http://www.farminguk.com/News/UK-free-range-egg-production-welfare-standards-amongst-some-of-the-best-in-the-world_26826.html (accessed 10 October 2014).

FSA (2014) *Campylobacter*. Available at: https://www.food.gov.uk/science/microbiology/campylobacterevidenceprogramme (accessed February 2015).

Garnett, K., Delgado, J., Lickorish, F., Medina-Vaya, A., Magan, N., Shaw, H., Rathe, A.A., Chatterton, J., Prpich, G.P., Pollard, S.J.T. and Terry, L. (2014) Plausible Future Scenarios for the UK Food and Feed System - 2015 & 2035. Cranfield University, Bedfordshire (UK) report to the Food Standards Agency on behalf of the Defra Futures Partnership project. Available at: https://www.food.gov.uk/sites/default/files/FINAL_FFS_Report-June-2014.pdf (accessed 27 April 2015).

Harrabin, R. (2014) *Greenhouse Gas Fear over Increased Levels of Meat Eating*. Available at: http://www.bbc.co.uk/news/science-environment-29007758 (accessed September 2014).

Houston, S. (2012) *Market Impact of EU Regulations on Group Housing of Sows*. Available at: http://www.bpex.org.uk/media/2357/market_impact_of_eu_regulations_on_group_housing_of_sows.pdf (accessed February 2014).

Koolman, L., Whyte, P. and Bolton, D.J. (2014) An investigation of broiler caecal Campylobacter counts at first and second thinning. *Journal of Applied Microbiology* 117, 876–881.

Lapola, D.M., Schaldach, R., Alcamo, J., Bondeau, A., Koch, J., *et al.* (2010) Indirect land-use changes can overcome carbon savings from biofuels in Brazil. *Proceedings of the National Academy of Sciences* 107(8), 3388–3393.

Lindquist, E.J., D'Annunzio, R., Gerrand, A., MacDicken, K., Achard, F., *et al.* (2012) *Global Forest Land-Use Change 1990–2005*, FAO Forestry Paper No. 169. Food and Agriculture Organization of the United Nations and European Commission Joint Research Centre. FAO, Rome.

Magdelaine, P., Spiess, M.P. and Valceschini, E. (2008) Poultry meat consumption trends in Europe. *World's Poultry Science Journal* 64, 53–63.

Mason, P. (2014) Dietary change important for climate mitigation. *The Pharmaceutical Journal.* Available at: http://www.pharmaceutical-journal.com/opinion/blogs/dietary-change-important-for-climate-mitigation/20066543.blog (accessed October 2014).

McDermott, N. (2013) Free range isn't better than factory farmed. *Daily Mail* 13 November 2013.

Ryan, C. (2013) Changing Face of Retail. Available at: http://www.meatinfo.co.uk/news/archivestory.php/aid/15973/Changing_face_of_retail.html (accessed February 2014).

Salatin, J. (2011) *Folks, This Ain't Normal: A Farmer's Advice for Happier Hens, Healthier People, and a Better World.* Center Street, Hachett Book Group, New York. ISBN: 0892968192.

Sani, P. and Gobbi, L. (2006) Thinning flocks to fatten profits. The Poultry Site. Available at: http://www.thepoultrysite.com/cocciforum/issue12a/3/thinning-flocks-to-fatten-profits (accessed February 2014).

SARE (2012) Profitable Poultry: Raising Birds on Pasture. A Sustainable Agriculture, Research and Education Bulletin. Available at: http://www.sare.org/Learning-Center/Bulletins/Profitable-Poultry (accessed 10 October 2014).

Sherwin, C., Richards, G. and Nicol, C. (2010) A comparison of the welfare of layer hens in four housing systems used in the UK. *British Poultry Science* 51(4), 488–499.

Silverman, R. (2013) Caged hens are 'happier than free range'. *The Telegraph* 13 November 2013.

Stein, K.F. (2013) *Rachel Carson: Challenging Authors.* Sense Publishers, Rotterdam. ISBN: 9462090688.

USDA (2014a) Field to Market Group. Available at: http://www.fieldtomarket.org (accessed 10 October 2014).

USDA (2014b) Understanding and Communicating Sustainable Agriculture. Available at: http://www.fieldtomarket.org/fieldprint-calculator (accessed 10 October 2014).

van Horne, P.L.M. and Bondt, N. (2013) Competitiveness of the EU poultry meat sector. *LEI Report 2013-068*, Project code 2273000568. LEI Wageningen UR, The Hague.

Xin, H., Ibarburu, M., Vold, L. and Pelletier, N. (2013) A Comparative Assessment of the Environmental Footprint of the US Egg Industry in 1960 and 2010. Available at: http://www.ans.iastate.edu/EIC/Media/50yrStudy/EggIndustryCenterReportEnv50yrStudy081613FINALWeb.pdf (accessed 10 October 2014).

PART III
People as a Sustainable Resource

CHAPTER 5
How to Attract, Retain and Develop Talent within the Industry

Colin T. Whittemore,[1]* Falko Kaufmann[2] and Robby Andersson[2]

[1]British Society of Animal Science, UK; [2]University of Applied Sciences Osnabrueck, Germany

INTRODUCTION

People choosing their profession or occupation usually ask themselves: 'Do I want to do that for the rest of my life?', 'What if I am not happy with that?', 'How about my career prospects?' The danger is that for the agricultural sector there may be an increasing tendency for the answer to be negative, resulting in the avoidance of careers in the agricultural sector. The perception of agriculture in general and of the poultry industry in particular, has steadily deteriorated over recent years. At the same time, the poultry sector has become more transparent and publicly available. However, because the general public and consumers necessarily lack professional knowledge, well-established production systems are viewed with suspicion. In addition, livestock industries are failing to promote the positive aspects of a poultry industry career, and at the same time are also failing to counter the lack of positive information available to the public at large. Certain fields such as animal welfare, animal health and risk-oriented food safety have moved to the centre of public attention, and thus political interest. It follows that the political and legal framework for all intensive husbandry systems, and in particular all stages of poultry production upstream and downstream, are placed under critical scrutiny. The 'negative image' of intensive livestock production has resulted in reducing numbers of young people deciding to opt for vocational and/or academic training in the poultry sector. This is, to say the least, unfortunate as the poultry industry offers a broad spectrum of highly diverse and challenging jobs. Moreover, the poultry sector is rapidly growing and therefore skilled employees are urgently sought, meaning career prospects are highly attractive.

Changes within the Animal Welfare Act and the Regulations for Productive Livestock will make poultry production more complex at every stage of the

*Corresponding author: colin.whittemore@btinternet.com

process. As a result, demands will be further heightened for well-educated and trained personnel to be present throughout the industry. Graduates entering their professional careers have aspirations which must first be recognized, then cherished and lastly rewarded. If a likelihood of this happening is not apparent, or if there is failure along the way, they will enter into a different career. Greater levels of emphasis therefore need to be attached to adult education, vocational and advanced training, and development (continuous education) in those disciplines that relate to the knowledge and skills needed for a successful poultry industry. Simultaneously, dual training and study courses are gaining importance for workers already employed in the industry as well as for lateral entrants.

General and specific educational options that are required to supply talent to the poultry industry in breadth and depth are not sufficiently well established. From the middle of the last century through to the end of the millennium, graduate expectations were largely satisfied through membership of National Societies of Animal Production (in GB and Ireland, WPSA and BSAP/BSAS). Government agendas for certification, societal demands for independent quality assurance and the need for recognition on an international scale, have however created an imperative for a further step beyond that of simple membership of a professional society. That step takes us into formalized accreditation schemes.

Professions are groups of like-minded people who offer services in a particular declared area of activity. A profession is regulated by its professional body, which oversees education, training and skills development, and governance of behaviour. A professional body can also offer a degree of protection against incursion by those who have lesser standards (are not members of the profession). Reassurance from fellow members within a profession also provides job satisfaction. In some cases, persons can only practise within a profession if that individual is recognized by their professional body by, for example, accreditation or certification. This restriction might be enshrined in law, or exist de facto by the will of society (the prospective customer base). The existence of a profession further implies that within it there will be areas of specialization for which particular (often post formal education) training will be required.

One may therefore see here three strands.

1. Governance – governance of standards of declared members by their professional body.
2. Accreditation – formal education, training and experience prior to full entry into the profession.
3. Continuing Professional Development (CPD) – continued training throughout career and in specialized areas of professional activity.

It is an important point of principle with regard to both accreditation and governance in science and technology that the needs of the 'academic' sector and those of the 'industry' sector are not differentiated. The need for the above three strands apply equally. Issues relating to governance, for example, appertain as much to those working in production industries as in research institutions.

GOVERNANCE

Governance of a profession has a value judgement element that is subject to the social mores of the times. For judgements to be safe and to reflect the general opinion and expectations of the group represented, a committee of elected members may be delegated to the task. Further, the decisions of those elected members should be overseen (audited) by a higher body, preferably of international repute.

Governance mainly concerns itself with two matters: the appropriateness of applicants to become members in the first case, and in the appropriateness of the conduct of the declared members. The 'appropriateness' of applicants to become members is a matter of the exclusion of those who profess expertise or knowledge in areas where they are neither expert nor knowledgeable. They are, in essence, those who claim to be that which they are not, and to do that which they cannot do. Such persons may either have never gained requisite qualifications and/or experience, or have allowed their skills/knowledge base to lapse through failing to keep active and up to date. This element of governance can be achieved by the responsible professional body administering an accreditation or certification scheme. Such schemes usually require both validated entry-level qualifications and/or experience, and also a validated diet of continuing professional development throughout the duration of the professional career. Structures that can be set up to achieve these goals are dealt with in greater detail later. The second matter, that of 'appropriateness of member's conduct' is more difficult to pursue as the necessary structures are not inherent. If inappropriate behaviour has occurred outwith a public forum, then an evidence base may be difficult to establish, relying on 'reporting' of misdemeanours by others (often clients), or upon 'whistle-blowing' by colleagues. Within professions, these elements remain fraught with difficulty and are often avoided or passed to other (statutory) authorities. For the responsible committees (especially the chair), there is a fervent hope that such matters do not arise 'on my watch'.

For science and technology, governance of conduct is usually restricted to bringing the professional society into disrepute in one way or another – failing to 'meet obligations to maintain the standards of the profession and of professional integrity in oneself and in other members'. The social construct within which science works has changed in recent years. First, funding has shifted away from 'independent' government and toward 'goal-oriented' industry. Second, scientists and technologists, both directly and indirectly funded by industry, may find themselves serving not so much the common public good by promulgating information gratis, as serving private individuals and corporations with interests in sales, revenue streams, confidentiality and patent.

The professional life of most scientists is subject to review by peers and peer review must necessarily fall within the orbit of governance and accreditation. Peer review is defined as the process whereby persons of equal status look keenly at each other's output. We should be alarmed at propositions that peer review is not appropriate for activities such as work with industry and work with lay clients. This is an issue with which a self-regulating professional society, such as that representing animal scientists and technologists, cannot avoid becoming involved, inconvenient though that may be. Peer review is, regrettably, often considered solely as part of the activity of the 'academic research' community. But the principle applies equally to all who have responsibility to the science and technology professions, especially perhaps those employed within industrial concerns. Unfortunately, objective evidence and sales promotion can become confused, a shortcoming not by any means limited to the commercial domain. Peer review should be common to all sectors. As such it plays an important part in attracting excellent young people to the community; and of accreditation. Persons of excellence wish to be members of communities that take it upon themselves to assure the quality of their own work.

In science peer review takes place primarily in two arenas: that of published works in scholarly and technical journals and that of oral presentation at meetings. In industry it takes place as a part of quality assurance for the flow of information. Both 'science' and 'industry' should perhaps pay more attention as to how diligently their 'peer review' is conducted, especially if the outcome of such review might be inconvenient. Those responsible for governance of a scientific profession such as animal science and technology might wish to ask themselves if they are currently satisfied on the following sorts of points:

- Are the 'peers' independent of both authors and funders, and do the 'peers' have predetermined expectations for the outcomes of investigations?
- Did the rigour of the interrogation by peers take any account of the standing of the person presenting the results?
- Are negative and non-conforming results acceptable?
- Were the results reported specifically those for which the investigational design was created; were the parameters to be tested chosen before the outcomes were known; were the number of parameters tested proportionate to the strength of the test?
- Are the outcomes presented justified by the results presented?
- Has experimental confirmation of anecdotal expectation been taken as an objective proof?
- Were the correct analytical techniques used, and did the author understand their proper interpretation?
- Were the parameters measured exactly and directly what was claimed as being measured?
- Was there realistic assessment of whether the same result would occur again in different places, times, circumstances etc.?
- In the case of a meta-analysis of numbers of experiments, has a thorough review been completed to establish results of unpublished works?
- Does the funder of the work (be it government or industry), or the researcher themselves, have a predetermined interest in a particular outcome?
- Was the work robust in terms of: (i) numbers of plots (animals); (ii) reliability of experimental (and laboratory) method; and (iii) the skills/competency base of experimenters (note – finding a statistical significance and/or demonstration of expected outcome is not a test for robustness)?
- Were the results internally consistent with each other, or were there dichotomies occurring within the results matrix?

- Have reasons/causations for outcomes been adequately argued?
- Is the strength of expression of commercial utility of any investigational outcome consistent with the strength of the primary finding?
- Was it evident that the hypothesis and interpretative structure were determined before the investigation began?
- Is the science coherent?

Talent in the animal science and technology communities will only be attracted, retained and developed if those that govern our sciences are content in relation to such issues as those exampled above.

ACCREDITATION

Schemes that recognize skills by maintenance of a register are common to most livestock industries. The UK Pig Industry Professional Register, for example, encourages life-long skills acquisition through practical training courses for stock workers arranged by the British Pig Executive. The veterinary profession has similar arrangements centred around species- and discipline-related programmes of study, which allow veterinarians to be registered as specialists in particular areas of endeavour, such as poultry or milk production. Recently the Agricultural Industries Confederation (a trade association) has set up their Feed Adviser Register in response to concerns that government would be seeking assurances on the quality of advice coming from the feeds trades (especially technical field salespeople) in respect to such matters as environmental protection, inappropriate emissions (greenhouse gases) and carbon footprint. The scheme centres around 'core competencies' and demonstration of their being kept up to date through CPD programmes. Sometimes accreditation and certification may mean no more than recognition of a 'pass' for a training programme offered for sale by the certificating body itself. The Advanced Training Partnership (run mostly by university departments and funded by government) concentrates on high-level training provision that contributes to the CPD required by independent accreditation bodies that rely mostly upon other bodies to provide. In this regard, accreditation and registration are quite separate functions to CPD provision (although depending upon them), as for example is the case for the scheme run by the British Society of Animal Science.

Usefully, an accreditation scheme should offer entry at more than one level. The BSAS scheme, for example, has both fully Certified and Associate levels. Associate is effectively a 'training' grade. Accreditation is not only about entry-level educational attainment, although, as for any profession, this is seminal. Accreditation assures ability to carry out declared functions: it is about high-level skills, competencies and knowledge, not just about ability to pass college examinations. Such skills are invariably learnt post-graduation, and have within them a large component of experiential learning. An accreditation scheme for animal scientists and animal technologists should therefore encourage early entry following graduation (Associate level), and then looks for evidence of career development experience in the following years before raising member's status to full certification level. This, of course, also induces motivation to up-skill. Those who can already demonstrate experience in the use and development of their competencies and special abilities will, naturally, enter directly to the Certified grade.

Benefits of becoming accredited

1. Assures professional competence and status.
2. Certifies special skills, knowledge and experience, giving international public recognition.
3. Provides support for career opportunities.
4. Fosters a culture of staff training and skills development within organizations.
5. Increases opportunities to provide expert advice and opinion.
6. Maintains up-to-date knowledge through CPD.
7. Demonstrates integrity through audit by a higher professional body.
8. Verifies responsible conduct and practice.
9. Entitlement (usually) to post-nominal designations.

Administrative structure

The core of an accreditation scheme is its public register of Accredited Members, which identifies for each member their area of professional activity, the level at which they are registered, the special competencies for which they have been assessed and found good, and a current e-mail address.

The professional accrediting body for the Register of Accredited Animal Scientists and Technologists is The British Society of Animal Science (BSAS). This is fitting because the BSAS serves the same community as the Register, namely professionals in industry, commerce, public service and academia who are involved in any discipline relating to the care, sustainability and productivity of animals.

The need for oversight – a specific example

However, it is essential that the executing body is overseen and audited by a higher body; in the case of the BSAS scheme, this is the Society of Biology, which has international standing across all the biological disciplines. In turn, the Society of Biology is monitored by the Science Council, within which it represents the sector. Within BSAS, there is an Accreditation and Governance Group which answers to the BSAS Council, an elected and formally constituted body originating in the middle of the last century to represent professionals in the industry. Accreditation is handled through a panel that scrutinizes individual applications to the register and decides upon appropriate action (acceptance, level, suitability of requested declaration of specialisms, referral, condition, rejection, etc.). In this task they are guided by the opinion of two independent assessors who review and comment upon the application in detail. It is a crucial point of principle that the expertise of both panel and assessors reflects that of the individual applicant. In this way parity is achieved across contrasting sectors such as research, technology and information transfer, and the production industries.

It has been particularly recognized by the UK's Science Council that professional status is as appropriate for those with industrial skills and competencies as those with academic flair. This should be fully reflected in any accreditation scheme seeking government support.

Continuing Professional Development

Encouragement of a formalized track for CPD is one of the main attractions of accreditation. Most accreditation schemes should therefore be 'CPD-driven'; maintaining one's position on a register being dependent upon the annual completion of a documented diet of CPD that can be shown to enhance skills, experience, performance and knowledge. An excellent example of a CPD scheme associated with a professional register is that of the Society of Biology for its Chartered Scientist Register. In the Society of Biology structure activities are grouped across five areas to ensure balanced development. These are formal training and education programmes (maximum 60% of points), work-based learning (maximum of 40% of points), professional activity (maximum 40% of points), self-directed learning (maximum 20% of points) and other (maximum 20% of points).

CONCLUSION

Globalization and the pursuit of a sustainable improvement of living conditions of a growing world population requires an increasingly efficient poultry industry. This can only be achieved with a well-qualified and career-ambitious work-force. The ancient presumption that upon graduation, persons entering their profession are fully competent is no longer appropriate. Accreditation for animal scientists and technologists, and proper governance of their activities by a recognized professional body provides a means to attract, retain and develop talent in the animal sciences, technologies and industries. Schemes are now patent that deal with all levels and types of competencies, but many of these are newly launched and still in a formative state. As these schemes develop, the responsibilities of accreditation authorities for the governance of science and science professionals – in academia, industry and the public sector – must become more apparent. Persons of excellence invariably wish to be members of communities that take it upon themselves to assure the quality of their own work.

A CASE STUDY IN GERMANY

A substantial part of the German poultry industry is located in the state of Lower Saxony. This makes the ongoing activities there with regard to attracting talent into the industry a particularly apt area of study. Germany functions as federal states. Primary responsibility for legislation and administration in fields of education, science and culture lies within the different federal states. However, the Standing Conference of Ministers of Education and Cultural Affairs manages and monitors a comparable and proper development of education across states. The government is also responsible for parts of legislation in the field of continuous education, non-school training, grants and the labour market. After the reform of Germany's federal system in 2006, education policy has largely transferred to the states. The government is still regulating higher education and qualifications but the states are able to deviate, especially with regard to vocational and in-service training.

Vocational training in Germany

There are various vocational training pathways in Germany. In secondary level II (upper secondary), vocational training can either take place in a combination of company-based training and part-time school, or a full-time vocational school. For poultry, the 'Tierwirt Geflügel' offers a special vocational training programme specifically tailored for the needs of the industry, especially at production level. The programme is supported by national and global companies as well as several education institutes for advanced and tertiary training. The training lasts for 3 years and can be divided into two parts. After an initial basic course, covering the broad spectrum of related professional fields, apprentices receive specific knowledge and skills in the disciplines of:

• husbandry and herd management;
• animal nutrition;
• product recovery and marketing;
• reproduction, breeding and hatching; and
• recycling and disposal of residues.

A vocational qualification in 'Tierwirt Geflügel' enables graduates to both directly join companies actively involved in the poultry industry and in the scheme, or to progress academically with an entrance qualification for the tertiary system (such as Universities of Applied Sciences), even without having previously gained higher qualifications at school.

The tertiary system covers education programmes at colleges and universities. Depending on the different states in Germany, there also exist certain vocational academies and other non-university institutes assigned to the tertiary level. Entrance to the tertiary level demands a qualification out of the upper secondary level. After the reform in 2006, those regulations regarding entrance qualification are the responsibility of the federal state. However, what all different education pathways in the tertiary have in common is their lack of poultry-specific training programmes in relevant breadth and depth.

In this context, an exception should be made for the University of Applied Sciences in Osnabrueck where agricultural students can focus their studies on

'Applied Poultry Sciences (StanGe)'. The major study course is accredited and also mentioned in the graduate certification, which is rather unique in the German landscape of higher education. Similar to the 'Tierwirt Geflügel' on the secondary level, the Osnabrueck poultry-specific study course was initiated in collaboration with the poultry industry to fulfil their need for well qualified personnel on an academic level. Those study courses with strong industry collaboration benefit the students, the companies themselves, and not least, the scientists. During study, students are in contact with the industry from an early stage. There are, for example, various project modules throughout the course, and importantly the theses on projects launched in collaboration with the industry itself. The companies themselves are able to select research questions of direct interest to them, and to work with potential candidates for recruitment into their businesses. However, there remains the challenge to integrate in a systematic way vocational training and continuous education.

Continuous education in Lower Saxony

The continuing education sector may be seen as the fourth level in the education system. Characteristics are:

- a large variety of providers;
- market-oriented nature;
- soft governmental regulations;
- broad spectrum of participants (voluntary and mandatory); and
- multi-functional.

According to paragraph 1(4) of the German Vocational Act – BBiG, continuous further training intends to serve the maintenance, adaptation and expansion of vocational competence, whilst advanced further training may act as an expansion of vocational competence in order to promote the career. Continuous education is aimed at different target groups. On the one hand, people are 'required' to participate in further training to approve and renew their knowledge, and on the other hand people participate voluntarily in order to improve their competences with focus on personal development and career setting. Both scenarios can be illustrated in more detail using the framework found in the German state of Lower Saxony.

Lower Saxony is characterized by its large agri-food sector, which is the second most important economic sector and the most important employer in rural areas. Including the upstream and downstream production stages, every fifth workplace is associated with the agri-food sector. Nationwide, Lower Saxony is the most relevant location for animal production: 35% of laying hens and 60% of poultry for meat are located in Lower Saxony. We would contend that whilst Lower Saxony makes a good example to show what can be done in agricultural education, there is no reason to believe that the political drivers could not be usefully followed in other German states and in other countries.

Currently, Lower Saxony administrators and working groups are developing the so-called 'Tierschutzplan'. This is a state-based animal care act dealing with all animal production systems in order to improve animal welfare and animal health. In poultry for example, beak trimming in layers and management-dependent stocking density in broilers and turkeys are discussed and reviewed. However, the focal point in all species and production systems is the general knowledge of the responsible animal keepers. It is not that the knowledge or ability of people running those systems is questioned but that they have to prove production system-related knowledge and skills via a certificate. The knowledge and skills sought can be obtained in vocational training. However, what is new is the fact that once obtained, knowledge and skills have to be renewed at regular intervals. Continuing education and the certificates to validate that it has happened are therefore mandatory for those involved in the poultry industry. For example, farmers who run broiler production systems need to prove skills and knowledge regarding:

- anatomy, physiology and species-specific ethology;
- needs-based animal nutrition;
- animal handling and transport of animals;
- indicators for disturbance of general condition and appropriate countermeasures;
- proper killing and slaughtering of productive poultry;
- measures to prevent outbreak and spread of diseases; and
- regulatory frameworks.

Besides the 'Tierschutzplan', the newest modification in the German Animal Welfare Act requires *every* animal keeper to commit to a self-monitoring programme based on livestock health-related indicators and parameters. In order to assess and evaluate appropriate indicators, animal keepers require state of the art knowledge and skills, which may be obtained via continuing education. Self-monitoring programmes will be audited by the animal welfare authority in the framework of regular operational control. This will inevitably lead to the demand for further education increasing *even more*. This demand opens the discussion for a closer integration of vocational training, continuous education and advanced further education (such as Dual Course studies and Open Universities).

Other countries have in place elements of the Lower Saxony framework, but these tend to be less formal, less 'state-driven' and more voluntary. For example, in the UK, Bright Crop targets 15/16 year olds with a wealth of information and contacts across the range of agricultural production and support industries. Skills training is often included as a required part of Livestock Quality Assurance Schemes. For graduates, several companies have graduate schemes to attract tertiary education leavers, while many companies in the agricultural supply and support industries contribute to industry scholarships through trade associations. However, it is evident that these programmes, in contrast to those for the poultry industry in Lower Saxony, lack a coherent framework and the impulsion such as would come from a legislative imperative.

CONTINUING EDUCATION IN VETERINARY AND HUMAN MEDICINE – A ROLE MODEL FOR THE POULTRY SECTOR?

According to the Medical Professional Code of Conduct, veterinary and medical doctors are obligated to undertake continuous education. Depending on the qualification level or the current job, veterinarians have to attend a defined number of courses. Participants collect points or 'hours', which are credited on a yearly basis. Every veterinarian has to collect a certain amount of credits per year. For example, 'normal' veterinarians on the job need to collect 20 h in continuous education per year, whereas specialized veterinarians need to collect 30 h/year, with 15 h/year in their respective field of specialization. Education programmes need to be accredited by the Academy of Veterinary Improvement (ATF). The respective methods of continuous education can vary largely in terms of their means of presentation, but as long as they are accepted by the ATF, the following methods are possible:

- media-based self-study, distance learning;
- participation at congresses, seminars, courses, colloquies etc.;
- sit in on lectures, case studies; and
- curricular programmes.

Within the veterinary model, the intention of an obligate continuous education is to maintain and develop competence. Usually this takes the form of absorbing the newest scientific knowledge, methodology and technology. However, the same model also encourages lateral learning; the development of different skill sets and the accumulation of groups of competencies whose whole is greater than the sum of their parts. The assurance of quality is seminal to all such schemes, and necessarily the prime focus for a scheme's management. Such is also pre-requisite from a legal point of view. Such a continuous education, obligatory or voluntary, may be usefully transferred to the poultry sector. Such a model would also offer quality assurance in the whole process chain, thus dealing with the sensitive issues of food safety, animal welfare and not least the public image of the poultry industry.

Integration of vocational training, continuous education and modular advanced further education: Continuous Education in Poultry Science (CEPS)

The University of Applied Sciences worked out a concept that may meet all demands and requirements by integrating vocational training, continuous education and modular advanced further education. This is illustrated in Fig. 5.1.

This concept combines different levels of training and education for the poultry science sector. It also enables a larger group of persons to obtain access to education at universities:

- graduates of vocational training;
- individuals with professional experience; and
- employees who want or need to qualify further alongside their job.

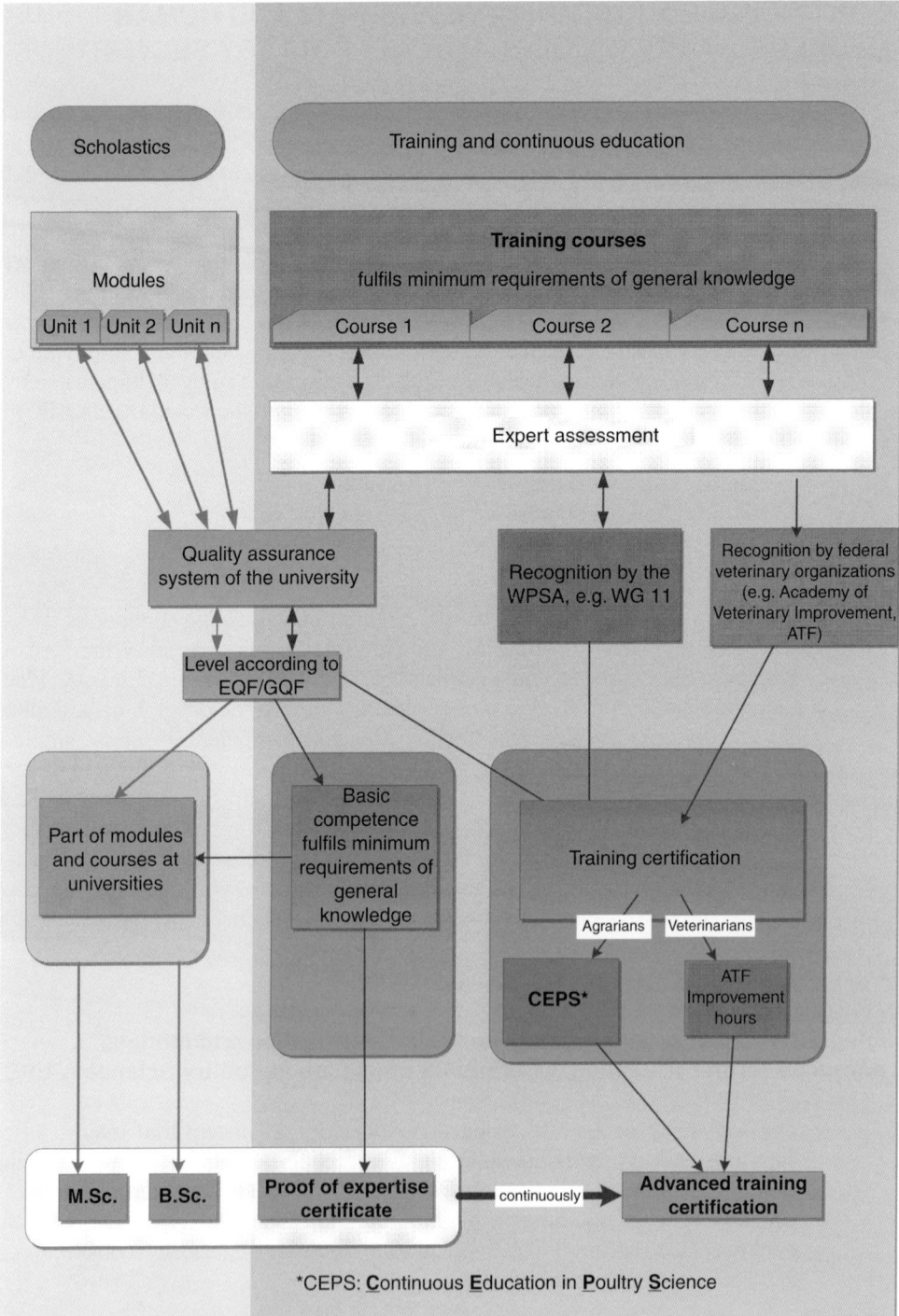

Fig. 5.1. Concept of a modular structured education system for the poultry sector integrating vocational training, continuous education and study.

On the one hand such a concept enables the industry to recruit well-qualified personnel and on the other hand enables individuals to maintain their employ-ability. Overall, the skilled personnel pool is augmented. The same concept covers the legal requirements for the attainment of knowledge and skills.

The keystone of the concept is its modular structure. It enables individuals or groups of persons to participate in certain modules or courses and collect credit points (CEPS) and/or to approve their knowledge and skills via certificates. Due to the modular structure, single units can be studied, certified and credited, and then assembled into a complete study programme. Quality assurance is handled by the assurance system of the university (accreditation) on the study side and may be the responsibility of bodies such as the WPSA (agrarians) or ATF (vet-erinarians) on the CPD side. Much of the relevant framework for this concept is already worked out and present. However, there are still uncertainties to be over-come before we have a fully operational education system for the poultry sector.

CONCLUSION

Vocational and advanced training and development programmes via continuous education have become a vital necessity for future-oriented companies and life-long learning is demanded. The crisis in attracting, retaining and developing talent within the poultry industry is being directly addressed in Lower Saxony. The framework developed there can serve as a useful model for others. Amongst its useful attributes are a specialist focus on the poultry sector, the combining of formal and continuing education, the direct involvement of industry as well as the 'legislature', and the imperative of the holding of required certificates by those responsible for the care of poultry and poultry production systems. Funda-mental to the whole framework are mechanisms for the assurance of quality and the continuous – lifetime – development of professional knowledge and skills.

PART IV
The Role of Nutrition in Sustainability

Chapter 6
Which Feedstuffs Will Be Used in the Future?

P.J. van der Aar,* J. Doppenberg and C. Kwakernaak

Schothorst Feed Research B.V., Lelystad, the Netherlands

INTRODUCTION

Livestock production and thus the poultry industry will be facing different challenges in the near future – challenges that require adaptation to a changing world. Many of these cannot be foreseen at this moment. Current ideas and trends may not become true and others will; however, a few trends seem to be persistent. In this chapter the consequences of the currently foreseen trends in the consumer market, the genetic development in livestock animals, international trends in consumption of poultry products and their consequences for the demand and supply of feedstuffs for poultry will be discussed. This discussion will be based on the most dominant trends at this moment in time. Since the feedstuff market is highly internationally oriented, where production and demands in one continent affect the prices and availability in others, developments should be viewed in a global perspective.

INTERNATIONAL TRENDS

In the next decades the most dominant trend is that the human development index (HDI) will increase. This index represents a combination of life expectancy, education level and buying power. On a global level this is a positive trend. However, the regions with the highest HDI also have the highest ecological footprint. The higher the HDI, the larger the demand for resources, which increases almost exponentially. Regions that have reached a high or very high development index have a higher footprint than is potentially available. For example, Western Europe and northern America need respectively 2.5 and 3.5 times more resources to support their needs than is (on average) available worldwide. Even less

*Corresponding author: pvdaar@schothorst.nl

developed regions such as Eastern Europe, the Middle East and Latin America are requiring more resources than are on average available (Global Footprint Network, 2011). The current ecological footprint associated with high develop-ment is not sustainable. Therefore, the major challenge will be to reduce the eco-footprint for highly developed populations. A 'business as usual' approach is not an option in this scenario, and the livestock industry has to contribute to this alteration in thinking.

Besides the improved living standards, the world population will increase further to approximately 9 billion in the year 2050, which puts more pressure on the available resources per inhabitant.

When populations increase their HDI, food consumption will change to more balanced and higher quality diets. The consumption of animal proteins will increase. On a global scale it is predicted that the demand for animal protein will have been increased by 50–70% in 2050. This growth will be mainly in Asia and Africa. The rapid expansion of livestock in South-east Asia has resulted in a rapid shift in the feedstuff streams. In less than 20 years Chinese soybean imports have increased from almost none to 60 million t (Mt) in 2013, whereas the rest of the world imported approximately 32 Mt. The role of Europe in the global feedstuff market is therefore reduced.

The competition for raw materials between food, fuel and feed has exerted an important impact on the feed industry. The growing use of starch-rich feed-stuffs for the production of bioethanol together with the rise of oil prices has elevated the cost price of energy in feed. The world cereal production has to reach record crops year after year to match the demand. Worldwide growing conditions have large consequences on the volatility of the feedstuff markets. The development of shale gas production and second generation bioethanol production will reverse this trend. It is only speculation about when and if this affects the feedstuff market. However, the existing investments in bioethanol plants, the ratio between the production costs and fuel prices, will warrant a demand for starch-rich crops from the biofuel industry.

Another geopolitical trend is the desire of the EU to become more self-sufficient for protein. The EU imports approximately 70% of protein-rich feedstuffs. More than 60% of these imports were soybeans or soybean meal (Fefac, 2012).

Sustainability issues, like the use of antibiotics, animal welfare, environmen-tal issues, product safety etc., will play an important role in the relation between livestock producer and society. It will not only be restricted to developed coun-tries but will influence the global production and eventually will be integrated globally in legislation.

The development of production oriented and applied knowledge in animal science will shift from the Western world to the areas where animal production continues to grow. European governments have cut back on the financial support; larger companies may reduce their research and development (R&D) facilities and their marketing support research budgets will be shifted towards expanding countries. In Asia the reverse trend can be observed. This trend has considerable consequences for the European livestock industry. The expertise and the research to maintain the existing knowledge diminishes, but more important, the shift from publicly available knowledge towards private knowledge will eventually hamper the rate of innovation.

TRENDS IN THE FEED TO FOOD SUPPLY CHAIN

During the first 25 years after World War II the demand for animal products rose and production was focused on the volume of food production. The consumer ate what the farmer produced. Over the years this has changed. The food chain has had more and more demands on the production of animal products. The requirements occurred at different levels.

1. Legislative level: public concern about several aspects of animal production has resulted in legislative measures either on an EU or national level. This includes regulation on the use of antibiotics, product safety, environmental regulations, building permits, etc.

2. Consumer demands: the changing society and the changes in lifestyle have also altered the consumer's attitude towards food. Health concerns and the demand for more convenience food has created different markets. The growing public concern about production systems focusing on high production and efficiency has created a market for alternative production systems. Nevertheless, a quality to price ratio will remain the dominant driver for consumers.

3. Retail: retailers play a major role in the chain. They are the interpreter of the consumer demands in the food chain. Their position is strong; not only from a buying power point of view, but also they decide whether new products can be successfully introduced or not. They stimulate product diversification to fulfil the demands of a variety of consumers. An example can be the variation of eggs on sale, whereas in the past this was limited. Retailers are very sensitive to criticism from action groups that question the way animal products are produced, especially if they follow name-and-shame tactics. This sensitivity results in additional demands on the producers.

4. Processing industry: the processing industry will exert their influence on the production in various ways. They also require more diversification. They are stimulating the production of agricultural products that meet their production processes the best. The optimal production of milk will depend on whether the milk is used for consumption or for processing to cheese. Furthermore, to meet the demand of the retailer for diversification they will ask for different products from the producers.

The producers have to fulfil these requirements. The shift has been from bulk production to product diversification and quality. The next step in production development will be the increasing focus on sustainability, stimulated by the growing awareness that the demand for crops will be doubled in 2050, and that we have to reach this with 50% of the resources currently used. For all partners in the feed-to-food chain sustainability, and especially the reduction of eco-footprints, will be important in the next decennium. Future developments in poultry feed will have to take these developments into account.

EXPECTED SCENARIOS IN BROILER PRODUCTION

A consequence of the diversification in consumer demand is that one should consider that different production systems, each with its own characteristics, will exist next to each other. For the sake of simplicity, future production systems can be divided into three groups.

1. Maximal production performance.
2. Slow-growing animals that have a minimum age at slaughter.
3. Regional production with feeds grown in the region.

Production system 1

Maximal performance: in these systems the broilers are grown to the maximum of their genetic potential with low feed conversion and the lowest production cost. The housing and management will be at the level that is required by legislation. In these systems breeds with the highest genetic potential for production will be used.

Production system 2

This system is based on slow-growing breeds which have a minimum age at slaughter. Chain demands will determine the management and housing conditions and may place additional requirements with regard to environmental aspects.

Production system 3

This type of production is following the trend that the consumer has a growing need to get reconnected to the origin of the food. Locally, organically produced food, often in small production units, appeals to this trend.

CONSEQUENCES FOR FEEDSTUFF AVAILABILITY AND USE

The feed industry will be forced to search for alternative feedstuffs. Over the last few decades the emphasis on production efficiency and, as a consequence, the development of broilers with a high growth rate and low feed conversion, has led to a situation in which good digestible feedstuffs are the first choice in formulating feeds with a high nutrient density. In least-cost formulations the energy component is the most expensive. In the low cost price scenario it means that for alternative feedstuffs the cost per energy unit will always have to compete with the currently used raw materials. For alternative protein sources it also means that not only the cost for the essential amino acids is important but also the ratio energy:essential amino acids will be decisive as to whether they will be incorporated in feeds. Since soybean meal is the major amino acid source in broiler feeds the energy:amino acid ratio of any alternative should be at least comparable with soybean meal. Other evaluation criteria for alternative feedstuffs will be the presence of anti-nutritional factors, which determine their incorporation level in diets, their effect on the quality of the products, the processing required, animal welfare and health. But for the next decades the more dominant criterion will be their effect on the different eco-footprints, like land and water use, CO_2 footprint and the potential use as food.

PROTEIN SOURCES

Bioethanol co-products can provide valuable sources of proteins that are inexpensive and sustainable. Examples include brewery co-products, rapeseed meal from biodiesel production and maize gluten meal. The yeast constituent of bioethanol co-products is of particular interest because it has very high protein content and the resulting fibre, from the separation of the yeast from distillers dried grains with solubles (DDGS), can be fed to ruminants. Yeast protein concentrate (YPC) is derived by separating the yeast-containing high protein fraction from distillery stillage by a continuous-flow process (Williams *et al.*, 2009). The resulting concentrate can then be dried into a powder so it can be fed to monogastrics as a protein source or feed additive. Positive performance effects have been seen in poultry and fish fed YPC, and its digestible amino acid content has been shown to be similar to that of soya (Scholey *et al.*, 2011). Legumes can also yield high protein levels and the current worldwide use of lupines, peas and beans is substantial, particularly in developing countries. Although they have high protein levels, the amino acid profiles of plant protein is poor. Another protein source of interest is algae from biofuel production; Lei (2012) estimated that algae could replace one-third of soya protein in monogastric diets.

Van Krimpen *et al.* (2013) produced a list of 62 ingredients containing a wide range of protein sources. A short list was then produced of the potentially most interesting protein sources to increase EU feed protein production. The criteria applied to select these protein sources were: (i) the protein source should be able to perform well in the climate conditions of north-west Europe; (ii) the cultivation of the protein source in Europe is currently not common practice; and (iii) in the long term (after 2020) the protein source will still be applied in feed and not food. The shortlist included oilseeds (proteins of defatted soybeans, rapeseed and sunflower seed), grain legumes (e.g. peas, chickpeas and lupines), forage legumes (lucerne), leaf proteins (e.g. grass and sugarbeet leaves), aquatic proteins (e.g. algae and duckweed), cereals and pseudo-cereals (proteins from oats and quinoa) and insects (e.g. mealworms and house flies). These protein sources differ substantially in terms of their environmental sustainability; products with low dry matter content, such as leaves and aquatic proteins, are considered to be less sustainable due to the high energy costs required for drying. More research is required to determine if protein extraction processes are sustainable. European-produced soybean meal is thought to be the most promising alternative to soybean meal from beans imported from South America as it has comparable protein yield. It could be improved further by selecting varieties with a very short growth season. Peas appear to be the most promising grain legume as an alternative to soybean meal as the protein yield is high, although techniques for extracting the protein need to be improved. Leaf and aquatic proteins could definitely be used to reduce soybean imports, mainly as they are not in direct competition with the land use of other crops (such as potatoes and sugarbeet), but again more research is required into protein separating techniques.

There is currently increased interest in the potential of insects to produce animal feed protein. The nutritional aspects of insects have been reviewed by Veldkamp *et al.* (2012) and Van der Poel *et al.* (2013). The use of insects or insect protein fractions as a sustainable protein-rich feed ingredient is technically feasible; insects can efficiently turn low-grade bio-waste into high quality protein for use in poultry diets. Cultivation and processing insects seems a promising innovation and it is expected that insect protein will be used as a feed ingredient in the poultry industry within the next 5 years. It is generally expected that the use of insects as a feed material in aquaculture is the nearest future application. Insects have a well-balanced nutrient content; they have the same or an even better amino acid profile compared to soybean meal and fishmeal. A rich content of polyunsaturated fatty acids, micronutrients and vitamins can also be attained, and the chitin in insects has many beneficial properties. The insect species most suitable for use in poultry diets, due to their high amount of protein and ability to degrade organic waste, are the black soldier fly (*Hermetia illucens*), the common housefly (*Musca domestica*) and the yellow mealworm (*Tenebrio molitor*). Additionally, a low carbon footprint facilitates insects such as mealworms and black soldier flies to be used as feed ingredients. Replacing 5% of compound feed with insects in broilers would mean that 72,000 t of insects a year would be required. Insects need to be further processed to get them into a form in which they are usable in the feed industry. Shelf-life of insects is increased significantly by processing methods like freezing and freeze-drying, but these methods are expensive. Additional research is needed into the feeding value and functional properties of insects, and hence the inclusion levels in poultry diets, before they can be introduced as a feed ingredient in the poultry feed chain. Further research is also needed into safety when using bio-wastes as a rearing substrate and the use of left-over or extracted residue products after production. Large scale production of insects is hindered by legislation (both environmental and in the poultry industry); in the EU the use of insect protein in feed for pigs and poultry is not allowed due to the Transmissible Spongiform Encephalopathies (TSE) regulation. Further risk assessment of use of insects as a feed ingredient is required to develop new regulations. The use of insects is also hindered by concerns over product safety and guarantee that a low cost price as a result of automation of the production process can be achieved. Currently, the production volume of insects on rearing companies in the Netherlands is low and the market is mainly focused on zoos and pet shops.

Most oilseed meals and grain legumes have a well-known composition and nutritive value. The feed industry is familiar with them and if the price is right uses them in poultry feeds. In many situations the relatively low energy concentration of protein sources is limiting for broiler feeds. Their future use will not change in the low cost scenario unless political pressure stimulates the production of these crops in the EU. If reduction of carbon footprint becomes a criterion, European sources may have an additional benefit. Protein isolated from grain legumes and co-products of cereals might meet the nutritional demands, but the required processing will determine their potential as a feed ingredient. Owing to their low nutritional value leaf proteins probably will not be used. The same will apply to aquatic proteins such as seaweed and duckweed. More promising might be protein from single cell organisms, such as algae, yeasts etc. Their high production potential per hectare promises a protein production that meets future eco-footprint demands. However, the nutritional value is still not well known, mainly due to the wide variation in type of these products. In the 1970s single cell proteins had been marketed in Europe. One of the limiting factors of these compounds was the content of nucleic acids. Although some consider nucleic acids as semi-essential nutrients, too high levels may cause negative effects and therefore their incorporation in feeds may be limited. At the time they were popular with nutritionists because of the uniformity of the amino acid content and their high digestibility. The production ceased when the product became uneconomic because the cost of the feedstock, i.e. methane and ammonia, rose in price and the cost of soya which it aimed to replace dropped in price. This is an illustration of the difficulty of predicting the ingredients to be used in the future. It will be heavily dependent on the relevant costs of all ingredients when put into least-cost formulation programmes.

In scenarios in which lower growth rates are acceptable, lower digestible feed ingredients might be economic as long as the cost per unit of energy and essential amino acids on a feed level are not affected. However, as a consequence the non-digestible protein content of these diets will increase and thus the N-emission to the environment. In this scenario the slower growth rate also will result in negative effects on eco-footprints. The Dutch organization of retailers (CBL) has launched a programme that in the near future they will only sell chicken meat from slow-growing animals. The aim is that broilers will be slaughtered at 56 days at a weight of 2200 g. Currently this weight is reached at 35 days. In an experiment performed by the Hubbard company, it was found that the lower growth rate reduced the mortality, but that the feed conversion ratio (FCR) increased from 1.60 to 2.00, mainly due to higher maintenance requirements. For the Dutch broiler production it would mean that approximately 120,000 ha of additional land is needed to produce the feedstuffs required with all consequences for the different eco-footprints per kilogram chicken.

If diets are diluted with coarse insoluble fibre there might be a benefit. Rougière and Carré (2010) showed that coarse particles from fibre stimulated the development of the upper parts of the digestive tract. Furthermore, Carré *et al.* (2010) observed in a meta-analysis from eight studies that the larger was the ratio gizzard:small intestine, the higher were the protein digestibility and the apparent metabolizable energy (nitrogen corrected) (AMEn). This effect was the largest for feeds with a low digestibility. This might reduce some of the negative effect of higher maintenance requirement. It may create possibilities for feedstuffs that have a low energy concentration owing to high levels of insoluble fibre.

A group of feed ingredients hardly discussed as alternative feedstuff are pure amino acids. In an evaluation of the environmental implications of the incorporation of amino acids in broiler feeds, Mosnier *et al.* (2011) concluded that utilization of amino acids reduced the need for protein-rich ingredients and that the costs of the feed with three amino acids were competitive with those in which no or hardly any amino acids were used. In most situations the amino acid incorporation was associated with reduced impacts on eutrophication, terrestrial ecotoxicity, cumulative energy demand and land occupation. However, the effect on greenhouse gases was variable and highly dependent on the nature of the diet, but the largest for cereal soybean diets.

In the short term, the mechanical separation of EU-produced legumes such as peas and field beans into high and lower protein products would seem to offer the most potential to replace some soya imports. One method is by fine milling and separation of particles on the basis of differences in the density of protein and starch granules in an air flow. Van der Poel *et al.* (2013) showed that with peas a product with circa 55% protein could be separated. Nixey and Little (2013) demonstrated that the removal of hulls from field beans significantly improved the nutrient content of the seed remaining. The process has the added advantage that the anti-nutritional factors, being concentrated in the hulls, will be markedly reduced, enabling the nutritionist to have the confidence to increase inclusion levels in poultry diets. The economics of both these processes is greatly influenced by the income that can be obtained from the co-product.

If it is expected that sustainability will become one of the leading factors for the evaluation of feedstuffs in the future, it is likely that more locally produced protein sources, especially rapeseed meal, will be used at the expense of imported soybean meal. The effect on ecological benefit is limited and mainly caused by less transportation. Of the alternative protein sources the single cell proteins are from this point of view the most promising. Their production potential per hectare is large but the development as feedstuff requires still much effort. There will be an increasing role for the use of pure amino acids, especially if they can be produced with less associated greenhouse gases. Since energy concentration will remain an important factor in the formulation of feed, methods to increase the energy content of low quality feedstuffs and co-products will have to be developed. This can either be reached by new processing methods or by an increased use of fermentation techniques or by a new generation of enzymes. The trend towards more slowly grown animals allows the use of lower quality feedstuffs. The question is, however, how persistent this trend is if sustainability is the dominant trend.

REFERENCES

Carré, B., Rougière, N., Bastianelli, D., Lafeuille, O. and Mignon-Grasteau, S. (2010) Relationships between individual digestion efficiencies and gut anatomy in broilers from experimental D+ and D- digestion lines or from commercial strains. In: *Proceedings of the XIIIth European Poultry Conference*, Tours, France, p. 192.

Fefac (2012) The Compound Feed Industry in the EU Livestock Economy. Available at: http://www.fefac.eu/files/55172.pdf (accessed 3 July 2015).

Global Footprint Network (2011) 2011 Annual Report. Available at: http://www.footprintnetwork.org/images/article_uploads/2011_Annual_Report_RF.pdf (accessed 3 July 2015).

Lei, X. (2012) Algae as Sustainable Protein Alternative for Animal Feed. Available at: http://www.allaboutfeed.net/news/algae-as-sustainable-protein-alternative-for-animal-feed-12701.html (accessed 18 June 2015).

Mosnier, E., van der Werf, H.M.G., Boissy, J. and Dourmad, J.-Y. (2011) Evaluation of the environmental implications of the incorporation of feed-use amino acids in the manufacturing of pig and broiler feeds using life cycle assessment. *Animal* 5(12), 1972–1983.

Nixey, C. and Little, A. (2013) Making organic poultry feed more sustainable: Dehulling homegrown protein crops. Report published by Organic Centre Wales, IBERS, Aberystwyth University, UK.

Rougière, N. and Carré, B. (2010) Comparison of gastrointestinal transit times between chickens from D+ and D− genetic lines selected for divergent digestion efficiency. *Animal* 4, 1861–1872.

Scholey, D.V., Williams, P. and Burton, E.J. (2011) Potential for alcohol co-products from potable and bioethanol sources as protein source in poultry diets. *British Poultry Abstracts* 7(1), 23–24.

Van der Poel, T., van Krimpen, M., Veldkamp, T. and Kwakkel, R.P. (2013) Unconventional protein sources for poultry feeding: opportunities and treats. In: *Proceedings of the 19th European Symposium on Poultry Nutrition*, Potsdam, 26–29 August, pp. 1–14.

Van Krimpen, M.M., Bikker, P., van der Meer, I.M., van der Peet-Schwering, C.M.C. and Vereilken, J.M. (2013) Cultivation, processing and nutritional aspects for pigs and poultry of European protein sources as alternatives for imported soybean products. *Report 662*, Wageningen UR Livestock Research, Lelystad, the Netherlands.

Veldkamp, T., Van Duinkerken, G., Van Huis, A.C., Lakemond, M.M., Ottevanger, E., *et al.* (2012) Insects as a sustainable feed ingredient in pig and poultry diets-a feasibility study. *Report 638*, Wageningen UR Livestock Research, Lelystad, the Netherlands.

Williams, P., Clarke, E. and Scholey, D. (2009) The production of a high concentration yeast protein concentrate co-product from a bioethanol refinery. In: *Proceedings of the 17th European Symposium on Poultry Nutrition*, Edinburgh, UK, 326 pp.

Chapter 7
Limiting Factors for Nutritional Efficiency

Brett Roosendaal[1]* and Annsofie Wahlstrom[2]

[1]RCL Foods, South Africa; [2]Zinpro Corporation, the Netherlands

INTRODUCTION

What do we mean by nutritional efficiency and how does this differ from traditional feed efficiency? Feed efficiency is defined as the amount of feed required to produce a kilogram of live weight gain. This definition is being challenged as to its usefulness due to it being a general term and affected by many factors. More specific and meaningful measures for the efficiency of energy and nutrient utilization of the major-cost components in poultry diets are required. The large contribution of energy to total feed and production cost requires that our focus be on improving energy utilization. Nutritional efficiency can be defined as improving the proportion of dietary nutrients into carcass lean tissue or egg mass that meets the biological targets of the genetic potential being used to produce the animal protein. Economic measures for efficiency such as feed cost per kilogram live weight gain or feed cost per kilogram egg or even feed cost per kilogram meat per square metre of floor space are already commonplace. Pertinent sustainability measures today would be nitrogen and phosphorus excretion per bird or CO_2 equivalent per kilogram meat produced. Given that definition of nutritional efficiency, this chapter focuses on how we can feed and measure broilers more efficiently with minor references to other poultry.

PREDICTING FEED INTAKE

Gous (2013) eloquently captured the essence of food intake in the following paragraph:

*Corresponding author: brett.roosendaal@rclfoods.com

Birds attempt to consume sufficient of a given food to enable them to meet their nutrient requirements for maintenance and growth. Maintenance requirements are the first priority followed by growth to meet its genetic potential. Food intake is governed by potential performance of the bird, the limiting nutrient in the feed and any constraints that may reduce the desired feed intake. The accurate prediction of food intake by a given bird on a given food when housed in a given environment makes it possible to define the optimum economic levels of energy and essential nutrients in feeds for poultry and consequent design of feeding programmes.

The above theory for predicting food intake appears straightforward; however, is the broiler capable of consuming the amount of food required to maximize growth? It is unlikely that feed digestion and metabolizability have changed over the last 50 years, so the continuing increases in growth are a consequence of increased daily feed intake (Leeson, 2012). Improved feed efficiency is a result of this ever-increasing feed intake being used proportionately more for growth and less for maintenance as days to reach a defined live weight continue to decrease.

Some of the constraints to feed intake would be feed texture, environmental temperature and the ability of the bird to dissipate the heat of digestion and stocking density. Birds are meal eaters and will eat for about 8 minutes each hour, preferably as one meal, although this is often interspersed with voluntary pauses, so feed intake is not a biological limit to productivity. After 25 to 28 days of age, broilers rarely have the luxury of finishing a meal at a single sitting as competition for feeder space intensifies. This means that an obvious interrelationship exists between stocking density, feeder pan space, feed texture, lighting programme and nutrient density as it affects feed intake (Leeson, 2012).

Predicting and measuring feed intake accurately under commercial conditions is absolutely fundamental to poultry production and the measurement of nutritional efficiency.

NUTRIENT RESPONSES

Dietary energy and amino acids are the largest components of broiler diets and constitute the majority of the cost. The significance of the relationship between energy and amino acids is their effect on controlling feed intake. The manipulation of energy and amino acid constraints in feed formulation via linear programming has a major impact on nutritional efficiency.

Protein

The concept of ideal protein is widely established and describes the amino acid profile of a diet that maximizes the utilization of all the essential amino acids. In combination with adequate levels of amino acids, applying an ideal amino acid profile maximizes the efficiency of nitrogen retention. Lysine is chosen as the reference amino acid because its major utilization in the body is for protein deposition. The methods used to determine the responses of broilers to an essential amino acid have been classified into either empirical or factorial categories. Neither of the methods is without its limitations and a full review of the methodologies can be found in D'Mello (2003).

A further extension of the 'ideal' concept is the feeding of dietary balanced protein (BP), where ratios of all essential amino acids are kept constant in relation to lysine as described in Wijtten *et al.* (2004), Eits *et al.* (2005a, b), Plumstead *et al.* (2007), Aftab (2009), Lemme *et al.* (2009) and Madsen *et al.* (2010). When this approach is used, broilers can respond to a higher level of lysine than those obtained with the conventional dose-response techniques alluded to above because of an absence of an imbalance leading to the higher estimates for the amino acid requirement.

More recently the emphasis has been on addressing the impact of amino acid levels on optimizing profitability (De Beer, 2010; Tillman, 2011, 2012; Aftab, 2012; Tillman and Dozier, 2013). The response criteria are typically body weight gain, feed conversion, carcass weight and breast meat weight. Analysis of the digestible lysine levels which maximized the amount of feed per gain (feed conversion ratio) typically shows a higher requirement than that for body weight gain. Tillman and Sriperm (2011) concluded that the lysine requirement for Cobb 700 males from 28 to 42 days were 0.95%, 0.99%, 1.00% and 0.97% for body weight gain, feed conversion, carcass weight and breast weight, respectively. Kidd *et al.* (1999) noted that the optimal level of digestible threonine which maximized profitability was near the digestible threonine level which also maximized broiler performance (feed conversion) and processing (carcass composition) parameters. It can be postulated that is likely to be the case for all the essential amino acids – that feeding near their requirement for performance is also close to the point that maximizes profitability. In the overwhelming majority of studies, there is a positive response in growth and feed utilization efficiency to increasing BP levels and the relationships are stronger in the faster growing broiler strains.

Energy

Approximately three-quarters of the cost of broiler feed is made up of the energy specification. The primary response to dietary energy concentration is seen in feed consumption and in productive efficiency rather than in production level (Fisher and Wilson, 1974). Fisher and Wilson (1974) also demonstrated that sex, age and broiler strain influenced the response of broilers to dietary energy as well as other factors such as feed density.

Classen (2013) summarized the effect of the confounding factors that affect the response of broilers to dietary energy and are listed as broiler genotype, diet composition and digestible nutrient content, feed form and processing, bird age, energy to protein ratio, environment and disease. Environmental factors of significance include temperature, environmental contaminants such as ammonia, stocking density, water and feed availability, lighting programmes, disease challenge and immunological stress.

Lemme *et al.* (2005) fed diets to male Ross 308 broilers from 1 to 46 days of age with increasing levels of BP and apparent metabolizable energy (AME).

The recommended AME levels for the starter (1–10 days), grower (11–32 days) and finisher diets (33–46 days) were 3010, 3175 and 3225 kcal/kg, respectively. These levels were reduced to 95% or 90% in maize–soybean diets. BP levels were adjusted to 100%, 90%, 80% and 70% of the ROSS recommendations, which resulted in digestible lysine levels of 12.7, 10.8 and 8.8 g/kg in the starter, grower and finisher diets, respectively. Weight gain generally increased non-linearly with increasing dietary amino acids (Fig. 7.1). When related to the amino acid intake (methionine plus cysteine as reference) all the weight gain data points described one response curve ($r^2 = 96\%$). The authors conclude that the responses on weight gain and breast meat yield appeared to be a function of amino acid intake rather than (or only indirectly of) energy intake. Decreasing dietary energy actually improved amino acid intake due to an increase in feed intake, which resulted in improved weight gain.

Cho (2011) conducted two trials with diets containing 11.3, 11.86, 12.42 and 12.97 MJ/kg of AME on Ross 308 mixed sex broilers. The digestible amino acids were at the recommended requirements of Aviagen (2007). The four levels of energy were fed from 0, 10 or 26 days of age until the end of the trial at 35 days of age. Body weight gain increased and feed to gain decreased with dietary energy. Dietary energy did not affect feed intake. The second trial used the same energy levels, but three levels of amino acids were used that approximated 90%, 80% and 70% of Aviagen's (2007) recommendations. At the highest level of dietary amino acids, body weight, carcass and breast yield increased and feed to gain ratio decreased with dietary energy, but feed intake was not affected. With intermediate levels of amino acids, dietary energy did not affect response criteria. With low levels of dietary amino acids, weight gain, feed intake and carcass yield decreased with dietary energy content.

When describing and analysing the effects of increasing energy content of a broiler diet, one has to take cognizance of the confounding factors when interpreting the results. The factors are well documented, but suffice to say in general terms that increasing energy content of diets in growing broilers results in a

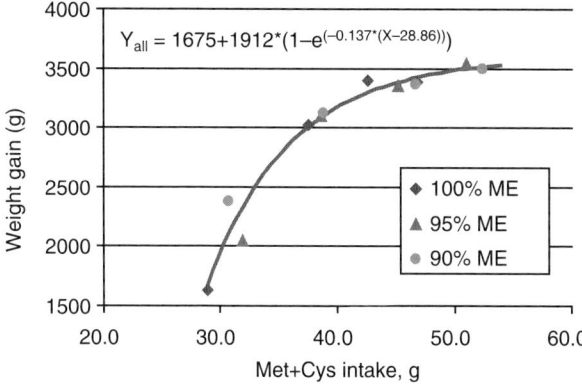

Fig. 7.1. Weight gain responses of 1–46-day-old broilers fed increasing levels of BP at very low, low and adequate dietary energy plotted against methionine plus cysteine intake.

decrease in feed intake. When the next nutrient becomes limited (usually amino acids in typical commercial diets), broilers attempt to try and increase intake and would use protein (as in this example) for muscle deposition at maximum efficiency and surplus energy is deposited as fat. Efficiency of energy utilization is affected at this point. An important criterion to use in the evaluation of nutrient responses should be the efficiency of energy conversion to the broiler operation's product objective to ascertain where the optimum lies, be this body weight gain, carcass or breast meat yield.

Starch

Starch is highly digestible in poultry and generally comprises more than half of the AME content of diets for broiler chickens. Weurding et al. (2001) observed differences in performance of broiler chickens receiving diets that were iso-energetic, but different in terms of starch digestion rate. Feeding diets containing slowly digestible starch (SDS) resulted in an improved feed conversion for broilers and an interaction was observed between starch digestion rate and amino acid content. Weurding et al. (2003) fed iso-energetic broiler diets (2975 kcal/kg) containing 34% of SDS (peas and maize as starch sources) or rapidly degradable starch (RDS) (tapioca and maize as starch sources) with a low (8.5 g/kg) and a high (11.0 g/kg) digestible lysine content. Intermediate lysine contents (9.13, 9.75 and 10.38 g/kg) were obtained by mixing the low and high lysine diet with the same starch sources. The diets with SDS showed consistently better production performances than the diets with the RDS (Fig. 7.2). SDS improved protein and energy utilization in broilers. The effect of starch sources was more pronounced at the lower digestible lysine contents indicating that amino acid and glucose supply might be unbalanced in the RDS diets, which implies that initially after ingestion of a diet more glucose conversion into glycogen or fat is needed in the rapidly digestible starch diet because of the absence of protein (Van Der Klis and Fledderus, 2007). Synchronizing glucose and amino acid supply might therefore stimulate energy efficiency.

Fibre

The inclusion of additional fibre in the diet improved nutrient digestibility and growth in broilers (Gonzalez-Alvarado et al., 2007; Kalmendal et al., 2011). When moderate amounts of coarse fibre (<5%) are fed, fibre accumulates in the gizzard and slows feed passage rate in the proximal part of the gastrointestinal tract (Jimenez-Moreno et al., 2011, 2013b; Svihus, 2011). Large, well developed gizzards improve gastrointestinal tract motility, favour gastro-duodenal refluxes and stimulate the secretion of pancreatic enzymes. The improved grinding activity together with the increase in reverse-peristalsis, facilitates the mixing of the digestive juices with the digesta, which may explain the positive impact of dietary fibre on the digestibility of dietary components (Jimenez-Moreno et al., 2009).

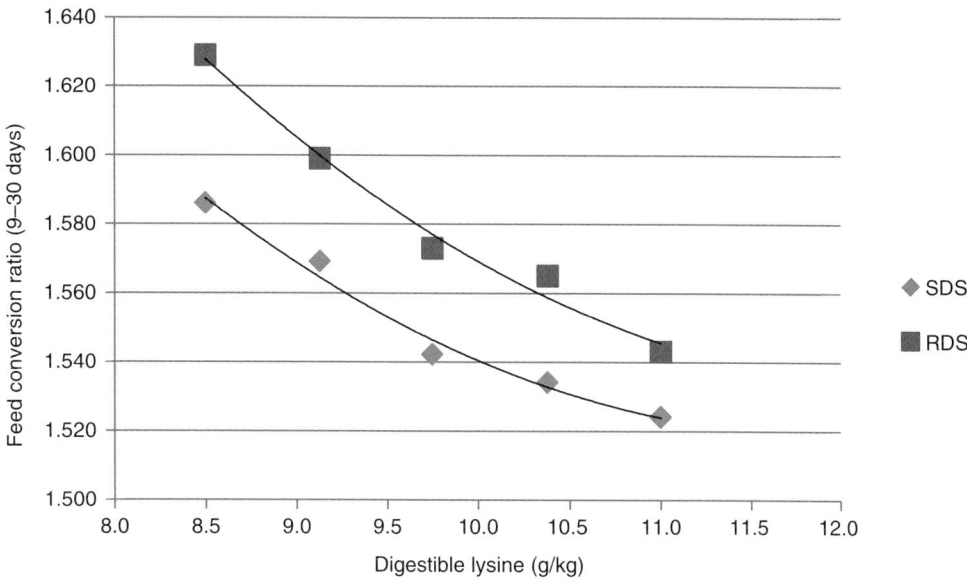

Fig. 7.2. The effect of dietary digestible lysine content in broilers fed iso-energetic diets with slowly degradable starch (SDS) or rapidly degradable starch (RDS) on feed conversion ratio from 9 to 30 days of age.

Dietary fibre is fermented by bacteria within the gut, producing lactic acid and short chain fatty acids thus reducing pH, which is beneficial from a bacterial point of view in the gut. Fibre also tends to decrease the number of goblet cells in the intestine villi and hence improves the digestibility. Esmail (2012) stated that the effect of fibre depends on its molecular weight: high fibre weight tends to increase the number of goblet cells.

More recently Jimenez-Moreno *et al.* (2013a, b) studied the effects of oat hulls and sugarbeet pulp on total tract apparent digestibility (TTAD) and apparent ileal digestibility of nutrients in 18-day-old broilers. They found improvements in performance and increased digestibility of nutrients in broilers fed 2.5% or 5% oat hulls but no improvement when 7.5% oat hulls were included. The effects of fibre inclusion on TTAD were more evident with oat hulls than with sugarbeet pulp and more importantly oat hull inclusion and not sugarbeet pulp improved crude protein and starch apparent ileal digestibility. Body weight was unaffected by dietary fibre inclusion, but feed conversion was improved in broilers of 1 to 18 days of age.

Jimenez-Moreno *et al.* (unpublished), as referenced by Mateos *et al.* (2013), studied the effects of feed form and the inclusion of fibre (2.5 and 5% oat hulls, rice hulls or sunflower husks) on productive performance and digestive traits in broilers from 1 to 21 days of age. The fibre sources were introduced on a weight for weight basis to the control diet. No interactions between fibre source and feed form were detected. Fibre inclusion improved feed conversion ratio (Table 7.1) and tended to improve body weight gain. However, there were no differences

Table 7.1. Influence of feed form and fibre inclusion of oat hulls (OH), rice hulls (RH) or sunflower husks (SFH) in the diet on growth performance of broilers from 1 to 21 days of age (Jimenez-Moreno *et al.*, unpublished).

Feed form[c]	Fibre (%)	BWG (g)	ADFI (g)	FCR (g/g)	Energy efficiency[a]	AME$_N$ (Mcal/kg)[b]
Mash		29.1[2]	37.3[2]	1.277[2]	3.97[1]	3.27[1]
Pellet		38.6[1]	48.1[1]	1.238[1]	3.85[2]	3.25[2]
Diet						
Control	0	32.9	41.9	1.237	4.07[1]	3.21[1]
OH	2.5	34.0	42.9	1.264	3.95[2]	3.26[2]
OH	5	33.7	42.2	1.257	3.85[3]	3.26[2,3]
RH	2.5	34.2	42.8	1.256	3.92[2]	3.29[1]
RH	5	34.7	43.0	1.264	3.85[3]	3.24[3]
SFH	2.5	33.8	42.1	1.249	3.90[2]	3.26[2,3]
SFH	5	33.7	42.2	1.252	3.83[3]	3.28[1,2]
SEM		0.49	0.58	0.01	0.02	12

BWG, bodyweight gain; ADFI, average daily feed intake; FCR, feed conversion ratio; AME$_N$, nitrogen corrected apparent metabolizable energy; SEM, standard error of the mean
[a]kcal AME$_N$ ingested/g body weight gain
[b]AME corrected to zero nitrogen balance
[c]$P < 0.001$ for all variables, superscript numbers denote significant differences within columns

between 2.5 or 5% fibre inclusion. The inclusion of fibre increased the empty gizzard weight, gizzard contents and reduced gizzard pH.

Dietary fibre utilization as a functional ingredient in poultry nutrition appears to be limited at present. Numerous factors such as hygiene status, the presence of various antinutritional factors such as non-starch polysaccharides in feed and management practices need to be taken into consideration. Mateos *et al.* (2013) concludes that the inclusion of moderate amounts of coarse insoluble fibre sources such as oat hulls or sunflower husks, at levels between 2% and 4%, improves growth performance of birds fed low fibre diets, especially broilers at young ages.

FEEDSTUFF EVALUATION

Knowledge of the nutritional value of feedstuffs and the nutrient response of poultry is crucial for the nutritionist when designing feed specifications. The most important tools are an extensive, robust feedstuff list and sound nutrient responses. Determining the nutritional value of each individual feedstuff for each poultry category requires a continuous research effort that is combined with accuracy.

Energy

The nutritional value of a feedstuff is highly dependent on the energy value given and the three most important nutrients contributing to energy are starch, fat and

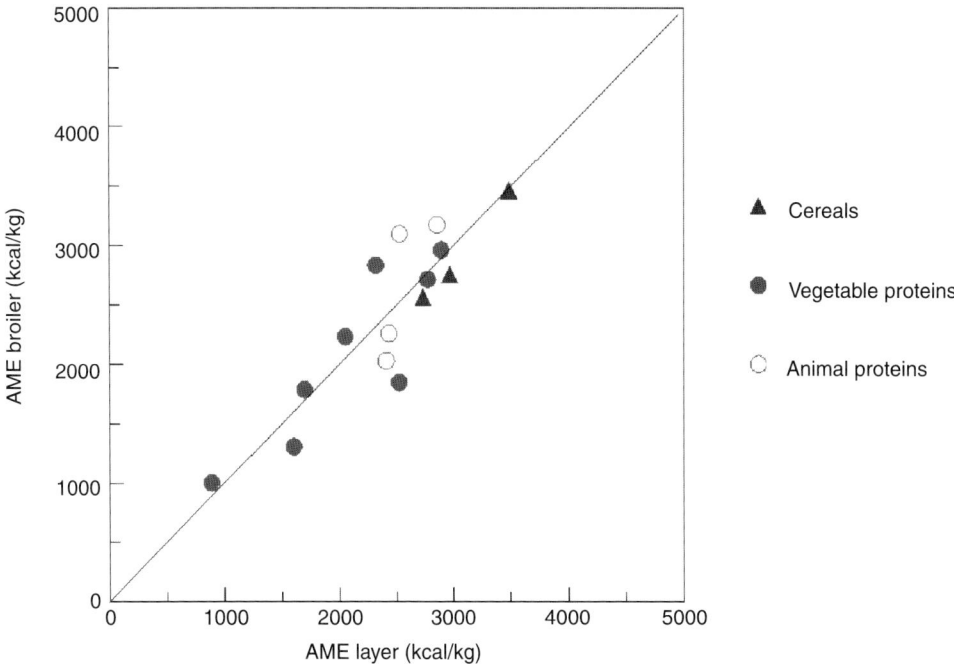

Fig. 7.3. The AME_N value in feedstuffs for laying hens and broilers. Line x=y is indicated to illustrate that the two systems are not equivalent for various raw materials.

protein. Bird type also has a profound effect on the prediction of nutritional values of feed ingredients for poultry as indicated by Van Der Klis *et al.* (2014) in Fig. 7.3, which shows that feedstuff AME_N values as determined in layer hens can be both higher and lower than in broilers. Van Der Klis *et al.* (2014) further conclude that rooster data are not only questionable for broilers, but also for laying hens.

The energy content in and the ranking between feedstuffs will be dependent on the energy evaluation system used. Several energy evaluation systems can be used, such as apparent and true metabolizable energy (corrected or uncorrected for zero nitrogen balance) with growing broilers or adult cockerels. Using adult cockerels, force-feeding methods are used, feeding the test material as such or as part of a complete diet, whereas broilers are generally fed complete diets to nearly *ad libitum* intake.

Several systems are also used to predict the AME_N value from digestible nutrient concentrations in feedstuffs as summarized by Van Der Klis and Fledderus (2007)(CP: crude protein; CFAT: crude fat; NfE: 1000-moisture-crude ash-CP-CFAT-crude fibre; NFR: 1000-moisture-crude ash-CP-CFAT-STARCH-SUGARS; d: digestible; values in g/kg unless stated otherwise):

1. The European Community equation:

$$AME_N \text{ (MJ/kg)} = 0.3431 \text{ \%CFAT} + 0.1551 \text{ \%CP} + 0.1301 \text{ \%TOTAL SUGAR}$$
$$\text{(Luff-Schoorl method)} + 0.1669 \text{ \%STARCH}$$

This equation is based on coefficients determined with adult cockerels, ignoring differences due to species and age and differences in nutrient digestibilities between feedstuffs.

2. The CVB (2004) equations:

$$AME_N \text{ adult (kJ/kg)} = 18.03 \text{ dCP} + 38.83 \text{ dCFAT} + 17.32 \text{ dNfE}$$

$$AME_N \text{ broiler (kJ/kg)} = 15.56 \text{ dCP} + 38.83 \text{ dCFAT} + 17.32 \text{ dNfE}$$

In the Netherlands two different general equations are used: one for adult poultry and one for growing broilers, which are based on measured nutrient digestibilities in target animals for each feedstuff.

3. The Rostock equation:

$$AME_N \text{ (kJ/kg)} = 18.8 \text{ dCP} + 39.8 \text{ dCFAT} + 17.3 \text{ dSTARCH} + 16.0 \text{ dSUGARS} + 17.2 \text{ dDNFR}$$

In the Rostock equation the NfE fraction of the CVB equation is split into starch, sugars and NFR, which implies that the variation in the starch content of a feedstuff is directly related to its AME_N value.

The effect of the different feed evaluation systems on the AME_N value and on the ranking of feedstuffs is shown in Table 7.2.

It can be seen from Table 7.2 that the variation in energy values differ between the prediction systems used, except for the cereals. More importantly, the ranking between feedstuffs varies, which implies that the feedstuff composition based on linear programming of the diets will change due to the energy evaluation system. The system used will impact on the nutritional efficiency of the poultry stock being fed.

Table 7.2. The effect of different feed evaluation systems on the AME_N value and on the ranking of feedstuffs.[a]

Feedstuff	EU AME_N MJ/kg	%	INRA AME_N MJ/kg	%	Rostock AME_N MJ/kg	%	CVB[b] AME_N MJ/kg	%
Maize	13.4	100	13.4	100	13.5	100	13.6	100
Wheat	12.6	93	12.5	93	12.6	94	13.0	95
Barley	11.2	83	11.5	86	11.4	84	11.6	85
Oats	9.4	70	9.8	73	9.7	72	10.4	77
Soybean meal	9.0	67	9.9	74	8.2	61	9.3	68
Rapeseed meal	7.0	52	6.1	46	5.4	40	7.0	51
Sunflower seed meal	6.5	48	6.3	47	6.6	49	6.1	45
Peas	11.5	86	11.5	86	10.4	77	11.3	83
Tapioca	12.1	90	12.6	94	–	–	12.2	89
Wheat bran	7.6	57	7.0	52	8.2	61	6.4	47

[a]Values have been recalculated based on the feedstuff composition as published by Sauvant *et al.* (2004)
[b]AME_N Poultry (no correction for higher efficiency of fat utilization)

Amino acids

Not all amino acids found in the digesta originate from the diet; some are of endogenous origin. Endogenous amino acids at the ileal level can be divided into a basal fraction, assumed to be independent of the raw material and to occur in any diet, and a specific fraction, which is considered a characteristic of the single raw material. The relative contribution of those endogenous losses depends on the amino acid intake. Due to this, digestibility coefficients are distorted for low protein ingredients like grains. To overcome this, Lemme *et al.* (2004) corrected or standardized for basal endogenous loss using an average digestibility coefficient from published data. Adedokun *et al.* (2014) review the methodology and comment on the repeatability and consistency of endogenous amino acid losses from different laboratories. Surgical procedure, age of the animal, type of diet/substances tested, feeding method, gut stimulation, microbial fermentation and the contribution of urinary amino acids all contribute to differences in digestibility coefficients. Table 7.3 presents the differences between the true faecal digestible system (Sauvant *et al.*, 2004) and the standardized ileal digestible system (Lemme, 2005) for some selected raw materials and amino acids.

Differences vary from −14 to +10 points. The overall lack of consistency between the two systems prevents the use of any systemic and easy correction factor. There is no doubt that the ileal digestible system is more reliable and should predict animal performance more accurately, but due to the lack of available data, commercial implementation has been slow.

Table 7.3. Differences between true faecal digestible coefficients (F) and standardized ileal digestible coefficients (I) for selected raw material and amino acids.

	Lysine			Methionine			M+C			Threonine		
	F	I	Diff	F	I	Diff	F	I	Diff	F	I	Diff
Grains												
Barley	78	88	10	80	88	8	82	89	7	76	85	9
Maize	85	92	7	94	94	0	93	90	−3	88	85	−3
Sorghum	87	90	3	90	89	−1	88	84	−4	89	83	−6
Triticale	85	85	0	89	90	1	85	88	3	89	87	−2
Wheat	84	86	2	90	91	1	91	91	0	83	87	4
Plant protein												
Maize gluten meal	90	76	−14	98	88	−10	96	83	−13	93	79	−14
Cotton seed meal	63	65	2	75	72	−3	71	73	2	69	68	−1
Lupines	92	87	−5	91	89	−2	94	85	−9	94	83	−11
Peas	79	85	6	–	73	–	–	68	–	81	78	−3
Rapeseed meal	78	80	2	87	84	−3	84	80	−4	84	73	−11
Soybean meal	91	90	−1	91	91	0	88	86	−2	89	85	−4
Sunflower meal	83	87	4	93	92	−1	89	87	−2	87	82	−5

Minerals

In poultry, phosphorus is the main concern for growing chickens and calcium for laying birds. Mineral nutrition and mineral profile is about satisfying specific mineral requirements and also balancing their mutual interactions. Mineral requirements, e.g. calcium, are influenced by many different factors such as genetic progress, co-dependences between calcium and other minerals, calcium source and particle size (Roland, 1986).

Phosphorus evaluation in poultry has become increasingly important due to environmental reasons and the lowering of feed costs. Currently, three different feed evaluation systems are used in poultry:

1. Available phosphorus (NRC, 1994).
2. Slope-ratio technique for measuring available phosphorus giving a relative bioavailability (Sauvant *et al.*, 2004).
3. Dutch retainable phosphorus system (Van Der Klis and Blok, 1997).

There are differences between the systems due to the principles employed in their determination, ranging from ratio of calcium to phosphorus, response parameters, and differences between feed phosphates. Phosphorus plays a vital role in energy metabolism and the optimum for maximal tibia ash is higher than for maximal body weight gain. Phosphorus requirements for poultry must be based on animal performance, animal welfare, environmental impact and cost.

Calcium is an important mineral with regard to feed conversion ratio. Calcium may interact with protein, reducing the solubility (Gifford and Clydesdale, 1990) or forming calcium soaps with dietary lipid in the gut reducing the utilization of energy (Leeson, 1993).

FEED STRUCTURE

Particle size

Gizzard function is enhanced when structural components such as whole grain (wheat), coarse particles (mean particle distribution >1 mm) or fibre sources are included in broiler diets. The coarse structure will improve AME_N content and starch digestibility in the ileum and excreta, reduce feed cost and improve gut health. Broilers fed whole grains have better feed conversion ratios and reduced water intake relative to broilers fed pellets with ground wheat. Broilers fed whole grains have better gizzard development and the addition of whole grains affects the microbial population in the intestine. The usage of whole grain may also have an economic benefit in the feed mill, due to the reduction in energy and time necessary for grinding. Whole wheat combined with a protein-rich concentrate is widely used in broiler production. The next step is to evaluate the addition of a calcium source (limestone) to the diet mix. The optional addition of calcium makes it possible for the bird to balance the calcium concentration in the gut and may also improve the feed efficiency (Cowieson and Selle, 2012).

Recommendations regarding optimal particle size are contradictory mainly due to confounding factors such as: feed physical form, complexity of the diet, grain type, endosperm hardness, grinding method, pellet quality, access to litter and particle size distribution (Amerah *et al.*, 2007, 2008; Jacobs *et al.*, 2010; Chewning *et al.*, 2012; Pacheco *et al.*, 2013). The majority of broilers are fed crumbles in the starter period followed by pellets. The particle size has to be measured in the pellet and results indicate that coarsely ground grains had better feed efficiency than finely ground grains. The particle size distribution is known to be affected by the grain hardness and appears to be more important in wheat-based rations. Coarse particles that persist after pelleting affect gizzard development and the positive attributes associated with this have been discussed already.

Pellet quality

McKinney and Teeter (2004) and Quentin *et al.* (2004) produced experimental pelleted diets and simulated different pellet qualities with graded proportions of fines (20%, 40%, 60%, 80% and 100%). Both papers have mash as a treatment as well as 100% pellets. In both trials the increasing proportion of fines in the diets resulted in impaired feed intake of broilers and weight gain. McKinney and Teeter's (2004) data indicate that at 80% fines, animal performance was almost as low as that of the mash-fed birds. Quentin *et al.* (2004) also observed a reduced eating frequency and increased resting frequency with a lower proportion of fines resulting in more net energy used for production.

Lemme *et al.* (2006) conducted two experiments with 14–35-day-old Ross male broilers to investigate the interactions between increasing levels of BP (9.7, 10.7, 11.7 and 12.7 g digestible lysine/kg of feed) (CVB, 2004) and physical feed form. In this trial a good pellet contained 17% fines and a bad pellet 80% fines. Highest weight gain was achieved with good pellets; to achieve similar performance with poor pellet quality, higher levels of BP were needed. Arce-Menocal *et al.* (2009) compared feeding crumbles versus pellets to 42 days of age. They found a 14.6% increase in feed intake from using pellets versus crumbles was associated with an improvement in growth of 22%.

Pelleting broiler diets improves growth performance and therefore increasing the proportion of intact pellets in front of the bird is economically advantageous from a feed efficiency point of view.

GASTROINTESTINAL TRACT OPTIMIZATION

The objective is to establish an appropriate balanced microbiota biomass that provides immunity to its host and withstands any environmental or dietary challenges whilst supporting efficient feed conversion.

The intestinal ecosystem

The intestinal microbial community is shaped by many different factors, including the host itself. In poultry the intestinal microbiota are primarily composed of two beneficial phyla: Firmicutes and Bacteroidetes (Lu *et al.*, 2008). The ratio between these two groups shifts in relation to the chemical composition of the host's mucosal surface. The Mollicutes, a division of the Firmicutes phylum, are normally present in low abundance in the gut of lean animals, but they expand dramatically when fed high-density diets. The Mollicutes-enriched community not only facilitates transfer of calories from the diet to the host, but also affects the response of the host to absorbed calories. This finding supports the hypothesis that the gut microbiota of obese animals are more efficient in releasing nutrients from the diet than the microbiota of lean animals (Turnbaugh *et al.*, 2006). The gut profile could positively affect the energy balance of the animal, stimulating the possibility that microbiota manipulation may lead to better animal performance. The study of the gut microbiota of animals with superior performance and better feed conversion could give us biomarkers for manipulation and early implantation of gut microbiota (Pedrosa *et al.*, 2012).

Exogenous enzymes

The benefits of exogenous feed enzymes that are pertinent to nutritional efficiency are: (i) reduced variability in the nutritive value between batches of ingredients; (ii) improved digestion and lesser amounts of undigested nutrient reaching the lower gut as well as a shift in gut flora toward favourable bacterial species (Bedford and Cowieson, 2012); and (iii) improved intestinal morphology and integrity, which results in enhanced digestion and absorption of dietary components.

The supplementation of animal feed with exogenous enzymes improves the nutritional value of some feed ingredients through increasing the efficiency of digestion. Exogenous feed enzymes are chosen on the basis of substrates in the ingredients used in feed formulation. However, birds are not fed substrates but ingredients with substrates in complex matrices (Ravindran, 2013). Due to these reasons, the potential nutritive value of ingredients is not realized at the bird level and no common feed ingredient is 100% digested. Even the digestion of substrates (starch, protein and lipid) for which birds produce sufficient amounts of endogenous enzymes is incomplete, with 10% to 20% of these substrates being undigested and excreted. The need to improve the digestion of these undigested substrates is the principal rationale behind the use of exogenous enzymes. Bao *et al.* (2013) attempted to quantify the energetic loss in young broiler chickens in a typical maize–soybean diet. The energetic loss contributed by undigested fat and starch was calculated to be about 210 kcal/kg and the energetic loss associated with undigested protein is around 190 kcal/kg (Cowieson, 2010). Bao *et al.* (2013) concluded that there is a potential loss of around 400 kcal/kg from undigested substrates (Table 7.4). It is clear therefore that the potential for enzymes to deliver value is significant.

Table 7.4. Undigested fat fraction in different diets (from Almirall *et al.* (1995), Slominski *et al.* (2006), Cowieson (2010), Liu *et al.* (2010).

Fat sources	Dietary concentration g/kg	Ileal digestion coefficient %	Undigested fraction %	Gross energy loss kcal/kg[1]
Maize–soy with flaxseed	80.0	56.4	43.6	325.8
Wheat–fish meal	–	66.1	33.9	–
Barley–soy diets	56.0	76.4	23.6	123.4
Maize–soy with cottonseed	35.5	82.7	17.3	57.4
Maize–soy diets	80.0	85.1	14.9	111.3

[1]The gross energy value in fat was calculated as 39.04 MJ/kg or 5450 kcal/kg (Carre *et al.*, 1995)

Extensive reviews of the effect of phytate on the digestibility of carbohydrates, amino acids, proteins and minerals can be found in the literature (Selle and Ravindran, 2007; Adeola and Cowieson, 2011). More recently, the use of phytase at higher levels than recommended doses has become popular. The aim is to dephosphorylate the phytic acid as quickly and comprehensively as possible to less-reactive inositol phosphate esters during the early gastric phase of digestion and to reduce its anti-nutritional effects. Phytate degradation was correlated with marked improvements in bird performance, nutrient retention, tibia ash and AME and these increases were most pronounced at the highest phytase inclusion rate (Ravindran, 2013).

Mitochondria

Mitochondrial conversion of energy as NADH and FADH to ATP is an important contributor to energy supply accounting for approximately 20–30% of resting energy requirements (Zurlo *et al.*, 1990). Due to this, changes in mitochondrial efficiency will have large impacts on energetic and feed efficiency (Bottje and Carstens, 2008). Broiler chickens have more efficient muscle mitochondria than laying chickens and this is correlated with their higher efficiency and increased growth rates (Bottje *et al.*, 2002). A hallmark of low feed efficiency is greater oxidative stress that includes higher mitochondrial reactive oxygen species production, extensive protein damage and the up-regulation of stress-responsive genes. Currently it is unknown how diet (creatine, L-carnitine, B vitamins, vitamin A, vitamin C, vitamin E, trace minerals, antioxidants and polyunsaturated fatty acids) affects mitochondrial efficiency in chickens (Acetoze *et al.*, 2013). Fariss *et al.* (2005) demonstrated protection from oxidative stress by administration of antioxidants such as vitamin E and ubiquinone. Copper and zinc are known as minerals that have antibiotic and antioxidant properties by reducing the effects of secondary bacterial infection and macromolecular damage by free radicals through the action of superoxide dismutase (Nonn *et al.*, 2003). Understanding the role of mitochondrial efficiency in feed efficiency will aid in feeding decisions and selection of broilers that are more energetically efficient to optimize nutrient use.

CONCLUSION

The culmination of diet design, integrated with nutritional experience and prejudice is predicated on the knowledge of a given feed intake. The definition of feed intake is seemingly self-evident, but in practice its measurement is elusive and difficult to predict. Nutrient response research will continue as geneticists continue to select for traits that enhance bird performance. Maximum production is attainable. However, measuring optimal production under the given business objective requires continual research and reappraisal of specifications and nutritional systems as prices of raw ingredients change and nutritional technology advances. Efficiency of energy conversion based upon the product objective and the optimization thereof should supersede feed conversion ratio as a response criterion. New developments in nutritional thinking and interactions between various nutrients lead to greater efficiency and are easily measured in terms of product output and economic advantage. Feedstuff evaluation systems are by nature creating more sustainable solutions that lead to more accurate prediction of animal performance and optimal usage of available raw materials. The selection of feedstuff evaluation systems and the nutrient responses derived thereon are strategically key to the nutritionist. Raw material buyers need to understand the implications of the nutritionists' systems in their raw material selection and quality criteria to achieve the common business objective. Sustainable production of the major raw ingredients, grains and soya oilcake, need to be ensured/entrenched with the opportunistic use of alternatives when they are available. Alternative raw materials such as insect protein will be evaluated against soybean meal in terms of cost, nutritional value and availability. Feed additives that enhance animal performance have become endemic and require careful and accurate product evaluation to ensure repeatable bird performance and sustainability before being used.

REFERENCES

Acetoze, G., Kurzbard, R., Ramsey, J.J., Klasing, K.C. and Rossow, H.A. (2013) Influence of mitochondrial function on feed efficiency of broilers with and without growth enhancing levels of minerals supplementation during coccidiosis challenge. In: Oltjen, J.W., *et al.* (eds) *Energy and Protein Metabolism and Nutrition in Sustainable Animal Production.* EAAP Publication No. 134. Wageningen Academic Publishers, the Netherlands, pp. 393–394.

Adedokun, S.A., Jaynes, P., Abd El-Hack, M.E., Payne, R.L. and Applegate, T.J. (2014) Standardized ileal amino acid digestibility of meat and bone meal and soyabean meal in laying hens and broilers. *Poultry Science* 93, 420–428.

Adeola, O. and Cowieson, A.J. (2011) Opportunities and challenges in using exogenous enzymes to improve non-ruminant animal production. *Journal of Animal Science* 89, 3189–3218.

Aftab, U. (2009) Responses of broilers to practical diets with different metabolizable energy and balanced protein concentrations. *Brazilian Journal of Poultry Science* 11, 169–173.

Aftab, U. (2012) Dietary amino acid optima: an economical appraisal. *Journal of Applied Poultry Research* 21, 738–743.

Almirall, M., Francesch, M., Perez-Vendrell, M., Brufau, A.M. and Esteve-Garcia, E. (1995) The differences in intestinal viscosity produced by barley and B-glucnase alter digest enzyme activities and ileal nutrient digestibility more in broiler chicks than cocks. *The Journal of Nutrition* 125, 947–955.

Amerah, A.M., Ravindran, V., Lentle, R.G. and Thomas, D.G. (2007) Feed particle size: implications on the digestion and performance of poultry. *World's Poultry Science Journal* 63, 439–455.

Amerah, A.M., Ravindran, V., Lentle, R.G. and Thomas, D.G. (2008) Influence of feed particle size on the performance, energy utilization, digestive tract development, and digesta parameters of broiler starters fed wheat- and corn-based diets. *Poultry Science* 87, 2320–2328.

Arce-Menocal, J., Avila-Gonzalez, E., Lopez-Coello, C., Garibay-Torres, L. and Martinez-Lemus, L.A. (2009) Body weight, feed particle size, and ascites incidence revisited. *Journal of Applied Poultry Research* 18, 465–471.

Aviagen (2007) *Broiler Nutrition Specification: Ross 308*. Huntsville, Alabama.

Bao, Y.M., Romero, L.F. and Cowieson, A.J. (2013) Functional patterns of exogenous enzymes in different feed ingredients. *World's Poultry Science Journal* 69, 759–774.

Bedford, M.R. and Cowieson, A.J. (2012) Exogenous enzymes and their effects on intestinal microbiology. *Animal Feed Science Technology* 173, 76–85.

Bottje, W.G. and Carstens, G.E. (2008) Association of mitochondrial function and feed efficiency in poultry and livestock species. *Journal of Animal Science* 87, E48–E63.

Bottje, W., Tang, Z.X., Iqbal, M.D., Cawthon, D., Okimoto, R., Wing, T. and Cooper, M. (2002) Association of mitochondrial function with feed efficiency within a single genetic line of male broilers. *Poultry Science* 81, 546–555.

Carre, B., Gomez, J. and Chageneau, A.M. (1995) Contribution of oligosaccharide and polysaccharide digestion, and excreta loss of lactic acid and short chain fatty acids, to dietary metabolisable energy values in broiler chickens and adult cockerels. *British Poultry Science* 36, 611–629.

Chewning, C.G., Stark, C.R. and Brake, J. (2012) Effects of particle size and feed form on broiler performance. *Journal of Applied Poultry Research* 21, 830–837.

Cho, M. (2011) The relationship between diet content of energy and amino acids and the impact on broiler performance. MSc thesis, University of Saskatchewan, Saskatoon, Canada.

Classen, H.L. (2013) Response of broiler chickens to dietary energy and its relationship to amino acid nutrition. In: *Proceedings of the Australian Poultry Science Symposium*. The Poultry Research Foundation, University of Sydney, Sydney, Australia, pp. 107–114.

Cowieson, A.J. (2010) Strategic selection of exogenous enzymes for corn/soy-based diets. *Journal of Poultry Science* 47, 1–7.

Cowieson, A.J. and Selle, P.H. (2012) The Environmental Impact of Low Feed Conversion Ratios in Poultry. Available at: http://Engormix.com (published 17 July 2012).

CVB (2004) *Tabellenboek Veevoeding 2004*. Centraal Veevoerderbureau, Lelystad, the Netherlands.

De Beer, M. (2010) Nutrition for maximum profit – do the math. *Aviagen Brief* (October), 5 pp.

D'Mello, J.P.F. (2003) Responses of growing poultry to amino acids. In: D'Mello, J.P.F. (ed.) *Amino Acids in Animal Nutrition*, 2nd edn. CAB International, Wallingford, UK.

Eits, R.M., Kwakkel, R.P., Verstegen, M.W.A. and Den Hartog, L.A. (2005a) Dietary balanced protein in broiler chickens. 1. A flexible and practical tool to predict dose–response curves. *British Poultry Science* 46(3), 300–309.

Eits, R.M., Giesen, G.W.G., Kwakkel, R.P., Verstegen, M.W.A. and Den Hartog, L.A. (2005b) Dietary balanced protein in broiler chickens. 2. An economic analysis. *British Poultry Science* 46(3), 310–317.

Esmail, S.H. (2012) Fibre plays a supporting role in poultry nutrition. *World Poultry News*. Available at: http://www.worldpoultry.net (accessed 10 February 2012).

Fariss, M.W., Chan, C.B., Patel, M., Van Houten, B. and Orrenius, S. (2005) Role of mitochondria in toxic oxidative stress. *Molecular Intervention* 5, 98–114.

Fisher, C. and Wilson, B.J. (1974) Energy requirements of poultry. In: Morris, T.M. and Wilson, B.M. (eds) *Energy Requirements of Poultry*. British Poultry Science Ltd, Edinburgh, UK, pp. 151–184.

Gifford, S.R. and Clydesdale, F.M. (1990) Interactions among calcium, zinc and phytate with three protein sources. *Journal of Food Science* 55, 1720–1724.

Gonzalez-Alvarado, J.M., Jimenez-Moreno, E., Lazaro, R. and Mateos, G.G. (2007) Effects of type of cereal, heat processing of the cereal, and inclusion of fibre in the diet on productive performance and digestive traits in broilers. *Poultry Science* 86, 1705–1715.

Gous, R. (2013) Predicting food intake in broilers and laying hens. In: *Proceedings of the Australian Poultry Science Symposium*. The Poultry Research Foundation, University of Sydney, Sydney, Australia, pp. 99–106.

Jacobs, C.M., Utterback, P.L. and Parsons, C.M. (2010) Effects of corn particle size on growth performance and nutrient utilization in young chicks. *Poultry Science* 89, 539–544.

Jimenez-Moreno, E., Gonzalez-Alvarado, J.M., De Coca-Sinova, A., Lazaro, R. and Mateos, G.G. (2009) Effects of source of fibre on the development and pH of the gastrointestinal tract of broilers. *Feed Science and Technology* 154, 93–101.

Jimenez-Moreno, E., Romero, C., Berrocoso, J.D., Frikha, M. and Mateos, G.G. (2011) Effects of the inclusion of oat hulls or sugar beet pulp in the diet on gizzard characteristics, apparent ileal digestibility of nutrients, and microbial count in the ceca in 36-day-old broilers reared on floor. *Poultry Science* 90(Suppl. 1), 153 Abstract.

Jimenez-Moreno, E., Frikha, M., De Coca-Sinova, A., Garcia, J. and Mateos, G.G. (2013a) Oat hulls and sugar beet pulp in diets for broilers. 1. Effects on growth performance and nutrient digestibility. *Animal Feed Science and Technology* 182, 33–43.

Jimenez-Moreno, E., Frikha, M., De Coca-Sinova, A., Lazaro, R.P. and Mateos, G.G. (2013b) Oat hulls and sugar beet pulp in diets for broilers. 2. Effects on the development of the gastrointestinal tract and on the stratic of the jejuna mucosa. *Animal Feed Science and Technology* 182, 44–52.

Kalmendal, R., Elwinger, K., Holm, L. and Tauson, R. (2011) High-fibre sunflower cake affects small intestinal digestion and health in broiler chickens. *British Poultry Science* 52, 86–96.

Kidd, M.T., Lerner, S.P., Allard, J.P., Rao, S.K. and Halley, J.T. (1999) Threonine needs of finishing broilers: growth, carcass and economic responses. *Journal of Applied Poultry Research* 8(2), 160–169.

Leeson, S. (1993) Recent advances in fat utilization by poultry. In: Farrell, D.J. (ed.) *Recent Advances in Animal Nutrition in Australia*. University of New England, New South Wales, Australia, pp. 170–181.

Leeson, S. (2012) Biological limits to productivity. *Journal of Applied Poultry Research* 21, 145–148.

Lemme, A. (2005) *Standardised Ileal Amino Acid Digestibility in Broilers*. Degussa AG, Hanau, Germany.

Lemme, A., Ravindran, V. and Bryden, W.L. (2004) Ileal digestibility of amino acids in feed ingredients for broilers. *World's Poultry Science Journal* 60, 423–437.

Lemme, A., Kemp, C., Fisher, C., Kenny, M. and Petri, A. (2005) Responses of growing broilers to varying dietary energy and balanced amino acid levels. In: *Proceedings of the 15th European Symposium on Poultry Nutrition*. Balatonfured, Hungary, pp. 465–468.

Lemme, A., Wijtten, P.A.J., Van Wichen, J., Petri, A. and Langhout, D.J. (2006) Responses of male growing broilers to increasing levels of balanced protein offered as coarse mash or pellets of varying quality. *Poultry Science* 85, 721–730.

Lemme, A., Kemp, C., Fisher, C. and Fickler, J. (2009) Economically optimal amino acid levels in broiler diets. *Amino News Special Issue* 13(3), 3–13.

Liu, N., Ru, Y.J., Wang, J.P. and Xu, T.S. (2010) Effect of dietary sodium phytate and microbial phytase on the lipase activity and lipid metabolism of broiler chickens. *British Journal of Nutrition* 103, 862–868.

Lu, J., Santo Domingo, J.W., Lamendella, R., Edge, T. and Hill, S. (2008) Phylogenetic diversity and molecular detection of bacteria in gull feces. *Applied Environmental Microbiology* 74, 3969–3976.

Madsen, T.G., Carrol, S., Kemp, C. and Lemme, A. (2010) Influence of energy and balanced protein levels in wheat-based diets on Ross 308 performance. In: *Proceedings of the Australian Poultry Science Symposium*. The Poultry Research Foundation, University of Sydney, Sydney, Australia, pp. 95–99.

Mateos, G.G., Guzman, P., Saldana, B., Jiminez-Moreno, E., Perez Bonilla, A. and Lazaro, R. (2013) Relevance of dietary fibre in poultry feeding. In: *Proceedings of the 19th European Symposium on Poultry Nutrition*, Potsdam, Germany, pp. 52–59.

McKinney, L.J. and Teeter, R.G. (2004) Predicting effective caloric value of non-nutritive factors: I. Pellet quality and II. Prediction of consequential formulation dead zones. *Poultry Science* 83, 1165–1174.

Nonn, L., Williams, R.R., Erikson, R.P. and Powis, G. (2003) The absence of mitochondrial thioredoxin 2 causes massive atopsis, exencephaly, and early embryonic lethality in homozygous mice. *Molecular Cell Biology* 23, 916–922.

NRC (1994) *Nutrient Requirements of Poultry*, 9th revised edn. National Academic Press, Washington, DC.

Pacheco, W.J., Stark, C.R., Ferket, P.R. and Brake, J. (2013) Evaluation of soybean meal source and particle size on broiler performance, nutrient digestibility, and gizzard development. *Poultry Science* 92, 2914–2922.

Pedrosa, A.A., Maurer, J., Cheng, Y. and Lee, M.D. (2012) Remodeling the intestinal ecosystem toward better performance and intestinal health. *Journal of Applied Poultry Research* 21, 432–443.

Plumstead, P.W., Romero-Sanchez, H., Paton, N.D., Spears, J.W. and Brake, J. (2007) Effect of dietary metabolizable energy and protein on early growth responses of broiler to dietary lysine. *Poultry Science* 86, 2639–2648.

Quentin, M., Bouvarel, I. and Picard, M. (2004) Short- and long-term of feed form on fast- and slow-growing broilers. *Journal of Applied Poultry Research* 13, 540–548.

Ravindran, V. (2013) Feed enzymes: the science, practice and metabolic realities. *Journal of Applied Poultry Research* 22, 628–636.

Roland, D.A. (1986) Eggshell quality IV: Oystershell versus limestone and the importance of particle size or solution of calcium sources. *World's Poultry Science Journal* 42, 166–171.

Sauvant, D., Perez, J.M. and Tran, G. (2004) *Tables of Composition and Nutritional Value of Feed Materials*. INRA Editions, Wageningen Academic Press, the Netherlands.

Selle, P.H. and Ravindran, V. (2007) Microbial phytase in poultry nutrition. *Animal Feed Science Technology* 135, 1–41.

Slominski, B.A., Meng, X., Campbell, I.D., Guenter, W. and Jones, O. (2006) The use of enzyme technology for improved energy utilisation from full-fat oilseeds. Part II: flaxseed. *Poultry Science* 85, 1031–1037.

Svihus, B. (2011) The gizzard: function, influence of diet structure and effects on nutrient availability. *World's Poultry Science Journal* 67, 207–224.

Tillman, P.B. (2011) Special considerations for amino acids in broiler nutrition. In: *III International Symposium on Nutritional Requirements of Poultry and Swine*, 29–31 March, Viscosa, Brazil, pp. 21–31.

Tillman, P.B. (2012) Current amino acid considerations for broiler diets: requirements, ratios, economics. In: *Proceedings of Ajinomoto Animal Nutrition Technical Seminar*, Amino Acid Nutrition in Broiler Diets, Bangkok, Thailand, pp. 24–56.

Tillman, P.B. and Dozier, W.A. (2013) Current amino acid considerations for broiler diets: requirements, ratios, economics. In: *Proceedings of Ajinomoto Animal Nutrition Technical Seminar*. Amino Acid Nutrition in Broiler Diets, Bangkok, Thailand, pp. 24–56.

Tillman, P.B. and Sriperm, N. (2011) Estimating amino acid requirements of broilers using static: production and dynamic: market based analysis. September 6. *Pre-symposium of Arkansas Nutrition Conference by Huvepharma*. Rogers, Arkansas.

Turnbaugh, P.J., Ley, R.E., Mahowald, M.A., Magrini, V., Mardis, E.R. and Gordon, J.I. (2006) An obesity associated gut microbiome with increased capacity for energy harvest. *Nature* 444, 1027–1031.

Van Der Klis, J.D. and Blok, M.C. (1997) *Definitief system opneembaar fosfor pluimfee.* CVB documentatierapport nr. 20, the Netherlands.

Van Der Klis, J.D. and Fledderus, J. (2007) Evaluation of raw materials for poultry: what's up? In: *Proceedings of the 16th European Symposium on Poultry Nutrition.* World Poultry Science Association, Strasbourg, France, 26–30 August.

Van Der Klis, J.D., De Lange, L. and Kwakernaak, C. (2014) Feed ingredients: assessment and enhancement of nutritional value. In: *Proceedings of the Australian Poultry Science Symposium.* The Poultry Research Foundation, University of Sydney, Sydney, Australia, pp. 31–37.

Weurding, R.E., Veldman, A., Veen, W.A.G., Van Der Aar, P.J. and Verstegen, M.W.A. (2001) *In vitro* starch digestion correlates well with the rate and extent of starch digestion in broiler chickens. *The Journal of Nutrition* 131, 2336–2342.

Weurding, R.E., Enting, H. and Verstegen, M.W.A. (2003) The effect of site of starch digestion on performance of broiler chickens. *Animal Feed Science and Technology* 110, 175–184.

Wijtten, P.J.A., Prak, R., Lemme, A. and Langhout, D.J. (2004) Effect of different dietary ideal protein concentrations on broiler performance. *British Poultry Science* 45, 504–511.

Zurlo, F., Larson, K., Bogardus, C. and Ravussin, E. (1990) Skeletal muscle metabolism is a major determinant of resting energy expenditure. *Journal of Clinical Investigation* 86, 1423–1427.

PART V

Avian and Human Health – Interactions, Opportunities and Threats

CHAPTER 8
Food Safety: Prevention is Better than Crisis Management

Patrick Wall* and Zhongyi Yu

University College Dublin, Ireland

INTRODUCTION

Poultry meat and eggs are affordable sources of valuable protein and will be a large part of the solution to meeting the needs of the growing world population. In addition, as human nutritionists increasingly address the requirements of people at different life stages, and different levels of activity, both poultry and eggs are bioavailable sources of high quality protein. Therefore the commercial opportunities for the sector are many, and the future is looking good, provided adverse publicity can be avoided associated with: (i) food safety; (ii) animal welfare; (iii) health and nutrition; and (iv) environmental impact. Public perception is often informed by sensational news coverage and items are placed higher on the agendas of policy makers as a result of the intensity of the media coverage of an issue. Policy makers and regulators are not consistent in how they address risk along the food chain, or in society at large, and often their response is in proportion to the media coverage rather than the risk to public health. At times the regulatory response can be disproportionate to the risk. In most instances there are several solutions to a crisis and the focus should be on delivering the optimum level of consumer health protection whilst doing the minimum damage to both commercial interests and consumer confidence. Sadly this is rarely the case, emphasizing the importance of focusing on prevention rather than crisis management.

WHAT BUSINESS ARE YOU IN?

In the poultry sector many of the stakeholders work within their confined areas without realizing the real objective of their activities (Fig. 8.1). Phenomenal

*Corresponding author: patrick.wall@ucd.ie

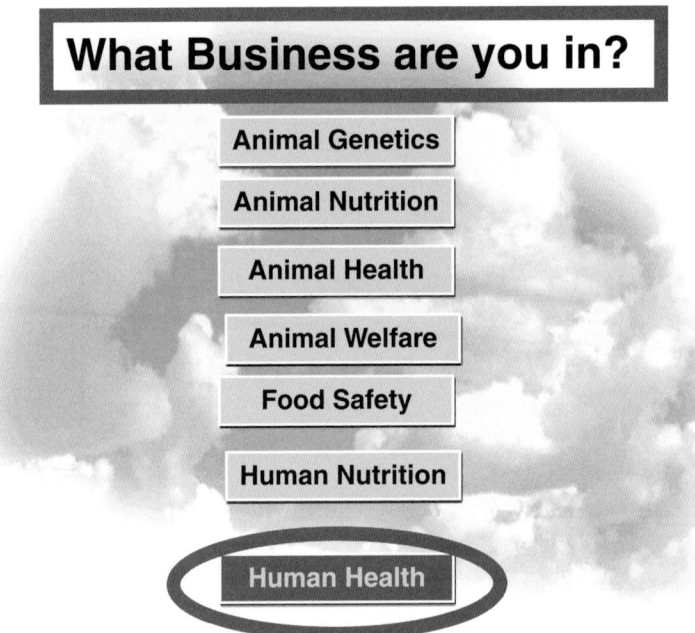

Fig. 8.1. No matter which part of the food chain you are working on, you are in the human health business.

advances have been, and are being, made in poultry genetics. Relentless selection of production traits has delivered us very different birds from those our forebears tended. Poultry nutritionists are far ahead of their human counterparts when it comes to diet formulation and performance. They control the totality of the diet, the rearing environment and have clear outcome measures such as food conversion efficiency, weight gain or egg production etc. Furthermore, breed types have dramatically reduced individual variability and for broilers and laying hens the stock are very closely related genetically. Genetics creates the potential and nutrition delivers on it, but suboptimal poultry health or welfare can undermine any gains from the former two. Good poultry health status is essential for safe food and stressed birds are more prone to disease, so the production of safe food should be a goal for all engaged in agri-food activities. Those working in animal genetics, in feed mills and on farms are as much in the food business as those operating processing facilities or hotels and restaurants. Outbreaks of food poisoning, recalls as a result of contamination incidents, and food fraud associated with poultry continue to occur, so there is no shortage of ammunition for the journalists to generate adverse publicity. Both human food and animal feed are globally distributed and so also are the media, which have an insatiable appetite for bad news. With the conventional media feeding off the social media and vice versa, often without taking time to verify the story, companies can suffer huge economic consequences if implicated in an incident, whether they are innocent or guilty. Therefore, it behoves all the stakeholders in the poultry sector to focus on prevention rather than crisis management. Consumer confidence is easily eroded and a scare associated with one company can often damage the entire sector.

Unfortunately the poultry sector still suffers from the legacy of the high incidence of *Salmonella enteritidis* outbreaks in the 1980s and early 1990s before the introduction of vaccines and other interventions. As a result, chicken and eggs are often the number one suspect in outbreaks of food poisoning where the food source has not been identified (Sharp, 1988; Rampling, 1993).

In addition the perception that antibiotics are overused in the sector puts chicken in the firing line every time the issue of 'superbugs' appears in the media.

However, producing safe food is not the end game. Food is the fundamental fuel for human health and 'you are what you eat' is a true dictum. Diet-related disease and obesity-related health problems in humans are major public health issues in both developed and developing countries. Primary agricultural output is coming increasingly under the spotlight in both the scientific and general media for contributing to human health problems. Headlines like 'red meat causes cancer', 'dairy products clog up your arteries' now join those of 'superbugs on farms', to undermine consumer confidence in some of the output from the agri-food sector. Poultry meat and eggs are well positioned to address many of the concerns about human nutrition.

In addition, it is now possible to modify the composition of the final output from farms by genetic selection of the poultry and altering the rations fed. Increasingly birds are being bred, and micronutrients are being fed to them, to deliver a healthier final product, whether that means less saturated fat, more vitamins or minerals, more omega-3 etc. Human nutrition is key to health so the final objective for most activities in the poultry sector should be improved human health and all engaged in activities along the food chain should consider themselves in the 'human health business'. Doctors and nurses are not in the health business; rather they are in the sickness business!

Once everyone in the poultry sector accepts that the end game is human health, consumer protection will become paramount and the rationale for biosecurity in mills and on farms, and increased attention to detail along the production cycle and supply chain will be more apparent. A greater understanding of why robust controls are essential at every stage will increase compliance.

COMPLEXITY OF THE FOOD CHAIN

The food chain is only as strong as its weakest link and from premix and animal feed forward to consumers' kitchens there can be no tolerance for substandard practices if adverse events are to be prevented.

The food chain is increasingly complex and consumers have been naively convinced that it is a straight line (Fig. 8.2), with the words 'farm to fork' written into the food legislation in many jurisdictions. The food chain is now more like a maze and even at farm level inputs such as agri-chemicals, animal remedies, animal commodity feed and micro-nutrients are globally sourced, e.g. many of the vitamins and minerals currently added to animal rations in the EU come from China. This illustrates that we truly live in a global village and a huge

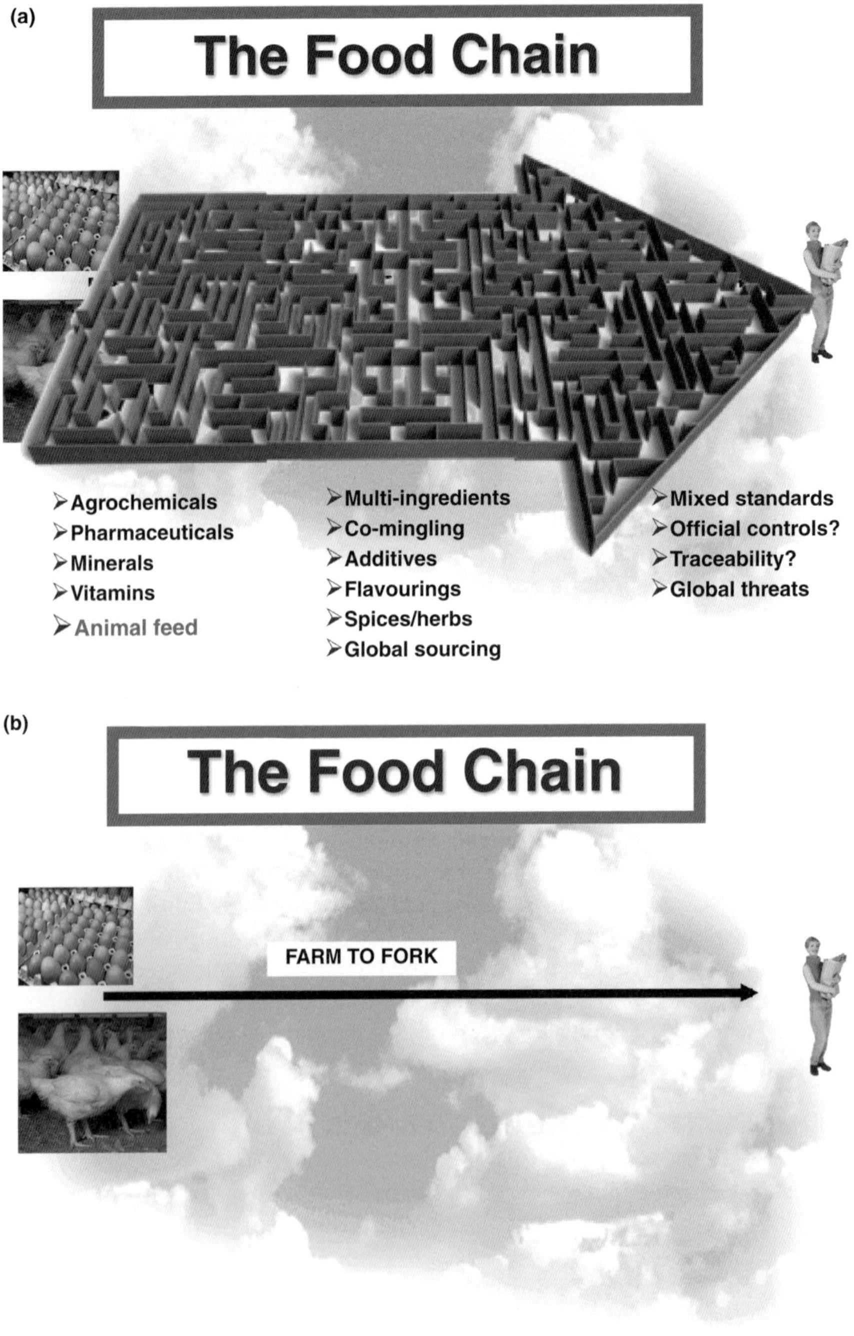

(a)

The Food Chain

- ➢Agrochemicals
- ➢Pharmaceuticals
- ➢Minerals
- ➢Vitamins
- ➢Animal feed

- ➢Multi-ingredients
- ➢Co-mingling
- ➢Additives
- ➢Flavourings
- ➢Spices/herbs
- ➢Global sourcing

- ➢Mixed standards
- ➢Official controls?
- ➢Traceability?
- ➢Global threats

(b)

The Food Chain

FARM TO FORK

Fig. 8.2. The modern food chain (a) is much more complicated than most people think it is (b).

interdependency exists between nations when it comes to protecting the food supply. The health of a country's citizens often depends on controls in operation in another jurisdiction completely.

Many foodstuffs can contain ingredients from many processors and manufacturers in different countries, or even different continents, and if one considers all of the ingredients in something like a pizza, and where they might come from, you could have the world on your plate! The flour, vegetables, cheese, processed meats, olive oil, spices and herbs are all globally sourced, posing a big challenge if you wanted to put a picture of the farm of origin on the packaging. The standards vary between manufacturers and processors within, and between, countries so the safety of the final product is governed by the standards of the weakest supplier of ingredients. The co-mingling of ingredients from several sources in finished products has made precise recalls nigh impossible. This was well illustrated in Ireland in 2008 when recycled bread, contaminated with dioxin, was fed on only ten pig farms and an inability to trace the contaminated pork after secondary processing necessitated a global recall of Irish pork (Casey et al., 2010).

The enforcement capabilities vary in different countries and many countries have strict regulations. Food laws are very similar across the EU and other single markets, and harmonized controls exist between trading partners; however, not every jurisdiction has the same resources for enforcement, and surveillance, leaving a far from ideal situation.

The more steps in the food chain the more opportunities exist for things to go wrong and the more people that are involved the more likely that one could be a shoddy operator or worse still a criminal. The recent EU horse meat scandal highlighted the vulnerability of our current supply chain and exposed weaknesses that can be easily exploited by those motivated to engage in criminal activity. While analytical chemists are now capable of detecting contaminants at parts per billion, or parts per trillion, and microbiologists are able to detect the presence of lower and lower numbers of microbes, lorry-loads of horse meat appeared to be moving around the EU disguised as beef, and horse meat was illegally included into a range of processed 'beef' products.

One might think that a food business operator purchasing inputs from a company registered or licensed by the authorities would have a legitimate expectation that it would be safe. However, a chronology of incidents has demonstrated that this is not the case. Therefore companies have to take responsibility for their own brands and reputations and for protecting the health of their customers and the health of their businesses. The legal requirements are like a pass level in an exam, but robust, accredited quality assurance schemes are like the honours paper.

FOOD FRAUD AND CONSUMER DECEPTION

In the global food chain a small percentage of the stakeholders have no regard for food safety legislation or consumer health and are prepared to engage in a range of shoddy practices and even criminal activity to increase their profits. A series of high profile food frauds on several continents, related to the poultry sector, continues to undermine consumer confidence in the safety of the food supply, in industries' commitment to produce safe food and in the regulatory authorities' ability to police the food chain.

In China in July 2014 American fast-food giants McDonald's and Yum, the owner of Pizza Hut and KFC, became embroiled in a food scare when under-cover video footage in one of their suppliers revealed falsified expiry dates on meat products, a range of substandard food hygiene practices and falsified records designed to mislead inspectors (Bloomberg, 2014). Nearly half of Yum's profits came from China in 2012; however, the 2014 scandal compounded dam-age done to their brands and reputation after a previous scandal in November 2012 when a supplier was alleged to have added 'industrial chemicals' to feed to accelerate the growth of the chickens (China Daily Mail, 2012).

In July 2014 an investigation by the Food Safety Authority of Ireland (FSAI) identified lamb kebabs on sale that were in fact 60% chicken (FSAI, 2014). This followed similar findings of an investigation by the Food Standards Agency in the UK (Food Standards Agency, 2014b). These revelations indicated that despite increased vigilance and awareness of food fraud after the 2013 horse meat scan-dal, some companies are still prepared to engage in food fraud (Premanandh, 2013).

In June 2013 the Chilean Ministry of Agriculture notified the US authorities of chicken contaminated with dioxin, which triggered a major recall of almost 200,000 lb of chicken in the USA (Food Safety News, 2013). In 2013, over 160 German farms were suspected of selling eggs labelled 'organic' but not adhering to the conditions required for this label (Global Post, 2013; Telegraph, 2013). In November 2006 the UK police had to initiate an investigation after over 30 mil-lion eggs supplied to shops and supermarkets may have been passed off illegally as free-range (Telegraph, 2006). In 2004 another fraud came to light when alle-gations began circulating in the egg industry that there were vastly more British free-range and organic eggs being sold in shops than could ever possibly be laid in UK farms (The Guardian, 2010).

In May 2002 the FSAI highlighted a practice where frozen, salted chicken-breast fillets, that attracted a lower import tariff, were imported into Holland from Thailand and Brazil and then defrosted and tumbled with water and other ingre-dients, including hydrolysed pig and beef protein, to assist in the retention of water. This resulted in as much as 50% of the fillet being composed of water. The fillets were then re-frozen, packed in bulk boxes and shipped to Ireland and the UK where they were predominantly sold to the catering trade. The FSAI high-lighted infringements of the labelling regulations and quality issues with the main concerns being the unacceptable practice of adding pig and beef protein to the product, incorrect declaration of meat content and the failure to disclose added collagen and water. Product descriptions suggesting whole, unadulterated chicken were misleading customers as to the content (FSAI, 2002a, b).

In 1999 in Belgium, animal feed contaminated with dioxins and polychlori-nated biphenyls affected more than 2500 poultry and pig farms. This incident led to the formation of the Belgium Federal Food Safety Agency and, with another scandal concerning BSE, resulted in a reform of EU food legislation. The loss to the Belgium economy was estimated at €1500–€2000 million (Van Larebeke *et al.*, 2001; Covaci *et al.*, 2008).

Traceability is key to maintaining consumer confidence and mislabelling country of origin undermines consumer confidence in the entire poultry sector. Initiatives are underway comparing stable isotopes from the poultry, eggs, and the water and poultry feed consumed by the birds to pinpoint where the birds were reared and hence the true origin of the products (The Ranger, 2013; Food-Integrity, 2014).

PRE-HARVEST FOOD SAFETY

Reducing the microbial load entering the food chain by implementing flock-health initiatives reduces the challenge on food-safety management systems and controls in food-processing plants, commercial catering establishments and in domestic kitchens. Intensification of farming systems can create increased opportunities for disease spread but can also present the opportunity to control the feed, water and environment to ensure disease incidence is reduced to a minimum. Local (on farm), regional, national and international issues can all impact on the health status of poultry.

There is a role for the use of antimicrobials and other pharmaceutical agents in poultry production, but they are not a replacement for good husbandry practices and stakeholders must be aware of when, and how, they should be used appropriately to avoid residues in the food chain and the generation of organisms resistant to antimicrobials (Tollefson and Karp, 2004). The Scandinavian countries have taken the lead on controlling *Salmonella* at farm level and also on reducing the use of antibiotics (Wierup *et al.*, 1995; Wegener *et al.*, 2003).

As a result of the adverse publicity associated with antibiotic resistant microbes, 'superbugs', and a perception that intensive agriculture is 'factory farming', there is a growing demand for antibiotic-free poultry in many jurisdictions, which is presenting a major challenge to producers (Bedford, 2000; Castanon, 2007; Daily Mail, 2013; NPR, 2014). Antibiotic residues are continually being detected in poultry meat from different parts of the world. Both banned and legal agents are being identified, which does nothing to inspire consumer confidence (Donoghue, 2003; Tajik *et al.*, 2010). In 2003 the banned veterinary antibiotics nitrofurans were found in chicken from Portugal and poultry from 43 farms was destroyed. Nitrofurans are banned from food because of concerns including a possible increased risk of cancer in humans through long-term consumption (University of Reading, 2003; Yahoo, 2014). China features highly in the list of countries where antibiotics are being inappropriately used in poultry production and it is not only food for human consumption that is being implicated. Since 2007, the US Food and Drug Administration has fielded over 3000 complaints of pet illnesses and deaths that apparently trace back to chicken jerky treats made in China (The Wall Street Journal, 2012; Veterinary News, 2013).

Campylobacter remains a major cause of food poisoning associated with poultry meat and it is disappointing that the many endeavours to understand fully how it persists and how to eliminate it from primary production units have failed (Adkin *et al.*, 2006). The UK Food Standards Agency carried out a survey of *Campylobacter* in chicken on retail sale in the UK between May 2007 and September 2008, and it reported that *Campylobacter* was present in 65% of the fresh chicken samples tested (Food Standards Agency, 2014a). In August 2014, the UK Food Standards Agency published results from the first quarter of the year, which revealed that '59% of whole chickens on retail sale were positive for *Campylobacter*' and 4% of the outside of the packaging tested positive. However, the presence or absence of *Campylobacter* is not the full story and the actual level is far more relevant in terms of public health and the setting of targets to achieve a one- or two-log reduction (Food Safety News, 2014). Only 16% of the birds had a colony count of over 1000 cfu/g, but this was not the figure that captured the media's attention.

An EU baseline survey carried out in 2008, and published by the European Food Safety Authority (EFSA) in March 2010, showed the UK estimated prevalence for *Campylobacter* in broiler birds (caecal contents) was 75.3% and 86.3% on broiler carcasses (skin samples). These results were above the weighted EU mean prevalences of 71.2% and 77%, respectively. There was a wide range of *Campylobacter* prevalence across EU member states varying from 4.9% to 100% on broiler carcasses and from 2% to 100% in broiler birds. It is estimated that there are approximately 9 million cases of human campylobacteriosis per year in the EU member states (EFSA Panel on Biological Hazards, 2011).

It appears that on-farm controls will not deliver the reduction needed to have a major impact on public health; initiatives at every stage in the food chain to deliver a sequential incremental risk reduction will be required. The EFSA opinion in 2011 suggests that long-term freezing (2–3 weeks) of carcasses could reduce the public health risk by more than 90%, while short-term freezing (2–3 days) would result in a public health risk reduction of between 50% and 90%. They also suggest that hot water carcass decontamination, or chemical carcass decontamination with lactic acid, acidified sodium chlorite or trisodium phosphate can have a role in reducing the level of *Campylobacter* contamination. New Zealand is one country that has succeeded in reducing the level of poultry-associated *Campylobacter* infections in humans by a range of regulatory and industry interventions applied along the entire food chain (Sears *et al.*, 2011).

SAFE FEED

Safe pre-mix, feed ingredients and finished feed are essential for safe poultry and eggs and there has been no shortage of recalls associated with contamination of these inputs. Everything from microbial contamination to mycotoxins to heavy metals and dioxins have contributed, and continue to contribute, to problems.

In the USA from 1 May to 30 November 2010, almost 2000 human illnesses, caused by *Salmonella enteritidis*, were associated with the consumption of contaminated chicken and contaminated feed was believed to have contributed to the problem (Centers for Disease Control and Prevention, 2010).

In 2012 more than a quarter of a million chicken eggs were recalled in Germany after in-house testing discovered 'excessive levels' of dioxin (BBC, 2011; Food Navigator, 2012).

A range of quality assurance schemes exists for feed and ingredients and harmonization is required to facilitate global sourcing. Strategic risk-based sampling has to be practised and the power of the analytical chemist can be used in the industry's favour by utilizing composite samples and only performing individual sample tests if a composite is positive. The ability to take a representative sample from a 20,000t or 50,000t shipload is often not possible and samples should be taken of subsets of the consignment before loading the ship; good agricultural practice needs to be adhered to in order to prevent untoward events occurring.

ANIMAL WELFARE

Concerns about animal welfare regularly trigger adverse publicity. In addition to a range of public perception issues, which may or may not contribute to compromised welfare, high stocking rates and stressed birds can be associated with avian health issues, increased disease transmission and a rise in zoonotic infection, which can lead to public health problems. The EU introduced a battery-cage ban in January 2012 and from that date laying hens are only allowed to be reared in enriched cages or in alternative systems such as barn or free-range. Enriched cages must allow at least $750\,cm^2$ per hen, and contain a nest, litter, perch and clawing-board. Farmers and operators were given a long transitional period to adjust to this measure (Europa Press Releases, 2008). Aligning the views of citizens and animal scientists as to what is good or bad welfare is particularly challenging (Miele et al., 2011). The perception that small scale is good and large scale is bad is not valid as the welfare of the birds depends not only on the system but also on the stockmanship and management. There is a disconnect between modern agricultural practices and consumer perceptions, as to how their food is produced, with large-scale efficient units often being perceived as factory farming or industrialized production. The Eurobarometer survey continuously demonstrates that, at least in the EU, consumers are concerned about animal welfare standards and they are deemed an integral attribute of total food quality. Producers, retailers and other food chain actors recognize that good animal welfare represents a business opportunity that can be profitably incorporated into their marketing strategies. An interdisciplinary EU project called Welfare Quality developed protocols for evaluating welfare to assist the European Commission adopt a more outcome-based approach to animal welfare legislation and recommendations for welfare improvements (Blokhuis et al., 2010). The welfare assessment of a chicken is, of course, independent of the country in which the bird is raised. A bird with foot-pad dermatitis, or one with cardiovascular problems, will have poor welfare no matter what language their caretaker speaks. The UN Food and Agriculture Organization (FAO) and the World Organisation for Animal Health (OIE) both have a huge role to address animal welfare on a global stage, which is important to facilitate global trade as farm-animal welfare is increasingly being included in bilateral agreements between trading blocs (Nielsen and Zhao, 2012). There are huge variations between countries and between subsets of populations within countries as to what is and what is not good welfare and consumer willingness to pay for enhanced welfare also varies greatly, with a positive relationship existing between respondent income and younger age in many jurisdictions (Lagerkvist and Hess, 2011).

TRADE-DISRUPTING DISEASES

Stakeholders need to be aware of the consequences of outbreaks of those non-zoonotic diseases, which, although they pose no risk to human health, disrupt the trade in poultry and eggs and damage both commercial interests and consumer confidence. In addition, outbreaks of the non-food-borne zoonosis avian influenza can have huge impacts on the global chicken supply chain (McLeod *et al.*, 2005; World Bank, 2005).

POSTHARVEST FOOD SAFETY AND PROCESS CONTROLS

The traditional role of meat inspection, both ante-mortem and post-mortem, is under the spotlight in many jurisdictions as food production and processing are becoming more complex. It is important to ensure that only clinically healthy poultry are slaughtered and diseased birds are not allowed into the food chain. The role of inspectors is two-fold here: they must protect consumer health and identify avian health issues that need to be addressed on-farm. In post-mortem inspections, the visual inspection approaches, that have been mandatory in many jurisdictions in the past, are now under review and are being enhanced, or replaced, by microbial monitoring to validate the hygiene measures and information from source farms demonstrating that the birds are healthy.

Enhanced surveillance

Enhanced surveillance capabilities are required to establish public health priorities, detect, delineate and investigate outbreaks, evaluate interventions and provide a detection service compatible with a modern food industry operating 'just-in-time' delivery systems in a global market place. Forensic microbiology including next-generation sequencing has spawned a new generation of 'disease detectives', hunting down dangerous microbes and tracking them back through the food chain to their source, or forward to identify foods in the market place that must be recalled to prevent human illness. In most jurisdictions multidisciplinary teams participate in outbreak investigations and, increasingly, international networks are facilitating the identification of globally distributed contaminated product. Companies producing contaminated product will eventually be uncovered with disastrous economic and reputational consequences (Swaminathan *et al.*, 2006).

Increasingly, cut-backs in the public sector in several jurisdictions mean that inspections and audits, and also surveillance, are not being scaled up to match the complexity of the food chain, which may lead to delays in identifying problems or some of them being missed. Therefore industry has to begin to use the same modern technology that is being used by the regulators, such as pulse field gel electrophoresis and next-generation sequencing, to protect their customers and their brands and reputations. Being able to distinguish between newly introduced and recurring microbial contamination by definitive typing may help differentiate sporadic isolates from those that may have become established in the production facility or are linked to one supplier. Being able to track microbes from the production facility to the farm of origin will facilitate control (Swaminathan et al., 2006; Tauxe, 2006; Lienau et al., 2011).

Robust food-safety management systems with adequate process controls are essential, but staff require knowledge of good manufacturing practice and hazard analysis and critical control points (HACCP). HACCP systems are not a replacement for other food hygiene requirements but part of a package of food hygiene measures that contribute to ensuring safe food. Prior to establishing HACCP, good food hygiene standards must already be in place, particularly in the following areas:

- infrastructural and equipment requirements;
- food safety specifications for raw materials;
- the safe handling of food (including packaging and transport);
- sanitation (cleaning and disinfection);
- water quality;
- maintenance of the cold chain;
- the health of staff;
- personal hygiene;
- training;
- food waste handling; and
- pest control.

Changing consumer lifestyles are creating a demand for more ready-to-cook and ready-to-eat meals, and the distance to the final consumer from the farms of origin is getting longer each year. The increasing competitive commercial environment is driving the need for efficiency, leading to consolidation and economies of scale that result in the mass production of increasing volumes at all stages of the food chain. In this environment, the consequences of a contamination incident can have devastating effects on health (with people often falling ill over large geographical areas) causing massive damage to the reputation of food companies and brand names. Reputations and brands that take years to build can be irreparably damaged overnight by being associated with a food scare or adverse health effects.

There are several factors that continually contribute to the occurrence of outbreaks of food-borne disease and often several of these occur simultaneously, thus amplifying outbreaks. These factors include: contaminated raw ingredients (including water), inadequate refrigeration or storage, insufficient cooking, cross-contamination between raw and cooked food, poor personal hygiene of staff, poor general hygiene on premises, and untrained staff. The tragedy is that, although these factors continually contribute to illness and deaths, they are all easily preventable.

In the education of stakeholders in the poultry sector so that they can see the relevance of appropriate food-safety management systems and process controls, examples of where process failures contributed to outbreaks of zoonotic disease should be highlighted (Knowles *et al.*, 2007).

Food safety in the boardroom

Often in global food companies the financial metrics dominate the discussions around the boardroom table, including: the next merger and acquisition to deliver double-digit growth, foreign exchange fluctuations, major competitors' strategies, performance of the pension funds, key executive remuneration and retention schemes etc. Risk management usually refers to financial risk and food safety only becomes a major issue when there is a crisis. Many companies have learnt, to their cost, the consequence of failing to give due regard to food safety and a culture of food safety must start at the top. Often one hears CEOs of major companies start off their speeches with 'in our company food safety is a given'; well, the question to ask is 'who has given it?'

THE 'TEN COMMANDMENTS'

There are ten points worth highlighting to food producers in bringing food safety centre stage.

1. Premises.
2. Plant and equipment.
3. Procurement.
4. Product.
5. Process controls.
6. Protocols.
7. Practices.
8. People.
9. Profit.
10. Paranoia.

Premises

The premises must be designed appropriately at the onset and have food-grade materials throughout. Basic issues need to be addressed including: being vermin-proof; being appropriately zoned with segregation of high- and low-care areas and their respective staff; having drainage systems that flow in the correct direction, i.e. from the high-care areas to the low-care areas and not the other way around; similarly the airflow must be in the correct direction, the water supply must be potable and the waste facilities must be adequate. The ideal is to have glass walls and windows to allow visitors to observe the process without having to enter the food production facilities. In summary, food safety must be part of the design of the building.

Plant and equipment

All plant and equipment must be such that it can be easily cleaned and maintained. Equipment that permits the build of biofilm and has inaccessible parts where microbes can hide and grow is a recipe for a food safety disaster.

Procurement

A company's reputation is only as secure as the standards of its weakest supplier and therefore whether it is a primary producer of eggs or poultry, or a supplier of other inputs from dry ingredients to packaging, a company may be relying on the compliance of others to protect its future. Therefore rigid supplier control has to be undertaken.

Product

The risk profile of both the raw ingredients and finished product has to be considered in deciding on the risk-management strategies. Ready-to-cook and ready-to-eat products require a different level of control than raw product that will be cooked by the consumer. Some consumers are more vulnerable than others, such as infants, the frail elderly and people who are ill, and a company's products must be safe for the weakest individual who may consume them.

Process controls

The process should be designed with the production of safe food in mind. The process flow should be from low to high risk whether in a slaughter facility, a packing station or a cooked-poultry operation. The risk of something going wrong should be engineered out of the process where possible. Technology can reduce the likelihood of human error with everything from compulsory handwashing, appropriate alarms and robotic and automatic process steps making a contribution. Appropriately placed closed-circuit cameras monitoring the facility can enable continuous monitoring of the process.

Protocols

Standardization of all activities ensures consistency, so documented protocols must exist covering the full range of production and quality assurance activities including prerequisites requirements, standard operating procedures, HACCP plans, cleaning and sanitation regimens and associated tests to verify their efficacy, traceability, internal audits, recall procedures, crisis management etc.

Practices

Often what happens in practice is at variance with what is outlined in protocols. The culture in the organization has to be one of compliance and companies must be doing the right thing because they believe it is the right thing to do and not out of fear of prosecution. Ensuring food is produced safely is everyone's responsibility not just those staff working in quality assurance. Practical protocols, and an understanding of the rationale behind them, make compliance more likely.

People

A company's staff are its greatest asset but untrained staff can be its greatest liability. Staff must receive training or a level of supervision commensurate with the risk associated with their activity. Often workers at certain levels in the poultry sector may not be literate, therefore appropriate educational material and approaches are required, tailored to the needs of the particular staff. Initiatives such as competitions between sections and rewards for high-performing staff in the areas of food hygiene and safety can drive standards up. However, there can be no tolerance for shoddy practices and the message has to be clear that contributing to a major non-compliance may be a sackable offence. Maintenance staff travelling, and bringing equipment, between low-care and high-care zones can contribute to cross-contamination so must be included in food safety training and protocols to ensure best practice.

Profit

Being highlighted by the authorities for breaking the law damages commercial viability and failed customer audits can result in companies losing valuable contracts. Recalls cost money and being associated with an incident can have huge economic consequences. Greed is a great incentive for some stakeholders to engage in shoddy practices or fraudulent activity.

Paranoia

With the increasing sensitivity of analytical chemists, now able to detect chemical contaminants at parts per trillion, and with the increasing power of forensic microbiologists to bar-code microbes and track them back from sick people to food outlets to production facilities and even to farms and feed mills, there is now no margin for error. Food-business operators need to be continuously vigilant, as the ability to identify them, if they are engaged in substandard practices, is now available to the authorities should they have the resources to do so.

CONCLUSION

Food safety is not rocket science, so while there are often many explanations for untoward events there is no excuse. These are not a result of bad luck, rather they are a result of bad management. Collaboration between all the stakeholders along the poultry supply chain has to be the way forward as everyone has a role to play if the sector is to realize its full potential. There can be no tolerance for shoddy operators as, with the intensity and speed of the global media, one player can damage the entire sector. If poultry and eggs are to capitalize on their strengths under the health and nutrition banner, food safety issues have to be resolved. The emphasis has to be on prevention rather than crisis management.

REFERENCES

Adkin, A., Hartnett, E., Jordan, L., Newell, D. and Davison, H. (2006) Use of a systematic review to assist the development of Campylobacter control strategies in broilers. *Journal of Applied Microbiology* 100(2), 306–315.

BBC (2011) More Farms Closed in Germany as Dioxin Scare Continues. Available at: http://www. bbc.co.uk/news/world-europe-12200619 (accessed 28 July 2014).

Bedford, M. (2000) Removal of antibiotic growth promoters from poultry diets: implications and strategies to minimize subsequent problems. *World's Poultry Science Journal* 56(4), 347–365.

Blokhuis, H.J., Veissier, I., Miele, M. and Jones, B. (2010) The Welfare Quality® project and beyond: safeguarding farm animal well-being. *Acta Agriculturae Scandinavica Section A* 60(3), 129–140.

Bloomberg (2014) McDonald's Food Supplier OSI Recalls Shanghai Products. Available at: http:// www.bloomberg.com/news/2014-07-27/mcdonald-s-food-supplier-osi-recalls-shanghai-unit-s-products.html (accessed 28 July 2014).

Casey, D.K., Lawless, J.S. and Wall, P.G. (2010) A tale of two crises: the Belgian and Irish dioxin contamination incidents. *British Food Journal* 112(10), 1077–1091.

Castanon, J.I.R. (2007) History of the use of antibiotics as growth promoters in European poultry feeds. *Poultry Science* 86(11), 2466–2471.

Centers for Disease Control and Prevention (2010) Multistate Outbreak of Human Salmonella enteritidis Infections Associated with Shell Eggs (Final Update). Available at: http://www.cdc. gov/SALMONELLA/ENTERITIDIS (accessed 27 July 2014).

China Daily Mail (2012) China: KFC Chickens Are Being Fattened with Illegal Drugs. Available at: http://chinadailymail.com/2012/12/19/china-kfc-chickens-are-being-fattened-with-illegal-drugs (accessed 28 July 2014).

Covaci, A., Voorspoels, S., Schepens, P., Jorens, P., Blust, R. and Neels, H. (2008) The Belgian PCB/dioxin crisis – 8 years later: an overview. *Environmental Toxicology and Pharmacology* 25(2), 164–170.

Daily Mail (2013) How Drugs Pumped into Supermarket Chickens Pose a Terrifying Threat to Our Health. Available at: http://www.dailymail.co.uk/news/article-2388444/How-drugs-pumped-supermarket-chickens-pose-terrifying-threat-health.html (accessed 28 July 2014).

Donoghue, D.J. (2003) Antibiotic residues in poultry tissues and eggs: human health concerns? *Poultry Science* 82(4), 618–621.

Europa Press Releases (2008) Animal Welfare: Commission Report Confirms the Potential Benefits of Banning Conventional Battery Cages for Laying Hens. Available at: http://europa.eu/rapid/press-release_IP-08-19_en.htm (accessed 28 July 2014).

European Food Safety Authority Panel on Biological Hazards (2011) Scientific opinion on Campylobacter in broiler meat production: control options and performance objectives and/or targets at different stages of the food chain. *EFSA Journal* 9(4), 2105.

FoodIntegrity (2014) Workshop: Determining the Geographical Origin of Food. Available at: http://secure.fera.defra.gov.uk/foodintegrity/index.cfm?pageid = 12 (accessed 28 July 2014).

Food Navigator (2012) Over 250,000 Eggs Recalled in Germany in Latest Dioxin Scare. Available at: http://www.feednavigator.com/Regulation/Over-250-000-eggs-recalled-in-Germany-in-latest-dioxin-scare (accessed 28 July 2014).

Food Safety Authority of Ireland (2002a) Food Safety Authority Finds Some Imported Chicken Fillets in Breach of Food Labelling Laws. Available at: http://www.fsai.ie/details.aspx?id = 7312 (accessed 28 July 2014).

Food Safety Authority of Ireland (2002b) Investigation of the Composition and Labelling of Chicken Breast Fillets from the Netherlands Imported into Ireland. Available at: http://www.fsai.ie/uploadedFiles/Monitoring_and_Enforcement/Monitoring/Surveillance/Poultry_Labelling_Report2.pdf (accessed 28 July 2014).

Food Safety Authority of Ireland (2014) FSAI Survey Finds no Horse DNA in Beef Products Tested – Presence of Chicken and Bovine DNA Found in Takeaway Lamb Dishes. Available at: http://www.fsai.ie/news_centre/press_releases/EU_horsemeat_testing_24072014.html (accessed 28 July 2014).

Food Safety News (2013) 188,000 Pounds of Chicken from Chile Recalled for Dioxin. Available at: http://www.foodsafetynews.com/2013/07/188000-pounds-of-chicken-from-chile-recalled-for-dioxin/#.U9X1icRDtyI (accessed 28 July 2014).

Food Safety News (2014) UK Survey Finds Campylobacter on 59 Percent of Chicken. Available at: http://www.foodsafetynews.com/2014/08/uk-finds-campylobacter-on-59-percent-of-chicken/#.U_tOusRDtyI (accessed 25 August 2014).

Food Standards Agency (2014a) *Campylobacter*. Available at: http://www.food.gov.uk/science/microbiology/campylobacterevidenceprogramme/#.U9YI1cRDtyI (accessed 28 July 2014).

Food Standards Agency (2014b) Local Authorities to Test for Lamb Meat Substitution. Available at: http://www.food.gov.uk/news-updates/news/2014/apr/testing#.U9X0Y8RDtyI (accessed 28 July 2014).

Global Post (2013) Probe Launched into German 'Organic' Eggs. Available at: http://www.globalpost.com/dispatch/news/afp/130225/probe-launched-german-organic-eggs (accessed 28 July 2014).

Knowles, T., Moody, R. and McEachern, M.G. (2007) European food scares and their impact on EU food policy. *British Food Journal* 109(1), 43–67.

Lagerkvist, C.J. and Hess, S. (2011) A meta-analysis of consumer willingness to pay for farm animal welfare. *European Review of Agricultural Economics* 38(1), 55–78.

Lienau, E.K., Strain, E., Wang, C., Zheng, J., Ottesen, A.R., Keys, C.E., Hammack, T.S., Musser, S.M., Brown, E.W., Allard, M.W., Cao, G., Meng, J. and Stones, R. (2011) Identification of a salmonellosis outbreak by means of molecular sequencing. *New England Journal of Medicine* 364(10), 981–982.

McLeod, A., Morgan, N., Prakash, A. and Hinrichs, J. (2005) *Economic and Social Impacts of Avian Influenza*. FAO Emergency Centre for Transboundary Animal Diseases Operations.

Miele, M., Veissier, I., Evans, A. and Botreau, R. (2011) Animal welfare: establishing a dialogue between science and society. *Animal Welfare* 20(1), 103–117.

Nielsen, B.L. and Zhao, R. (2012) Farm animal welfare across borders: a vision for the future. *Animal Frontiers* 2(3), 46–50.

NPR (2014) Americans Want Antibiotic-Free Chicken, and the Industry Is Listening. Available at: http://www.npr.org/blogs/thesalt/2014/02/14/276976353/americans-want-antibiotic-free-chicken-and-the-industry-is-listening (accessed 28 July 2014).

Premanandh, J. (2013) Horse meat scandal – a wake-up call for regulatory authorities. *Food Control* 34(2), 568–569.

Rampling, A. (1993) *Salmonella enteritidis* five years on. *Lancet* 342, 317–318.

Sears, A., Baker, M.G., Wilson, N., Marshall, J., Muellner, P., Campbell, D.M., Lake, R.J. and French, N.P. (2011) Marked campylobacteriosis decline after interventions aimed at poultry, New Zealand. *Emerging Infectious Diseases* 17(6), 1007–1015.

Sharp, J.C. (1988) Salmonella in eggs. *British Medical Journal* 297(6663), 1557–1558.

Swaminathan, B., Gerner-Smidt, P., Ng, L.K., Lukinmaa, S., Kam, K.M., Rolando, S., Gutiérrez, E.P. and Binsztein, N. (2006) Building PulseNet International: an interconnected system of laboratory networks to facilitate timely public health recognition and response to food-borne disease outbreaks and emerging food-borne diseases. *Foodborne Pathogens and Disease* 3(1), 36–50.

Tajik, H., Malekinejad, H., Razavi-Rouhani, S.M., Pajouhi, M.R., Mahmoudi, R. and Haghnazari A. (2010) Chloramphenicol residues in chicken liver, kidney and muscle: a comparison among the antibacterial residues monitoring methods of Four Plate Test, ELISA and HPLC. *Food and Chemical Toxicology* 48(8), 2464–2468.

Tauxe, R.V. (2006) Molecular subtyping and the transformation of public health. *Foodborne Pathogens and Disease* 3(1), 4–8.

Telegraph (2006) How the Egg Inspectors Spotted the Free-Range Label 'Fraud'. Available at: http://www.telegraph.co.uk/news/uknews/1534371/How-the-egg-inspectors-spotted-the-free-range-label-fraud.html (accessed 28 July 2014).

Telegraph (2013) German Farms in Egg 'Fraud'. Available at: http://www.telegraph.co.uk/news/worldnews/europe/germany/9895504/German-farms-in-egg-fraud.html (accessed 28 July 2014).

The Guardian (2010) Egg Boss Jailed for 'Free Range' Fraud. Available at: http://www.theguardian.com/uk/2010/mar/11/free-range-eggs-fraud (accessed 28 July 2014).

The Ranger (2013) BFREPA Take the Lead in the Fight against Egg Fraud. Available at: http://www.theranger.co.uk/News/BFREPA-take-the-lead-in-the-fight-against-egg-fraud_21968.html (accessed 28 July 2014).

The Wall Street Journal (2012) KFC Criticized Over Suppliers in China. Available at: http://online.wsj.com/news/articles/SB10001424127887324731304578189290798901754 (accessed 28 July 2014).

Tollefson, L. and Karp, B.E. (2004) Human health impact from antimicrobial use in food animals. *Médecine et maladies infectieuses* 34(11), 514–521.

University of Reading (2003) Contaminants – Illegal Drug Residues Found in Portuguese Chickens. Available at: http://www.reading.ac.uk/foodlaw/news/uk-03018.htm (accessed 28 July 2014).

Van Larebeke, N., Hens, L., Schepens, P., Covaci, A., Baeyens, J., Everaert, K., Bernheim, J.L., Vlietinck, R. and De Poorter, G. (2001) The Belgian PCB and dioxin incident of January–June 1999: exposure data and potential impact on health. *Environmental Health Perspectives* 109(3), 265.

Veterinary News (2013) Hartz Recalls Chicken Products Containing Antibiotic Residue. Available at: http://veterinarynews.dvm360.com/dvm/Toxicology/Hartz-recalls-chicken-products-containing-antibiot/ArticleStandard/Article/detail/804157 (accessed 28 July 2014).

Wegener, H.C., Hald, T., Wong, L.F., Madsen, M., Korsgaard, H., Bager, F., Gerner-Smidt, P. and Mølbak, K. (2003) *Salmonella* control programs in Denmark. *Emerging Infectious Diseases* 9(7), 774.

Wierup, M., Engström, B., Engvall, A. and Wahlström, H. (1995) Control of *Salmonella enteritidis* in Sweden. *International Journal of Food Microbiology* 25(3), 219–226.

World Bank (2005) Avian Influenza: Economic and Social Impacts. Available at: http://web.worldbank.org/WBSITE/EXTERNAL/TOPICS/EXTHEALTHNUTRITIONANDPOPULATION/0,,contentMDK:20663668~pagePK:64020865~piPK:149114~theSitePK:282511,00.html (accessed 28 July 2014).

Yahoo (2014) Concern over Bugs Resistant to Banned Antibiotic Found in Chicken. Available at: http://au.news.yahoo.com/world/a/24344268/concern-over-bugs-resistant-to-banned-antibiotic-found-in-chicken (accessed 28 July 2014).

Chapter 9
Endemic Disease – the Challenge to Reduce Antibiotic Use

Andrew Walker[1]* and Patrick Garland[2]

[1]Slate Hall Veterinary Services, UK; [2]Premier Nutrition, Rugeley, UK

INTRODUCTION

Antibiotics are one of the 20th century's great scientific discoveries (Swormink, 2014) and have impacted significantly on both animal and human health. However, they must be used in a responsible manner by all who prescribe and administer them. They are not a substitute for good management, biosecurity and hygiene practices.

Since the discovery of penicillin in 1928 by Sir Alexander Fleming antibiotics have become gradually more significant in human and animal health. The growth promoter effect of antibiotics on poultry was first discovered in the 1940s (Castanon, 2007). Since that point there has been a wide variety of antimicrobial active ingredients developed that are used in both humans and animals. The use of antibiotics as growth promoters in animal feeds has been permitted in the member states of the European Union (EU) since the 1960s. However, concerns about antibiotic resistance, especially associated with antibiotics that were used both in human patients and as growth promoters in livestock, led to the Swann Report (Swann *et al.*, 1969), in which it was recommended that antibiotics used in human medicine should not be used as growth promoters (Donaldson, 2007). The UK broiler industry largely dropped the use of antibiotics as growth promoters 5 years before the European Union ban as a result of Assured Chicken Production standards, to which 85% of the industry had signed up. On 1 January 2006 EU approval for the use of antibiotics as growth promoters was withdrawn (Castanon, 2007; Cogliani *et al.*, 2011) following concern about the transference of resistance genes from animal to human microbiota. With more recent evidence linking the possible use of antimicrobials, particularly in intensive livestock systems such as poultry production, to increased resistance via the transmission

*Corresponding author: andrew@slatehall.co.uk

of resistance genes into the human population, prophylactic, metaphylaxis and therapeutic antibiotic usage has come under closer scrutiny.

The poultry industry will play an increasingly large role, within both the animal health sector and the global health scheme, to help safeguard the future efficacy of the active molecules we have available to us and also to ensure continued efficacy of these molecules.

The poultry industry is an intensive production sector that has thrived through its dynamic ability to overcome the challenges raised by intensive farming systems. It has also faced many challenges from the legislation governing the industry. The reduction of antibiotic usage within the industry remains a key focus, however this must be achieved in a multifaceted and evidence-based approach, rather than based on the emotive perception of the consumer or retailer. Within the commercial poultry sector most medication is used on a flock basis, rather than treatment of individual animals. Therefore a thorough risk-based analysis and clinical assessment prior to the prescription of any medication is crucial.

One must ensure that when it comes to how much antibiotic is used it is as little as possible but as much as necessary (RUMA, 2005). However, the level of resistance to antibiotics is correlated to the level of consumption (Nordberg *et al.*, 2005). It is the authors' opinion that reduction in antibiotic usage should not be driven purely by a desire to reduce the total amount of antibiotic used, but rather that reduction should be based upon a targeted and risk-based reduction and refinement. The classification of antibiotic used and potential ramifications for usage of critically important human antibiotics must be considered. If the reduction is driven purely by a benchmarked key performance indicator such as milligram per kilogram of meat then there is a risk that critically important antibiotics with a lower therapeutic dosage could be used as a first-line treatment. For example, fluroquinolones used at 10 mg/kg live weight can reduce antibiotic usage figures when compared to amoxycillins or chlorotetracyclines used at 20 mg/kg.

Prior to prescribing antibiotics the veterinarian should have a full understanding of the case history, carry out a diagnostic investigation to ensure that the treatment is the most appropriate, is justified and that any alternative options are not appropriate. It is also essential to support and educate the client during these investigations on the best preventative health care and management of a disease once a diagnosis has been reached to reduce future disease outbreaks.

The strategies employed to help focus where these improvements can be made can be based on the three Rs of 'Reduction, Replacement and Refinement'.

REDUCTION

Legislation and voluntary bans

Veterinarians within some European member states have decided to impose a voluntary ban on those antibiotics that have a significant value in human medicine, mainly the third- and fourth-generation cephalosporins and fluroquinolones (Mevius and Heederik, 2014).

In the Netherlands in 2013 antibiotic use in livestock farming had dropped by 50% compared to 2009. By 2015 the aim is to reduce this by a further 20%. However, historical over or inappropriate usage can allow for a dramatic initial reduction. The Dutch government aims to achieve this by asking the sectors concerned to reduce their use of antibiotics and monitoring their response by:

- strengthening the powers of the Dutch Food and Consumer Product Safety Authority to supervise antibiotic use in livestock farming;
- strengthening the position of the Dutch Veterinary Medicines Authority (SDA), for example by ensuring that it gains access to company data about antibiotic usage;
- revoking the dispensing permit of veterinarians who break the rules;
- formulating an agreement with the farmers' unions limiting each livestock farmer to using only one veterinary practice; and
- publishing the names of livestock farmers who use an excessive amount of antibiotics and as a result are breaking the law (Government of the Netherlands, undated).

Early indications show that these strategies are having some beneficial effects in reducing antimicrobial resistance. The Dutch authorities have recorded that a decreased use of antibiotics leads to less antibiotic resistance at livestock farms. For example, the cefotaxime resistance in *E. coli* of broilers decreased from 20% to 5.8% in the period 2007–2012 (Swormink, 2014).

It has been suggested by some that the decoupling of the right to prescribe veterinary medicines from the right to sell these medicines might slow resistance. However, a report by Berenschot in 2010 (Beemer *et al.*, 2010) reviewed the effects of decoupling in the Netherlands and concluded that decoupling would not be effective and that the veterinarians' position should be strengthened in order to control the level of antibiotic usage. However, Denmark has not allowed prescribing veterinarians to supply medications for many years and in-feed medication has been avoided since the mid-2000s. All medications and vaccines are supplied by pharmacies who cannot employ prescribing veterinarians.

Hatching egg hygiene

Avian colibacillosis, omphalitis (navel infection) and yolk sac infections are the most common infectious diseases that the poultry industry faces worldwide (Calnek *et al.*, 1997). A full risk analysis should be carried out on any flock placement prior to being medicated at day old. This analysis should include all stages of egg production, through to the embryo development stages and right onto the growing farms.

To reduce the chance of bacterial infections in chicks one must evaluate the risk at every stage along the production process. This starts right from the point of the egg developing within the breeder and continues through to placement of the chicks. Adverse breeder intestinal health can have a negative impact on faecal contamination of the eggs and nestbox hygiene (Smith *et al.*, 2000). Nestboxes should be clean with good mats to ensure that egg hygiene, especially in the critical period as the egg is drying immediately post-laying, is optimal. There is also a multitude of other factors that when managed well can reduce the bacterial loading on eggs and ultimately bacterial contamination of the chicks (Cox *et al.*, 2000). These include egg belt maintenance and hygiene, regular egg collection, good egg handling and regular cleaning of the egg store. There is an inherent variation in the egg cuticle layer both between individual birds and between different strains of birds. This cuticle is involved in the first-line defence against bacterial penetration (Solomon *et al.*, 1994; Bain *et al.*, 2013). Any 'sweating' or condensation that forms on the eggshell surface, whilst in storage or incubation, can aid the bacterial passage through the shell pores, increasing the risk for bacterial contamination. It has been established that effective hatchery sanitation practices are necessary to ensure maximum hatchability and high quality chicks (Whistler and Sheldon, 1989; Bruce and Drysdale, 1994). Enteric microorganisms can penetrate the shell and result in decreased hatchability, poor chick and poult quality and perpetuation of infection (i.e. *Salmonella*, colibacillosis) in growing birds (Arcliicnbuwa *et al.*, 1980). Shell surface bacteria will have a lag phase before beginning an exponential growth phase, under the appropriate conditions for multiplication (Zwietering *et al.*, 1992). Therefore any interventions through egg fumigation will have their maximum effect on bacterial loading as soon after the egg is laid as possible. This makes early fumigation procedures on breeder farms critical in reducing bacterial numbers. The next risk for bacterial contamination is during the transportation and storage in the hatchery before being set for incubation. During transfer into the hatchery eggs can be exposed to variations in temperatures and humidity, which can influence moisture accumulation on the surface (Fromm and Margolf, 1958). Given the warm and humid environment within the hatchery there is a risk of increased bacterial growth, from the initial contamination from farm, on the eggshell surface. Hatchery hygiene is paramount in producing good quality, healthy chicks and minimizing bacterial infections later on in the growing phase. Therefore the hatcheries supplying chicks play a crucial role in the reduction strategy.

Table 9.1 demonstrates the impact of overall bacterial loading during incubation of a theoretical placement of eggs. If 0.5% of eggs set are dirty or floor eggs then this can contribute a significant proportion of the total bacterial loading

Table 9.1. A theoretical bacterial challenge of eggs during the incubation process (one incubator or hatcher (15,000 eggs) with 99.5% 'clean' eggs and 0.5% 'dirty eggs').

Process	Percentage of eggs set	cfu/egg	Bacterial numbers	Percentage of total egg surface bacteria	cfu/egg	Bacterial numbers	Percentage of total egg surface bacteria
		Pre-incubation			Post-incubation to hatcher		
Clean eggs	99.5%	500–10,000	$7.4625 \times 10^6 -$ 1.4925×10^8	80%	5–100	$7.4625 \times 10^4 -$ 1.49250×10^6	6%
Dirty eggs	0.5%	20,000– 2,000,000	$1.5 \times 10^6 -$ 1.5×10^8	20%	2,000,000– 20,000,000	$1.5 \times 10^8 -$ 1.5×10^9	94%

cfu, colony forming units

through the hatchers. It is therefore advisable to minimize the use of dirty and floor eggs through improved grading and selection or to incubate these eggs separately.

When deciding on whether to medicate day-old broilers there are also other factors that can influence the potential risks for bacterial infections. For example, breeder flock history, breeder flock age where older flocks' eggs become more porous with increased cracks, egg storage age, as well as the broiler farm factors such as site disease history and cleaning and disinfection procedures, will all influence the decision to medicate or not. Some vertically transmitted bacterial diseases can also be controlled through day-old medication.

Water hygiene

Microbial growth within poultry house drinking water systems can be a significant source for bacterial challenge. The widespread use of nipple drinkers, compared to open fonts or bell drinkers, will reduce contamination of the drinking water from environmental sources. If bell drinkers are used then they should be cleaned at least once daily with a good hygiene protocol in place for both the bells and the cleaning brushes and buckets. Good water hygiene should be part of a routine flock health programme and can be an important factor in reducing colonization of broilers with *Campylobacter* (Kapperud *et al.*, 1993) and *Salmonella* (Vandeplas *et al.*, 2010).

Water hygiene can often be overlooked, but given that chickens, for example, can drink between 1.6 and 2.0 times the amounts of feed consumed, by weight, it constitutes a significant proportion of potential bacterial intake into the intestinal tract. Water is a key nutritional and growth medium for most bacteria; combining this with often high environmental ambient temperatures, potential times of low flow rates through drinker lines, the ability of some bacteria to form protective biofilms and frequent introduction of nutraceutical supplements, vaccinations and medications means that there are many factors that can influence drinker system hygiene. Given the potential for biofilm formation (Zimmer *et al.*, 2002; Cox, 2005; Kalmokoff *et al.*, 2006) in drinking systems, regular cleaning and sanitation is crucial (Cox and Pavic, 2010).

The bacterial levels within drinking water should be below 100 cfu/ml total bacteria and 50 cfu/ml coliforms (Carter and Sneed, 1987). The routine testing of water from the drinker system at various points is recommended. Regular testing can help establish a baseline for the current management of the system and identify whether changes in the water sanitation process are required. Since biofilm can hold bacteria it is advisable to swab the inside of pipelines as well as take running-water samples.

Immunity

The immune system of the bird is able to deal with many infections, whether that is through the innate immune system or through cell-mediated or humoral defences. A robust immune system is crucial for ensuring that chickens are able to mount an appropriate immune reaction when challenged by a multitude of endemic diseases. Innate immunity can be affected by stressful physiologic events related to hatching and to environmental factors during the first week of life (Hoerr, 2010). To maximize the immunity it is necessary to not only ensure that the quality of the chick is good but that the vaccination status of the breeders is tailored to the risks in the progeny, as well as themselves. Minimizing dehydration, narrowing the hatch window and reducing bacterial challenges at the time of hatch will all help to lower physiological stresses and can improve chick quality (Aviagen, 2009). Optimal brooding conditions at placement, to maximize early food intake, gut development and consequently yolk absorption will also help to build a more robust chick (Dibner *et al.*, 1998). Control of immunosuppressive diseases, such as Marek's disease, chicken anaemia virus, infectious bursal disease and adenovirus infections will allow the bird to develop a full immunological response both to vaccinations and also to field challenges.

Vaccination

Vaccination of poultry, for a multitude of viral and bacterial pathogens, is widespread and has been a crucial aid for the industry's success during intensification. However, electing against which pathogens to vaccinate is often a difficult decision and requires consideration of:

- risk of challenge;
- consequence of a challenge;
- cost of a challenge;
- regulations or accreditation schemes;
- export implications;
- cost of vaccination;
- route of administration;
- maternally derived antibodies and protection of progeny;
- risk of spread of live vaccines to naive flocks; and
- differentiation tests between vaccinated and unvaccinated flocks.

The vaccination programme for any flock should be reviewed regularly to ensure that it is providing optimal protection at minimal risk. All vaccines will have an impact on the bird through the energy cost in stimulating the immune response. This cost can vary significantly between vaccines and vaccine types and increase feed conversion ratios (FCRs) and/or reduce live weight. For example, in Sweden where vaccination of broilers is rare, the performance achieved on relatively lower energy feed regimens compares favourably with the UK, despite the UK's higher energy feeds. Similarly in New Zealand the food conversion ratio is incredibly low and the negligible disease background and lack of vaccination no doubt contributes to this achievement.

The live attenuated bacterial vaccines tend to mimic the route of infection of the natural field challenge and as such stimulate a local cellular and humoral response, often increasing innate immune pathways in the process. However, the response with live vaccination alone can be short lived without boosting the immune system with either an inactivated vaccine or natural field challenge.

The inactivated or killed bacterial vaccines are injected either intramuscularly or subcutaneously and are designed to stimulate a systemic immune system, as they bypass the natural route of infection. However, the protection they offer is often longer lived through the aid of a vaccine adjuvant such as aluminium hydroxide or mineral oil. Due to the economic constraints in the UK such vaccines tend to be limited to breeders and longer lived egg layers. However, in countries where there are high endemic disease challenges they can be used in shorter lived birds. These vaccines can also increase the level of antibodies that are transferred into the yolk and subsequently the developing embryo when employed in breeding stock.

Inactivated vaccines are safe but costly and are, therefore, essentially used to produce high, uniform and long-lasting antibody titres prior to lay in hens that have been vaccinated with a live vaccine (Van den Burg, 2008).

Biosecurity

Biosecurity is also a key element in the control of endemic diseases and ultimately reducing antibiotic usage. Biosecurity is the measures that are taken to protect a population against a harmful biological or biochemical substance (Cox, 2005; Stevenson and Waite, 2011). Therefore optimal biosecurity requires a holistic approach encompassing hygiene procedures, vaccination strategies, maximizing bird health and immune status whilst minimizing environmental or nutritional stressors.

Endemic disease challenges can be transmitted either vertically or horizontally/laterally. Given the structure of the commercial poultry industry, vertically transmitted diseases are more readily controlled as the breeders have a clear segregation from the progeny via the hatchery process. Vertically transmitted diseases, such as mycoplasmosis, can be controlled by one of three approaches: (i) maintenance of flocks that are free of infection; (ii) medication; or (iii) vaccination (Kleven, 2008). This allows for endemic diseases, such as salmonellosis, mycoplasmosis and leucosis, to be screened for in the breeders and interventions put in place to either reduce the risk to the progeny, or remove breeding stock from the production system. This will help prevent infection of the direct progeny, but also horizontal infections through both the hatchery and growing farm environments. For mycoplasmosis a potential intervention can be strategic medication of the progeny of an infected flock for the first 3 days to minimize vertical or pseudo-vertical transmission and again later in the crop to reduce lateral transmission. Maintaining disease-free breeding stock will clearly help to reduce antibiotic usage in the progeny.

When establishing the disease status of a flock, whether it be layers, breeders or broilers, a regular screening programme should be implemented. This can incorporate regular post-mortem screening, serological testing, bacteriology and molecular screening with the use of polymerase chain reaction (PCR) testing. Working closely with your veterinarian will enable a bespoke screening programme to be implemented taking into account the local, national and global disease risk. It is also worth defining clear veterinary intervention points (VIPs), upon which defined actions are taken. For example, if the mortality of a broiler flock in the first 5 days is greater than 0.3% per day then post-mortems are carried out to establish the cause of mortality; or if egg production in a laying flock drops by a defined percentage then veterinary investigations are instigated. Veterinary intervention points can also be established for broiler processing data whereby if certain predefined factory reject parameters are exceeded then this triggers investigations into the growing farms or if egg packing facilities highlight issues with egg quality such as increased seconds or downgrades, pale eggs, cracked eggs or reduced Haugh Unit scores. With the implementation of VIPs a system of screening, reviewing, investigating and ultimately implementing actions can be established to give a fully integrated and consistent approach to disease investigation.

There has been an increase in free-range production within the UK over recent decades, both in the egg laying and meat sectors alike. Implementing a thorough disease prevention and biosecurity programme can often be difficult in these production systems due to the increased risks associated with wild bird contact, potential exposure to dirty stagnant water through access to ditches and puddles, as well as exposure to more adverse weather conditions and the associated increased stresses. However, there is still a significant number of critical control points that can be managed to ensure a robust biosecurity protocol in free-range production. Vertically transmitted diseases can be controlled as discussed above and a good risk-based vaccination programme can be used to ensure a good chick/pullet quality upon arrival to the farm. On-farm biosecurity should encompass the usual visitor restrictions to priority personnel only and interventions put in place upon entry onto the site (boot changes, overalls, wheel wash, foot dips and hand sanitizers). Implementing good range management will also reduce the risk for some diseases and can reduce general bacterial loading through contaminated water sources.

In the authors' opinion, this increase in free-range egg production is consistent with a rise in some diseases, such as mycoplasmosis, pasteurellosis and erysipelas (Wang *et al.*, 2010) due to these organisms' survival in the environment and difficulties in maintaining biosecurity. Free-range production has an inherent risk of increased wild bird contact. These endemic diseases are seen more rarely in colony systems, where the risks for disease transmission are less and tighter biosecurity procedures can be implemented. However, since the changes in European legislation and the ban on conventional cages since the beginning of 2012, epidemiology and disease transmission of some diseases would appear to be significantly different from conventional cages. This is thought to be due to increased bird to bird contact within and between individual colonies and increased opportunity for faeco-oral transmission through scratch areas, nestboxes and access to muck belts (depending on colony design).

Given the shift in production systems as shown in Fig. 9.1 and the increased risks for disease one must question whether meeting the current consumer demands, with respect to production systems, is a sustainable option for longer term global food production and pressure on resources.

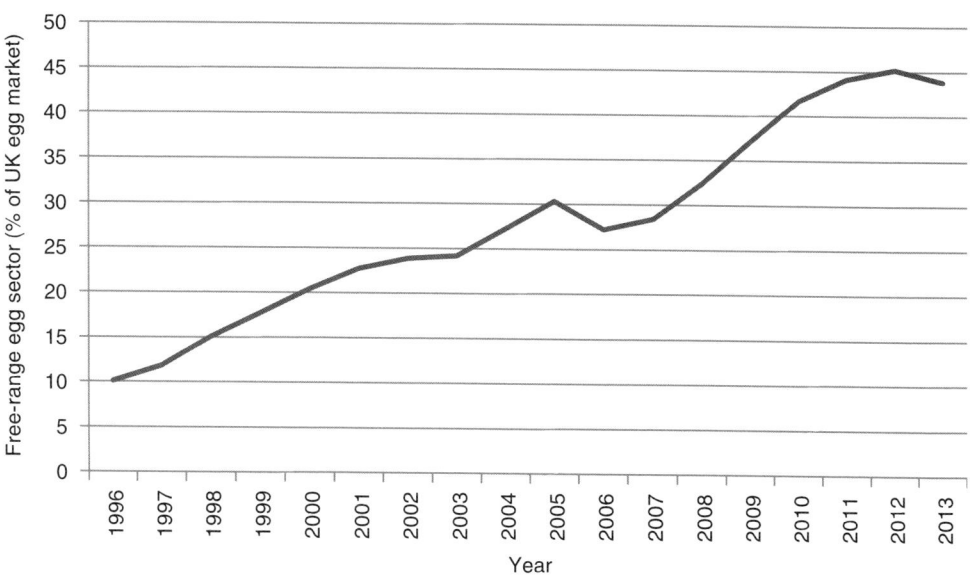

Fig. 9.1. Growth of the free-range egg sector (figures courtesy of the British Egg Industry Council).

Treating of primary or secondary pathogens

Antibiotics are often used to control secondary bacterial infections, which can be due to either primary bacterial or viral challenges. Virus infections tend to dominate in poultry and bacterial infections are frequently secondary invaders. *Escherichia coli* infections, following infectious bronchitis (IB), infectious bursal disease (IBD), Newcastle disease (ND), and avian pneumovirus (AP), especially in turkeys (TRT), and *Mycoplasma* infections, account for much of the antimicrobial medication used in poultry production. *Salmonella*, *Pasteurella multocida*, *Ornithobacterium rhinotracheale* and *Haemophilus paragallinarum* in warmer climates regularly occur and respiratory and systemic infections play a very important role in mixed disease challenges.

When treating a mixed bacterial infection, for example a *Mycoplasma galli-septicum* or *Mycoplasma synoviae* challenge combined with either *E. coli* or *O. rhinotracheale*, consideration must be given to sensitivity of both the primary and secondary pathogenic bacteria involved in the challenge. Does one look to treat the primary pathogen to reduce the risk for secondary bacterial challenges developing or is the priority to treat the secondary bacterial challenges that could be the main cause of mortality, poor egg production or increased processing rejects? Often there will be a good antimicrobial sensitivity to the primary pathogen if using a narrow spectrum antimicrobial, with poor activity against the secondary pathogens. For example, in the UK tylvalosin (Aivlosin™) is licensed for the treatment of *Mycoplasma*, however tylvalosin has no activity against Enterobacteriaceae including *E. coli* and *Salmonella* spp. (they are naturally resistant) (European Medicines Agency, 2009). In such a situation the synergistic effects of certain antimicrobials can be of benefit, for example the use of tiamulin combined with oxytetracycline (Stipkovits and Kempf, 1996; Valks and Burch, 2002) will give a better clinical response for *M. synoviae* or *M. gallisepticum* and *E. coli* mixed infections, with this synergistic combination often being used in commercial layers.

Genetics

With recent advances in avian genomics the possibilities for genetic selection of traits that can aid in the reduction of antibiotic use are endless. The chicken genome was first sequenced in 2004 (Burt, 2005). Selection of traits that can aid reducing antibiotic usage can vary from improving breeder eggshell quality and innate natural defences such as cuticle structure (Board and Fuller, 1994) to improving the chick's innate immunity and natural defences to specific diseases (Burnstead and Millard, 1987; Lamont, 1998). The heritability observed with cuticle thickness means that genetic selection should be possible to increase cuticle deposition in commercial poultry. This would reduce trans-generational transmission of microorganisms and reverse the lack of selection pressure for this trait during recent domestication (Bain *et al.*, 2013). Since genetic selection has been in place within the poultry breeding pyramid there has been a primary focus on production characteristics for growth and feed conversion. However, given the demands on the industry there is a shift towards more welfare-oriented characteristics such as leg health, liveability and intestinal health. This shift in focus is being embraced by the industry to meet the requirements of the consumer and is being guided by European regulation and legislation (Council Directive 2007/43/CE).

REPLACEMENT

There is no specific product that is a direct replacement for antibiotics, in terms of having the desired bactericidal or bacteriostatic effect. However, there are some live bacterial vaccines that are being used as a targeted therapeutic strategy in the face of a disease outbreak.

This competition for the same infection route and attachment sites can also be of benefit as certain live bacterial vaccines can be used as a targeted therapeutic in the face of a disease outbreak. This is not applicable to all live vaccines; although in the authors' experience the use of a live attenuated vaccine administered in the face of a coli septicaemia or *E. coli*-associated peritonitis outbreak can be beneficial in reducing mortality. The exact aetiology of this response is unclear, it is hypothesized that the attenuate non-pathogenic vaccinal *E. coli* strains are competing for the same cellular attachment sites, either in the respiratory tract or intestinal lumen, as the pathogenic strains, and therefore reducing the attachment and subsequent pathological pathways (J. Brown, United Kingdom, 2010, personal communication).

The concept of competitive exclusion (CE) has been used in poultry for some years. The process of seeding or colonizing a bird's intestinal tract with beneficial bacteria to outcompete or exclude enteric pathogens from colonizing can have significant impacts on disease control and shedding of pathogens (La Ragione and Woodward, 2003). This is a form of treatment with probiotics, one or more beneficial microorganisms derived typically from the gastrointestinal flora of an adult of the species to be treated. While single organism treatments have at times shown promise, the most efficacious CE preparations contain a large number and diversity of genera and species ranging from lactic acid bacteria to strict anaerobes (Cox and Pavic, 2010). Schneitz (2005) reviewed commercially available CE products and their efficacy in excluding *Salmonella* and other enteric organisms, concluding that diverse undefined caecal cultures offered the best protection followed by a defined consortium of many diverse strains. Competitive exclusion products, either used alone or as part of a wider therapy including prebiotics, can aid in controlling avian diseases that require antibiotic medication (Hofacre *et al.*, 2003).

There is a vast array of nutraceuticals used in the poultry industry. These nutritional supplements can include nucleic acids, prebiotics such as mannan oligo-saccharides (MOS) and fructo-oligo-saccharides (FOS) from yeast cell-wall extracts or herbal or plant extracts. These prebiotics enhance the natural intestinal flora, promoting the growth of probiotic species, such as lactic acid bacteria, thereby reducing the likelihood and persistence of colonization by enteric pathogens (Doyle and Erickson, 2006). Increased numbers of *Lactobacillus* and *Bifidobacterium* species correlated with reduced *Salmonella* prevalence (Fernandez *et al.*, 2002; Xu *et al.*, 2003). The variation in efficacy and level of clinical data on such products can be variable, in part due to the less well-regulated structure of the nutraceutical industry. Nevertheless this is an area of increasing interest and development with some good scientific data being generated.

In contrast to antibiotics, most alternative compounds do not reduce overall microbial loads in the gut and thus will not promote growth by a mechanism similar to antibiotics. Instead, they alter the gut microflora profile by limiting the colonization of unfavourable bacteria while promoting the fermentation of more favourable species. Consequently, alternatives to antibiotics promote gut health by several possible mechanisms, including: altering gut pH, maintaining protective gut mucins, selection for beneficial intestinal organisms or against pathogens, enhancing fermentation acids, enhancing nutrient uptake and increasing the humoral immune response. Strategic use of these alternative compounds will help optimize performance and reduce the need for antimicrobial therapy when they are used in a manner that complements their modes of action.

Although MOS have been shown to reduce some clostridial caecal levels (Biggs *et al.*, 2007; Benites *et al.*, 2008) their effects are limited and their use should be as part of a wider intestinal health control strategy. Given their limited antimicrobial effects such products are targeted at subclinical effects rather than as an adjunct to therapy in the face of a clinical disease outbreak.

The effects of the secondary metabolites of garlic, propyl thiosulfinate and propyl thiosulfinate oxide, have been shown to have a pharmacological effect on the cardiovascular system as well as exerting some immuno-modulating and antimicrobial effects (Chowdhury *et al.*, 2002; Choi *et al.*, 2010; Hanieh *et al.*, 2010). In addition to these effects they also have the ability to improve resistance to *Eimeria acervulina* infections (Kim *et al.*, 2013), hence could be an additional therapy in improving overall intestinal health and reducing necrotic enteritis. Coccidiosis is a major factor in clinical outbreaks of necrotic enteritis (Williams, 2005).

Organic acids, either as a single acid or as blended compounds, are widely used in poultry production for their antimicrobial effect. However, in addition to their antimicrobial effect, these short chain organic acids improve protein and energy digestibility by reducing microbial competition with the host for nutrients. They also reduce endogenous nitrogen losses, by lowering the incidence of sub-clinical infections and secretion of immune mediators and by reducing production of ammonia and other growth-depressing microbial metabolites (Dibner and Buttin, 2002; Griggs and Jacob, 2005). Organic acids can be used either as feed additives or via the drinking water, where they can also have a disinfectant effect on drinking water hygiene and biofilm reduction (Szewzyk *et al.*, 2000).

REFINEMENT

Once a bacterial disease has been diagnosed and a clinical decision to medicate with an antibiotic has been made, there are then a significant number of further factors to consider as to which antibiotic should be used:

- antimicrobial sensitivity of the organism (*in vitro* antibiogram, MIC);
- antimicrobial spectrum (narrow spectrum versus broad spectrum);
- route of administration (drinking water, in-feed medication, injection or topical);
- dose rate;
- duration of treatment;
- application period;
- licensed products;
- cascade use (UK);
- food residues – meat and egg withdrawals;
- voluntary industry restrictions;
- economics;
- retailer restrictions; and
- consumer pressure.

The list above is not in any order of priority and the weighting of the various factors can have many influences outwith the scientific evidence-based clinical decision.

There are various ways in which to measure the antibiotic sensitivity of a bacterial isolate. This can be done using a disc diffusion test. The use of the Iso-Sensitest™ has superseded the use of the DST (Direct Sensitivity Test) plates due to its more consistent zones of inhibition (Kuper *et al.*, 2009). The other commonly used method of ascertaining sensitivity is to measure the minimal inhibitory concentration (MIC) for an isolate. This method involves culturing the bacterial isolate in a serial dilution of the antibiotic to establish a 'break point' at which the isolate is deemed sensitive (Andrews, 2001). The break points are not fixed for an antibiotic and will vary depending on the pharmacokinetic (PK) and pharmacodynamics (PD) characteristics.

These methods of establishing antimicrobial sensitivities in general poultry practice are common. However, these methods have an inherent risk or potential low specificity as they are often done on a single bacterial isolate taken from a direct swab culture collected from necropsy. Therefore one could question whether the antibiotic sensitivity of a single bacterial isolate or colony forming unit (cfu) collected from an investigatory necropsy, potentially from a single bird, is representative of the causal pathogen within a given bird population.

Pharmacokinetic and PD information form the scientific basis of modern pharmacotherapy. Pharmacokinetics describes the drug concentration–time courses in body fluids resulting from administration of a certain drug dose, PD the observed effect resulting from a certain drug concentration (Meibohm and Derendorf, 1997). Understanding the pharmacokinetics of an antimicrobial, knowing its potential tissue concentration at the site of infection relative to its MIC for the organism to treat, can help the veterinarian to decide which antibiotic, which dose, route of administration and treatment period. This improves the therapeutic control of the disease and reduces the chance of developing antimicrobial resistance (Burch, 2002) whilst optimizing the clinical outcome.

Given the number of potential variables in deciding on a therapeutic regime there can often be more than one possible treatment option. For example, tylosin tartrate is indicated in outbreaks of necrotic enteritis associated with *Clostridium perfringens* and amoxicillin trihydrate also has an indication against *Clostridium* spp. Tylosin is a bacteriostatic, narrow spectrum macrolide, effective mainly against gram-positive bacteria (Goetting *et al.*, 2011), whilst amoxicillin is a bactericidal, broad spectrum β-lactam antibiotic active against both gram-positive and gram-negative bacteria. Macrolides are known to select for macrolide-resistant *Campylobacter* spp. in animals, especially *Campylobacter jejuni* in poultry. At the same time, macrolides are one of few available therapies for serious *Campylobacter* infections, particularly in children, in whom quinolones are not recommended for treatment. Given the high incidence of human disease due to *Campylobacter* spp., especially *C. jejuni*, the absolute number of serious cases is substantial (WHO, 2011). In addition, the use of the broad-spectrum amoxicillin also has the potential to select for antibiotic resistance (Editorial, 2013). It is the authors' personal preference to use amoxicillin as a first line treatment for *Clostridia*-associated enteritis in chickens due to the specific risks of resistance in *Campylobacter*.

Having a thorough understanding of the PK/PD of an antimicrobial, as well as the bird type and management of those birds, can influence the clinical efficacy. For example, colistin is often used in free-range egg layers within the UK to treat coli-septicaemia or egg peritonitis at the onset of lay. Although the condition is multifactorial in aetiology (Lister and Barrow, 2008), mortality associated with the condition can be controlled through the therapeutic use of colistin. Colistin is poorly absorbed from the intestinal tract of birds, with most of its activity remaining intra-luminal, with relatively low tissue concentrations achieved. Due to the management of egg layers, they tend to have water access over a 15 h period and assuming one can medicate over a similar time point there will be an activity over the required level for the required time period, sufficient to have a therapeutic effect. However, when used in broiler breeders, suffering from coli-septicaemia and egg peritonitis, this same antimicrobial has a significantly reduced efficacy, in the authors' experience. This is likely to be influenced by the management of broiler breeders, in the fact that they tend to consume feed in a single feed, over a short time period and drink most of their daily water intake over a relatively short period as well. As such there is unlikely to be sufficient colistin present, either intra-luminally or within the tissues to have a sufficient bactericidal effect (Botsoglou and Fletouris, 2001; Goetting *et al.*, 2011).

The use of sub-therapeutic doses of some antibiotic classes can lead to increased mutagenesis and resistance gene transfer, according to Patrice Courvalin of Institut Pasteur in Paris, France. Low concentrations of antibiotics induce random mutagenesis, he explains. If you use low concentrations of penicillin in pneumococci, you will select for resistance to other drug classes, while strains are still susceptible to penicillin, because it is just random mutagenesis. The increased mutation rate may promote horizontal gene transfer. Thus, low concentrations of fluoroquinolone and aminoglycoside resistance genes induce mutations, leading to transfers of these traits (Cogliani *et al.*, 2011).

CONCLUSION

To reduce antibiotic use within the poultry meat and egg production sectors the industry must embrace a multifaceted approach to reduce, refine and replace antibiotics in order to safeguard their use in the future. This approach must be scientifically sound and evidence-based and not reactionary to consumer demands and perceptions. There should be a strong focus on hygiene and bio-security with a vaccination programme based on the risk analysis. The supplementary nutraceutical, prebiotic and probiotic products can also be useful additions in a holistic one-health strategy. Continuing developments through improved diagnostics, genetic selection and further understanding of the bird's innate immunity will all help in the goal of antibiotic reduction. The authors have no doubt that the dynamic poultry industry will embrace this initiative and in so doing ensure a sustainable industry with a global health vision.

REFERENCES

Andrews, J.A. (2001) The development of the BSAC standardized method of disc diffusion testing. *Journal of Antimicrobial Chemotherapy* 48, 29–42.

Arcliicnbuwa, F.E., Adkr, B. and Wiggins, A.D. (1980) A method of surveillance for bacteria on the shell of turkey eggs. *Poultry Science* 58, 799–806.

Aviagen (2009) Broiler Management Manual. Available at: http://en.aviagen.com/assets/Tech_Center/Ross_Broiler/Ross_Broiler_Manual_09.pdf (accessed 1 October 2014).

Bain, M.M., McDade, K., Burchmore, R., Law, A., Wilson, P.W., Schmutz, M., Preisinger, R. and Dunn, I.C. (2013) Enhancing the egg's natural defence against bacterial penetration by increasing cuticle deposition. *Animal Genetics* 44, 661–668.

Beemer, F., of Velzen, G., of den Berg, C., Zunderdorp, M., Lambrechts, E., Gier, K. and Oud, N. (2010) What Would Be the Effects of Decoupling the Prescription and Sale of Veterinary Medicines by Veterinarians? Berenschot Report. Available at: http://www.fve.org/uploads/publications/docs/berenschot%20report_02_2010.pdf (accessed 16 June 2015).

Benites, V., Gilharry, R., Gernat, A.G. and Murillo, J.G. (2008) Effect of dietary mannan oligosaccharide from bio-mos or SAF-mannan on live performance of broiler chickens. *The Journal of Applied Poultry Research* 17, 471–475.

Biggs, P., Parsons, C.M. and Fahey, G.C. (2007) The effects of several oligosaccharides on growth performance, nutrient digestibilities, and cecal microbial populations in young chicks. *Poultry Science* 86, 2327–2336.

Board, R.G. and Fuller, R. (1994) *Microbiology of the Avian Egg*. Chapman & Hall, London.

Botsoglou, N.A. and Fletouris, D.J. (2001) *Drug Residues in Food*. Marcel Dekker, New York.

Bruce, J. and Drysdale, E.M. (1994) Trans-shell transmission. In: Board, R.G. and Fuller, R. (eds) *Microbiology of the Avian Egg*. Chapman & Hall, London, pp. 63–91.

Burch, D.G.S. (2002) *Antimicrobial Sensitivity Data for the Major Poultry Bacterial Pathogens*. Paper presented at 'Anti-Infectives: the way forward', 8–9 July 2002. Royal Pharmaceutical Society, London.

Burnstead, N. and Millard, B. (1987) Genetics of resistance to coccidiosis: response of inbred chicken lines to infection by *Eimeria tenella* and *Eimeria maxima*. *British Poultry Science* 28, 705–715.

Burt, D.W. (2005) Chicken genome: current status and future opportunities. *Genome Research* 15, 1692–1698.

Calnek, B.W., Barnes, H.J., Beard, C.W., McDougald, L.R. and Saif, Y.M. (1997) *Diseases of Poultry*, 10th edn. Iowa State University Press, Ames, Iowa.

Carter, T.A. and Sneed, R.E. (1987) *Drinking Water Quality for Poultry*. PS&T Guide No. 42, Extension Poultry Science, North Carolina State University, North Carolina.

Castanon, J.I.R. (2007) History of the use of antibiotic as growth promoters in European poultry feeds. *Poultry Science* 86, 2466–2471.

Choi, I.H., Park, W.Y. and Kim, Y.J. (2010) Effects of dietary garlic powder and (alpha)-tocopherol supplementation on performance, serum cholesterol levels and meat quality of chicken. *Poultry Science* 89, 1724–1731.

Chowdhury, S.R., Chowdhury, S.D. and Smith, T.K. (2002) Effects of dietary garlic on cholesterol metabolism in laying hens. *Poultry Science* 81, 1856–1862.

Cogliani, C., Goossens, H. and Greko, C. (2011) Restricting antimicrobial use in food animals: lessons from Europe. Banning nonessential antibiotic uses in food animals is intended to reduce pools of resistance genes. *Microbe* 6, 274.

Cox, B. (2005) *Biosecurity – The Economics and Benefits – Are We Fooling Ourselves*. Proceedings of the Poultry Service Industry Workshop 4–6 October.

Cox, J.M. and Pavic, A. (2010) Advances in enteropathogen control in poultry production. *Journal of Applied Microbiology* 108, 745–755.

Cox, N.A., Berrang, M.E. and Cason, J.A. (2000) Salmonella penetration of egg shells and proliferation in broiler hatching eggs – a review. *Poultry Science* 79, 1571–1574.

Dibner, J.J. and Buttin, P. (2002) Use of organic acids as a model to study the impact of gut microflora on nutrition and metabolism. *Journal of Applied Poultry Research* 11, 453–463.

Dibner, J.J., Knight, C.D., Kitchell, M.L., Atwell, C.A., Downsand, A.C. and Ivey, F.J. (1998) Early feeding and development of the immune system in neonatal poultry. *Journal of Applied Poultry Research* 7, 425–436.

Donaldson, L. (2007) Combating antimicrobial resistance: the role of the specialist advisory committee on antimicrobial resistance. *Journal of Antimicrobial Chemotherapy* 60, i27–i32.

Doyle, M.P. and Erickson, M.C. (2006) Reducing the carriage of foodborne pathogens in livestock and poultry. *Poultry Science* 85, 960–973.

Editorial (2013) Antibiotic resistance: long-term solutions require action now. *The Lancet Infectious Diseases* (published online) 13, 995.

European Medicines Agency (2009) Scientific Discussion. Available at: http://www.ema.europa.eu/docs/en_GB/document_library/EPAR_-_Scientific_Discussion/veterinary/000083/WC500061061.pdf (accessed 15 June 2014).

Fernandez, F., Hinton, M. and Van Giles, B. (2002) Dietary mannan-oligosaccharides and their effect on chicken caecal microflora in relation to salmonella enteritidis colonization. *Avian Pathology* 31, 49–58.

Fromm, D. and Margolf, P.H. (1958) The influence of sweating and washing on weight loss, bacterial contamination and interior physical quality of 12-day old shell eggs. *Poultry Science* 37, 1273–1278.

Goetting, V., Lee, K.A. and Tell, L.A. (2011) Pharmacokinetics of veterinary drugs in laying hens and residues in eggs: a review of the literature. *Journal of Veterinary Pharmacology and Therapeutics* 34, 521–556.

Government of the Netherlands (undated) Antibiotic resistance in the livestock industry. Available at: http://www.government.nl/issues/antibiotic-resistance/antibiotic-resistance-in-livestock-farming (accessed 15 June 2015).

Griggs, J.P. and Jacob, J.P. (2005) Alternatives to antibiotics for organic poultry production. *Journal of Applied Poultry Research* 14, 750–756.

Hanieh, H., Narabara, K., Piao, M., Gerile, C., Abe, A. and Kondo, Y. (2010) Modulatory effects of two levels of dietary alliums on immune response and certain immunological variables, following immunization, in white Leghorn chickens. *Animal Science Journal* 81, 673–680.

Hoerr, F.J. (2010) Clinical aspects of immunosuppression in poultry. *Avian Diseases* 54, 2–15.

Hofacre, C.L., Beacorn, T., Collett, S. and Mathis, G. (2003) Using competitive exclusion, mannan-oligosaccharide and other intestinal products to control necrotic enteritis. *Journal of Applied Poultry Research* 12, 60–64.

Kalmokoff, M., Lanthier, P., Tremblay, T.L., Foss, M., Lau, P.C., Sanders, G., Austin, J., Kelly, J. and Szymanski, C.M. (2006) Proteomic analysis of *Campylobacter jejuni* 11168 biofilms reveals a role for the motility complex in biofilm formation. *Journal of Bacteriology* 188, 4312–4320.

Kapperud, G., Skjerve, E., Vik, L., Hauge, K., Lysaker, A., *et al.* (1993) Epidemiological investigation of risk factors for campylobacter colonization in Norwegian broiler flocks. *Epidemiology and Infection* 111, 245–255.

Kim, D.K., Lillehoj, H.S., Lee, S.H., Lillehoj, E.P. and Bravo, D. (2013) Improved resistance to *Eimeria acervulina* infection in chickens due to dietary supplementation with garlic metabolites. *British Journal of Nutrition* 109, 76–88.

Kleven, S.H. (2008) Control of avian mycoplasma infections in commercial poultry. *Avian Diseases* 52, 367–374.

Kuper, K.M., Boles, D.M., Mohr, J.F. and Wanger, A. (2009) Antimicrobial susceptibility testing: a primer for clinicians. *Pharmacotherapy* 29, 1326–1343.

La Ragione, R.M. and Woodward, M.J. (2003) Competitive exclusion by *Bacillus subtilis* spores of *Salmonella enterica* serotype Enteritidis and *Clostridium perfringens* in young chickens. *Veterinary Microbiology* 94, 245–256.

Lamont, S.J. (1998) Impact of genetics on disease resistance. *Poultry Science* 77, 1111–1118.

Lister, S.A. and Barrow, P. (2008) Bacterial diseases: enterobacteriaceae. In: Pattison, M., McMullin, P.F., Bradbury, J. and Alexander, D.J. (eds) *Poultry Diseases*, 6th edn. Elsevier, Philadelphia, Pennsylvania, p. 140.

Meibohm, B. and Derendorf, H. (1997) Basic concepts of pharmacokinetic/pharmacodynamic (PK/PD) modelling. *International Journal of Clinical Pharmacology and Therapeutics* 35, 401–413.

Mevius, D. and Heederik, D. (2014) Reduction of antibiotic use in animals "let's go Dutch". *Journal of Consumer Protection and Food Safety* 10, 1007.

Nordberg, P., Monnet, D.L. and Cars, O. (2005) Antibacterial Drug Resistance: Options for Concerted Actions. World Health Organization, February 2005. Available at: http://apps.who. int/medicinedocs/documents/s16368e/s16368e.pdf (accessed October 2015).

RUMA (2005) *Responsible Use of Antimicrobials in Poultry Production*. Available at: http://www. ruma.org.uk/guidelines/antimicrobials/long/poultry%20antimicrobials%20long.pdf (accessed 14 July 2014).

Schneitz, C. (2005) Competitive exclusion in poultry – 30 years of research. *Food Control* 16, 657–667.

Smith, A., Rose, S.P., Wells, R.G. and Pirgozliev, V. (2000) The effect of changing the excreta moisture of caged laying hens on the excreta and microbial contamination of their egg shells. *British Poultry Science* 41, 168–173.

Solomon, S.E., Bain, M.M., Cranstoun, S. and Nascimento, V. (1994) Hen's egg shell structure and function. In: Board, R.G. and Fuller, R. (eds) *Microbiology of the Avian Egg*, 1st edn. Chapman & Hall, London, pp. 1–24.

Stevenson, A. and Waite, M. (2011) In: *Concise Oxford English Dictionary*. Oxford University Press, Oxford, UK, p. 1728.

Stipkovits, L. and Kempf, I. (1996) Mycoplasmoses in poultry revue. *Scientifique et Technique* (International Office of Epizootics) 15, 1495–1525.

Swann, M.M., *et al.* (1969) *Report of the Joint Committee on the Use of Antibiotics in Animal Husbandry and Veterinary Medicine*. Her Majesty's Stationery Office, London.

Swormink, B.K. (2014) Target reduction in antibiotics: 70 percent. *World Poultry Magazine* 30, 11 March 2014.

Szewzyk, U., Szewzyk, R., Manz, W. and Schleifer, K.H. (2000) Microbiological safety of drinking water. *Annual Review of Microbiology* 54, 81–127.

Valks, M. and Burch, D. (2002) The treatment and control of mycoplasma infections in turkeys. Paper presented at the Novartis animal health seminar, European Poultry Association Conference, Bremen, Germany, September 2002.

Van den Burg, T. (2008) Viral diseases: birnaviridae. In: Pattison, M., McMullin, P.F., Bradbury, J. and Alexander, D.J. (eds) *Poultry Diseases*, 6th edn. Elsevier, Philadelphia, Pennsylvania, p. 365.

Vandeplas, S., Dubois Dauphin, R., Beckers, Y., Thonart, P. and Théwis, A. (2010) Salmonella in chicken: current and developing strategies to reduce contamination at farm level. *Journal of Food Protection* 73, 774–785.

Wang, Q., Chang, B.J. and Riley, T.V. (2010) Erysipelothrix rhusiopathiae. *Veterinary Microbiology* 140, 405–417.

Whistler, P.E and Sheldon, B.W. (1989) Biocidal activity of ozone versus formaldehyde against poultry pathogens inoculated in a prototype setter. *Poultry Science* 63, 1068–1073.

WHO (2011) *Critically Important Antimicrobials for Human Medicine*, 3rd rev. edn. WHO Document Production Services, Geneva, Switzerland.

Williams, R.B. (2005) Intercurrent coccidiosis and necrotic enteritis of chickens: rational, integrated disease management by maintenance of gut integrity. *Avian Pathology* 34, 159–180.

Xu, Z.R., Hu, C.H., Xia, M.S., Zhan, X.A. and Wang, M.Q. (2003) Effects of dietary fructo-oligosaccharide on digestive enzyme activities, intestinal microflora and morphology of male broilers. *Poultry Science* 82, 1030–1036.

Zimmer, M., Barnhart, H., Idris, U. and Lee, M.D. (2002) Detection of *Campylobacter jejuni* strains in the water lines of a commercial broiler house and their relationship to the strains that colonised the chickens. *Avian Diseases* 47, 101–107.

Zwietering, M.H., Rombouts, F.M. and Riet, K.V. (1992) Comparison of definitions of the lag phase and the exponential phase in bacterial growth. *Journal of Applied Bacteriology* 72, 139–145.

Chapter 10
Human Nutrition and Health – Making Products More Desirable to Consumers

Dawn Scholey[1]* and Steve Pritchard[2]

[1]Nottingham Trent University, Nottingham, UK; [2]Premier Nutrition, Rugeley, UK

INTRODUCTION – THE HEALTH OF THE NATION

As a nation we are fat and getting fatter. Whilst the problems of being overweight or obese have traditionally been associated with developed western countries, the issue is spreading across the globe. Much has been made of the challenge for global agriculture to meet the demands of feeding a projected population of 9 billion people by 2050. These 9 billion people will have an appetite of 12 billion.

From the poultry industry perspective the projected population growth will, by default, increase demand for our products. However, for the European industry the increasing world demand for poultry meat and egg products is likely to be met by production from Latin America and Asia as these regions have a lower cost of production and more commercial approach to regulation. Again this may be of benefit to European producers as other regions become a more attractive marketplace for countries like Brazil and Thailand to export to. The opportunity for the European industry is to satisfy the growing demand in its local market for innovative products. There is a diverse range of criteria that will need to be met, including provenance, taste, convenience, welfare, safety and value. This chapter sets out one of the key challenges in terms of the health of the nation, namely obesity, and accepts that telling consumers to eat healthily does not appear to have any long-term impact on lifestyle habits. The industry has a tremendous opportunity to innovate both in terms of production and marketing to secure a sustained proportion of the nation's diet by positioning itself as an essential part of a healthy balanced diet.

*Corresponding author: dawn.scholey@ntu.ac.uk

OBESITY IN THE UK

The impact of obesity is both a burden on the individual and also on society. The European Commission policy on 'Nutrition and physical activity' states that in Europe, six of the seven biggest risk factors for premature death are related to how we eat, drink and move:

- hypertension;
- high cholesterol;
- high Body Mass Index (BMI);
- inadequate fruit and vegetable intake;
- lack of physical activity; and
- alcohol abuse.

The problems of an overweight population are well recognized. UK government policy specifically targets the issues of obesity and improving diet. In its policy document 'Reducing obesity and improving diet' (Department of Health, 2013), the Department of Health states that the majority of the population in England are overweight (61.9% of adults and 28% of children between the ages of 2 and 15). They describe the issue as increasing the risk of type 2 diabetes, heart disease and certain cancers. Obesity is the leading cause of cancer after smoking. Excess weight can also affect employment prospects, self-esteem and mental health. The government estimates that being overweight or obese costs the National Health Service (NHS) more than £5 billion per annum.

HEALTHY EATING ADVICE – IS IT WORKING?

It is UK government policy to encourage the population to eat and drink more healthily and to be more active. The action plan involves four main areas.

1. Advice to the general public on a healthy diet and physical activity through the Change4Life programme.
2. Improving labelling to allow consumers to make more informed choices.
3. Encouraging 'high street' businesses to include calorie information on menus.
4. Giving guidance on physical activity levels.

Whilst there might be general public awareness of the problem and some of the potential solutions there is very little evidence that the trend towards a more obese population is being reversed.

The main findings of a recent report published by the Health and Social Care Information Centre 'Statistics on Obesity, Physical Activity and Diet' (HSCIC, 2014) were that:

- Between 1993 and 2012 the proportion of adults in England with a normal BMI decreased from 41.0% to 32.1% in males and from 49.5% to 40.6% in females.
- Over the same time period the proportion of obese adult males increased from 13.2% to 24.4% and for females from 16.4% to 25.1% (Fig. 10.1).

   As a further illustration the UK government launched the '5 a day' campaign
in 2003, encouraging people to eat five portions of fruit and vegetables a day.
Figure 10.2 shows that despite an initial 6% increase in intakes between 2001
and 2006 this has then gradually decreased to 2009 and from there remained
relatively steady. The campaign appears to have had no lasting effect on the
mean fruit and vegetable intake of the population.

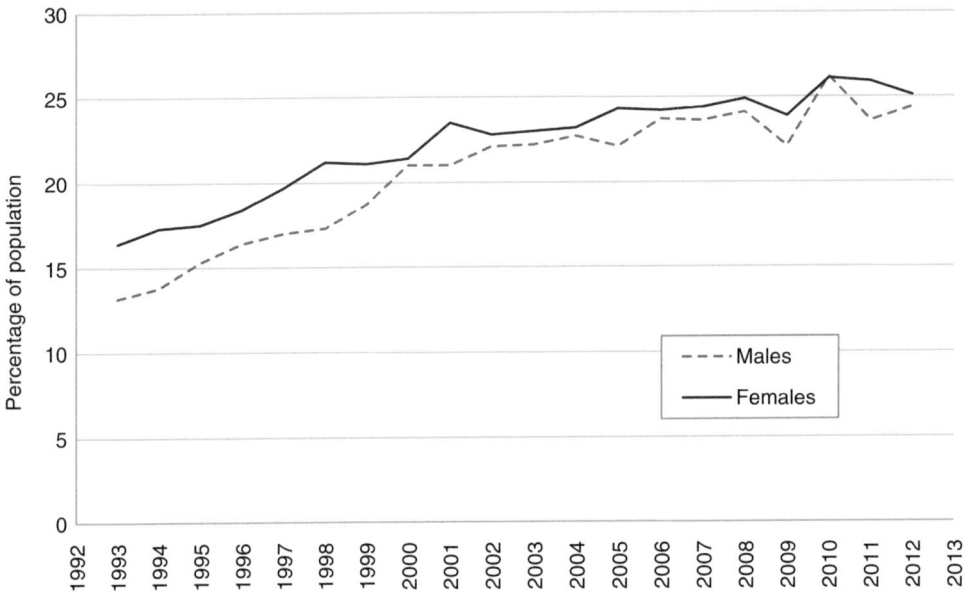

Fig. 10.1. Obesity prevalence in adults in the UK, 1993–2012 (Health Survey for England, 2012).

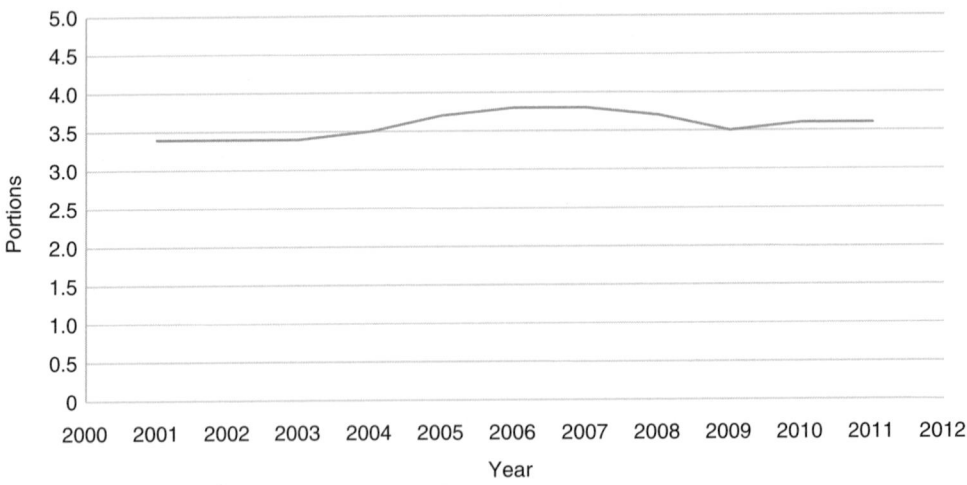

Fig. 10.2. Fruit and vegetable intake in the UK, 2001–2011 (Health Survey for England, 2012).

HEALTH BY STEALTH

Clearly the message is not achieving its goal in terms of changing the eating or exercise habits of the population. This dilemma has led to the suggestion of a new tactic: 'Health by Stealth' (Drummond, 2014). This involves improving the nutritional value or potentially reducing the calorific value of existing dietary components rather than fundamentally changing eating habits. The challenge for the food and agriculture industry is to develop healthier foods that consumers want to eat, but not to excess.

Taste and convenience are critical factors. Why would consumers eat more of something that is tasteless or that is difficult and time consuming to prepare? There are numerous press reports of the next new 'wonderfood'. Blueberries, for example, have a number of positive nutritional attributes, but rather than eat the fresh fruit as part of our five a day there is now a blueberry-flavoured crisp marketed by one of the big brands in Asia. This highlights the difficulties in getting the population to change their eating habits and to achieve a healthy balanced diet that tastes good and is convenient.

Tomatoes are generally accepted as good for us and can form part of our five a day. Would people eat more of them if they tasted better? The variety Tasti-Lee is an example of a collaborative effort in Florida to get a nutritious, tasty tomato to market. Further research is ongoing to define better the characteristics of a tasty tomato with the right balance of sugar, acid and volatile components that give a superior eating quality. By measuring these traits certain varieties have been shown to contain the right mix for the perfect tasting tomato that will encourage consumers to eat more of a healthy product.

This example illustrates the potential in the marketplace for the poultry industry to position products that are both tasty and healthy and that can form part of a balanced diet. Studies of consumer eating habits have shown that families typically eat a rotation of about ten meals in the home. If poultry products can establish themselves in this rotation, as a fundamental part of a tasty, healthy meal plan, then that should develop a consistent demand for our products.

PUBLIC PERCEPTION – CHALLENGES AND OPPORTUNITIES

Media coverage, particularly with cases like the dioxin and horsemeat contamination, has reduced public trust both in the food industry and the national bodies responsible for food safety. Food technologies are thought of as riskier than medical technological applications with DNA technologies perceived as particularly high risk, comparable with nuclear energy and radioactive waste (Fife-Shaw and Rowe, 1996). Biotechnology applications are all generally thought to be risky by the general public, but public perception of risk associated with food is often based on the type and level of media coverage (Miles and Frewer, 1999) and also on their level of trust. Beneficial claims can be correlated with high trust levels, particularly in people with little specialist knowledge.

Poultry meat has three risks associated with it by the public: microbiological, chemical (such as antibiotics and agricultural residues) and technological (i.e. genetic modification), with most consumers believing the microbiological risk to be the greatest, specifically from *Escherichia coli* (Yeung and Morris, 2001). Consumers also do not want their food 'messed around with'. This presents challenges to how the industry produces and markets products.

Eggs are still considered by some to be a risk in terms of high cholesterol and heart disease, although there is now strong evidence to the contrary. Moderate egg consumption does not increase the risk of cardiovascular disease (CVD) in healthy individuals (Fernandez, 2006) and there has been no correlation found between egg consumption and plasma cholesterol.

A study examining causal factors of CVD showed that one egg a day consumption was responsible for less than 1% of CVD risk compared with other lifestyle factors such as obesity, exercise and smoking, which accounted for up to 40% of the risk (Barraj *et al.*, 2009). Overweight men who eat three eggs per day have higher high density lipoprotein ('good' cholesterol) levels and less risk of metabolic syndrome (Mutungi *et al.*, 2008).

Despite these negative concerns, overall, poultry products are already desirable to consumers in their standard form. Poultry meat is considered to be low in saturated fat, has a high protein to calorific ratio, is convenient to prepare and supplies a range of essential nutrients as well as being good value. Similarly eggs, having come through the onslaught of both the *Salmonella* crisis and the cholesterol debacle, are now considered to be a nutritious ingredient providing good quality protein, vitamins and minerals (Ruxton *et al.*, 2010). Eggs are recognized as contributing valuable, high quality protein, which is important for low income families.

Chicken is the most popular meat in the UK, and whilst overall meat consumption fell by 13% between 2007 and 2013, chicken consumption rose by 7% over the same time period (DEFRA, 2013). Turkey and other poultry products also have a long-term increase in consumption, although levels are much lower than for chicken (18 g/person/week compared with 192 g/week for chicken). Egg consumption has been decreasing over the past 30 years, but the recent trend is a positive one (9% increase from 2009 to 2013; DEFRA, 2013), helped largely by the establishment of the Lion Egg Scheme and a number of positive health messages that have heralded eggs as a superfood.

The consumption of poultry meat does not appear to have a link with all-causal mortality unlike red meat and meat products, which have also been linked to increases in type 2 diabetes. High red meat intake can raise the risk of having a stroke, with people who ate chicken or turkey rather than red meat having a reduced risk (Bernstein *et al.*, 2012).

The potential benefits of eggs and poultry meat for the increasing proportion of elderly people in the population have also been recognized. In older people muscle deposition is often reduced, and a protein or amino acid supplement will often replace a meal with no calorific or nutrient intake. Therefore promoting muscle anabolism using protein-rich foods such as eggs and poultry has several advantages, including accessibility, palatability and low cost.

REQUIREMENTS FOR DESIRABLE PRODUCTS

What opportunities are there to further enhance the taste and nutritional value of poultry products, making products more desirable to consumers?

We need a balanced diet for good health and well-being. Unbalanced diets, be they as a result of poverty or overconsumption, can cause health issues. This is particularly relevant considering the poor nutritional quality of some diets in the developed world. Inadequately nourished children have a lower IQ, which can be improved with micronutrient supplementation (Schoenthaler *et al.*, 2000).

It is important to consider several factors when choosing a food type for enrichment (Table 10.1). Poultry meat and eggs are a good fit for all these factors, being low cost, with high acceptability across age/social groups and no religious considerations. Both eggs and poultry meat can be enriched via fortification of poultry diets to provide a source of vitamins and minerals for humans, and the potential is there for several nutrients to be included in an enriched product to add value to the consumer.

Under EU food labelling regulations, to make a nutritional claim a product must contain a minimum of 15% of recommended dietary allowance (RDA) in a 100 g portion to allow the use of the wording 'source of. . .'. To be able to claim to be 'high in' a particular nutrient a product must have a minimum of 30% RDA per 100 g portion (EU, 2006).

TACKLING OBESITY – THE POSITIVE EFFECTS OF POULTRY PRODUCTS ON SATIETY

Given the background of an increasingly overweight population the opportunity for poultry products to become established as an ingredient that can help control weight is a possibility. There is strong evidence that increased protein in diets increases satiety and reduces intake. High protein, low fat diets have been shown

Table 10.1. Food choice considerations for enrichment (Yaroshenko *et al.*, 2003).

The food needs to be:	Comments
Part of traditional meals	Cannot change cultural habits or religious prohibition
Consumed moderately and regularly	Regular consumption is required to ensure nutritional requirements are met, but moderation reduces the risk of over-supplementation
Consumed by all age groups	Old and young people are more likely to have nutrient-deficient diets and compromised immunity
Low cost	Consumer choice will be effected by price point
Enriched with more than one nutrient	Multiple nutrients will have an overall benefit on the nutrient balance of the diet – plus add value
Providing a meaningful amount of recommended daily allowance (RDA)	Minimum of 50% RDA reduces the risk of marketing hype for minimal enrichments

to improve weight loss (Skov et al., 1999). Protein intake has also been shown to limit weight re-gain after loss and improve overall body composition and reduce abdominal fat deposits (Due et al., 2004).

The industry needs to avoid promoting fad diets. Our customers should be encouraged to consume a healthy balanced diet that will help them, along with regular exercise, to maintain a normal body mass index (BMI). Research has shown that eating eggs for breakfast increases weight loss by 61% over cereal-based breakfasts of equal calories (Vander Wal et al., 2008). Eggs need to be positioned as a sensible part of a healthy diet.

Similarly from the poultry meat perspective, turkey breast has a high protein to calorie content and has been shown to keep post-meal insulin levels stable and help maintain blood sugar (Pal and Ellis, 2010). Poultry meat products in general have the potential to fulfil the role of first choice meat protein in the diet. With an obesity crisis looming both in the UK and further afield the growth of the 'slimming industry' is an inevitable consequence. Provision of good science-based information and ultimately healthy, tasty and convenient recipes to this sector should be a priority for the poultry industry.

IMPROVING TASTE

If we want consumers to eat more poultry products they need to be appealing in terms of taste. The factors affecting the appearance, taste and aroma of poultry meat have been reviewed by a number of authors (Berri, 2000; Fletcher, 2002; Jayasena et al., 2013). The key factors affecting poultry meat appearance and flavour were highlighted as follows:

- Breed/strain/sex of bird;
- Skin colour;
- Lipid class and fatty acid composition;
- Energy/protein content of the diet;
- Free amino acid and nucleotide content;
- Diet content;
- Pre-slaughter feed withdrawal and transportation;
- Stunning;
- pH post-slaughter;
- Post-slaughter ageing;
- Cooking process;
- Antioxidants; and
- Irradiation and high-pressure treatment.

However, influencing the taste of poultry meat through dietary means has proved challenging. Studies on alteration of the fatty acid profile (Lopez-Ferrer et al., 1999; Kiyohara et al., 2011) and using dietary amino acid manipulation to alter the free glutamate content of meat (Imanari et al., 2008) have shown it is possible to improve the sensory scores.

Whilst the bird itself may be resistant to significant changes in overall taste, poultry meat does have the advantage of being a good canvas on which to add post-slaughter flavours. Another challenge for the industry, given the consumer concerns over microbiological risk mentioned above, is making sure that product is not overcooked thus reducing the final taste experience. Innovation in packing does offer some novel solutions in this area such as roast in the bag products where handling of raw meat is eliminated and the enclosed packaging gives the potential for less drying out of the meat during the cooking process.

NUTRITIONAL ENRICHMENT

Poultry products have the potential for enrichment or enhancement of nutritional value through a number of nutritional strategies detailed below.

Amino acids

In addition to the influence of amino acids on taste, overall protein and individual amino acid content can be used to alter carcass fatness. Ebrahimi *et al.* (2014) showed that increasing dietary arginine levels in broiler diets by up to 83% gave benefits in terms of production parameters and also plasma thyroid levels. At the same time carcass fatness was reduced in terms of abdominal fat but levels of intramuscular fat were increased, which may help eating quality.

Turkey is anecdotally believed to contain a large amount of the amino acid tryptophan, which is an essential amino acid metabolized to melatonin and serotonin. Along with vitamin B6 (pyridoxine), tryptophan has been used as a treatment for depression as an adjunct to other therapies with some success (Volker and Jade, 2006). There is a public perception that turkey can be used as a sleep aid and causes drowsiness with some media sources recommending turkey sandwiches for insomniacs. Turkey, however, contains no more tryptophan than chicken (around 0.24 g/100 g) and less than pork, and insufficient to cause any of the reported effects. Although turkey as an aid for sleep has been discredited, the story created a great deal of media coverage, which highlights the potential of good marketing.

Long chain fatty acids

Nutrition agencies have made recommendations for n-3 intake as consumption has been linked to a reduction in all causal mortality, CVD and stroke. N-3 fatty acids may also have a role in cellular ageing and are essential for brain development and growth in infants. Inadequate consumption has been linked to atherosclerosis, hypertension, visual problems, cancer, asthma, diabetes and many other conditions (Simopoulos, 1991).

It has been suggested that the genetic selection of chickens for growth has altered the nutritional parameters, with the ratio of n-6:n-3 fatty acids increasing from 2:1 to 9:1 as n-3 fatty acids have declined (Wang *et al.*, 2009). The docosahexaenoic acid (DHA) to docosapentaenoic acid (DPA) ratio has also changed with traditional chick varieties having a ratio of DPA:DHA of around 0.3, which has now increased to around 1.2 (Wang *et al.*, 2009).

In large mammals, fast growth rates mean that sufficient DHA cannot be synthesized from muscle and liver, which results in a higher proportion of DPA, so the genetic selection of broiler for rapid growth may explain some of the reduction in DHA content. Whilst genetics may have played its part in this, the use of high n-6 fat sources in commercial diets will also have had a major influence on the fatty acid ratio of both poultry meat and eggs.

Chicken and eggs were one of the few sources of land-based DHA and the decline in DHA consumption has had adverse effects on human health, especially mental health (Hibbeln *et al.*, 2004). Mammals lack the enzymes required to convert n-6 to n-3 polyunsaturated fatty acids (PUFA), so rely on a dietary intake.

The n-3 enrichment of poultry meat could bridge the human diet gap as meat can contribute 75 mg per portion (Givens and Gibbs, 2006). Altering the fat profile and overall content has been achieved in both poultry meat and eggs (Hayat *et al.*, 2009). Poultry meat has been enriched with n-3 fatty acids to provide 90% of the RDA with some variation (Grashorn, 2007). Numerous poultry meat products enhanced with n-3 have been brought to market in recent years (Best, 2007).

The n-3 fatty acid content of eggs has also been enriched up to 65% of RDA and omega-3 eggs have higher n-3 PUFAs such as DHA and EPA. There has been a marked interest in this type of enrichment, with 28 new omega-3 enriched egg products launched worldwide in 2006 (Best, 2007).

By increasing the long chain PUFA content of poultry meat and eggs there will be an increase in the oxidation potential of these products, which needs to be considered both in terms of the health consequences but also taste. However, n-3 enriched egg yolks have been shown to be resistant to lipid oxidation (Meynier *et al.*, 2014).

Conjugated linoleic acid

Conjugated linoleic acid (CLA) refers to a group of isomers of linoleic acid present primarily in beef and dairy products. A number of health benefits have been ascribed to CLA intake (Roche *et al.*, 2001) including reduction in cancer risk and CVD risk. Given the relatively poor perception of beef and dairy products as part of a healthy diet there has been interest in improving the CLA content of poultry and eggs. Szymczyk *et al.* (2001) showed that it is possible to increase the CLA content of broilers and influence the amount of abdominal fat. However, high inclusion of CLA in the broiler diet did have an adverse impact on live weight and reduced the overall PUFA content of the bird.

Work on eggs by the same group (Szymczyk and Pisulewski, 2003) also showed that CLA in the diet is transferred into the yolk. The overall fatty acid composition of the yolk is significantly altered by increasing CLA supplementation, increasing the saturated fatty acid content and decreasing the non-CLA PUFA. The potential for CLA-enriched eggs is however limited by the impact on the eating quality of eggs, which have been described as having rubbery and elastic yolks when cooked (Ahn *et al.*, 1999).

Vitamins, minerals and trace elements

Standard eggs contain a number of vitamins and micronutrients that would allow them to make claims as being 'a source' of vitamin A, folate, pantothenic acid, choline and phosphorus. Additionally, they are 'high in' vitamin D3, riboflavin, vitamin B12, biotin, iodine and selenium (Department of Health, 2012). There is some variation in published work, with Hasler (2000) reporting that eggs naturally contain 15% of the RDA for riboflavin, 17% of selenium, and 31% of vitamin K per single egg. Eggs also provide the opportunity to enhance further their nutrient content by feeding increased levels of specific elements to the hen (Naber, 1993). Whilst eggs are generally well suited to enrichment with a range of nutrients, the potential for poultry meat is more limited.

Vitamin E

In human health, vitamin E supplementation, in conjunction with selenium enrichment can reduce the risk of some cancers, such as prostate and colon cancer. In poultry, vitamin E has been used to reduce ascites in broilers and alleviate heat stress. As vitamin E prevents oxidative stress from free radical formation, it is also used to reduce oxidative damage in meat (Grau *et al.*, 2001), especially in cases where there is potential for reduced meat quality, e.g. with omega-3 fatty acid enrichment (Gonzalez-Esquerra and Leeson, 2001). Vitamin E incorporation into poultry diets improves oxidative stability of both cooked and raw meat and can increase the vitamin E content of chicken muscle (Surai and Sparks, 2000). Wenk *et al.* (2000) suggested that dietary vitamin E levels were reflected in the content of the muscle and adipose tissue in broilers.

Alpha-tocopherol is easily deposited in the yolk and eggs can be enriched in a substantial way with no effect on egg quality (Grashorn, 2005). Eggs can be enriched to meet 50% of RDA, currently 12 mg (EC Directive, 2008), which increases as PUFA levels increase. Enrichment has been shown to be able to increase the vitamin E provided by a single egg from 16% of RDA to 60–80% (Schiavone, 2011).

Vitamin A

Vitamin A is involved in maintaining vision, immune functions and healthy teeth and skin. It acts as an antioxidant, by preventing free radical chain formation, similar to vitamin E (Grashorn, 2007). Vitamin A deficiencies in the developing world are responsible for approximately 2 million deaths and half a million cases of blindness every year. Retinol maintains epithelial cells and has importance for vision in humans. Hen diets have been supplemented with retinyl acetate, which is positively correlated with the retinol content of egg yolk (Mendonca *et al.*, 2002). Vitamin A has been increased in eggs from 10 to 24 iu/g (Squires and Naber, 1993). Supplementation of diets with beta-carotene has also been shown to increase egg retinol by around 20% (Jiang *et al.*, 1994). However, deleterious interactions have been suggested between vitamin A and vitamin D3 (Leeson and Caston, 2003).

Vitamin D

There has been interest in increasing the vitamin D content of eggs as vitamin D insufficiency has been reported in Europe as well as developing countries. There is evidence that vitamin D is involved in the aetiology of auto-immune disease and deficiency has been linked to an increased risk of several chronic conditions such as cancer, diabetes and CVD (Judd and Tangpricha, 2008).

It is possible to increase the vitamin D content of eggs, which are already 'high in' this vitamin (Yao *et al.*, 2013). However, commercialization of vitamin D-enriched eggs is currently limited by the legal restrictions on the maximum amount of vitamin D that can be added to feed in Europe.

Folate

Folate has an established role in the prevention of fetal neural tube defects; consequently, recommended levels are high for pregnant women. It also has a potential role in the prevention of heart disease. High folic acid doses in hen diets have been converted to increased folates in eggs (Hoey *et al.*, 2009), although a plateau is reached at a concentration of 30–50 μg/egg (Hebert *et al.*, 2005). There is the potential to enhance the natural folate content of eggs by two or three times the normal level achieving up to 75% of the RDA.

Poultry meats are poor sources of folate, but both hamburgers and sausages have been successfully enriched during processing with no detrimental sensory effects on the finished product (Cáceres *et al.*, 2008). Folic acid content does not appear to be affected by heating for sterilization purposes so there is potential for enrichment of processed foods, although no work to date has been carried out on poultry meat products.

Vitamins B6 and B12

Meat in general is an important source of most B vitamins, with chicken and turkey meat containing similar amounts of B vitamins, with the exception of niacin (B3), where chicken contains higher levels (up to 56% RDA per breast portion) compared to turkey.

Chicken and turkey meat are good sources of vitamin B6, with the breast meat containing more than leg for both species, with a 100 g serving providing 0.5–0.6 mg, which is almost 50% of the RDA for adults (1.3 mg/day). Vitamin B6 is important for the immune system, red blood cell and neural metabolism and low levels have been linked to depression (Hvans *et al.*, 2004). However, vitamin B6 deficiency is uncommon due to its ready availability in staple food sources. Eggs also contain some vitamin B6 (0.09 mg/egg), but attempts to increase this level by dietary manipulation of hen diets have been unsuccessful (Leeson and Caston, 2003). However, the feeding of a strain of *Lactobacillus* has been shown to increase the egg content of both vitamin B6 and vitamin B12 (to 0.115 mg and 1.13 µg per 100 g, respectively) (Al-Fataftah *et al.*, 2013).

Vitamin B12 deficiency can increase the risk of ocular vascular disease and both dementia and Alzheimer's disease. It is particularly prevalent in vegetarians (Antony, 2003) and this may be in part because vitamin B12 is less well absorbed from eggs and vegetable sources than from animal sources such as poultry meat (Doscherholmen *et al.*, 1976). Turkey contains a higher level of vitamin B12 than red meat, with 100 g providing 135% RDA for an adult (1.4 µg). Egg B12 content (0.5–1.4 µg/egg) has also been increased to over 100% of RDI by increasing vitamin supplementation to the bird (Leeson and Caston, 2003).

Choline

In humans, choline is important for memory and fetal/infant brain development. Choline is a natural emulsifier present in bile and is an important part of acetyl-choline and lecithin production.

Eggs are a good source of choline, which has been linked to reductions in CVD and dementia. The suggested intake of choline is 425–550 mg/day and egg yolk itself provides around 680 mg/100 g. Although there is no current RDA for choline, research has suggested a role for high choline diets in protecting against DNA damage, in cases where folate is restricted (Shin *et al.*, 2010). Choline supplements have been suggested for pregnant women, as large amounts are required for fetal development and adequate placental function, and many diets in both high and low income countries may be too low in choline (Zeisel, 2013). Adequate egg consumption would make supplements unnecessary.

Selenium

Selenium (Se) deficiency can cause a decrease in fertility, poor general health and compromised immunity (Surai, 2006). Selenium deficiency is a global problem, with the UK consumption estimated at 50% RDA (Fisinin *et al.*, 2009). There have been positive effects shown for both reduction of cancer risk and mortality with Se supplementation (Surai, 2006), and there is evidence that Se is anti-atherosclerotic and therefore can offer protection against CVD. Several studies have correlated Se deficiency with cognitive decline, therefore supplementation may be important for healthy ageing.

Increased Se content of eggs has been found in response to supplementation of hen diets (Surai and Sparks, 2001). Enriched eggs are capable of delivering 70% of the RDA per egg (Fisinin *et al.*, 2009) and are commercially available worldwide. Although Se is one of the most toxic of the trace minerals, it is safe to deliver Se in egg form as they would need to be consumed in great excess for an extended period for toxicity to occur in humans.

Generally meat is a good source of Se, but this is dependent on the area of production and any supplements used. A 95% uptake of supplemented Se has been shown in chicken meat and Se in chicken meat has been shown to have high bioavailability (77%; Wen *et al.*, 1997). There is a premium Se-enhanced chicken brand in Korea (Selen chicken) and Ukraine produces Se- and vitamin E-enriched meat with Se content of 284 mg/g (from 85 mg/g basal) (Yarashenko *et al.*, 2004), which would mean that a 100 g portion of chicken meat would provide 50% of the RDA. Turkey meat can provide 60% of the daily recommended intake (per 100 g) and Schubert *et al.* (1987) noted a high Se content in turkey meat in areas with high soil concentrations, which suggests an additional potential for enrichment in this market.

Iodine

Iodine is a constituent of thyroid hormones, which affect cell protein synthesis throughout the body. A lack of iodine is a major cause of brain disorders, with 740 million people suffering from goitre. Many people also have thyroid hypertrophy, which is of particular concern in women of childbearing age. Follicles in the chicken ovary concentrate iodine, so there is a relationship between dietary iodine and egg content (Klasing, 2000). Up to 6 mg/kg iodine can result in 26 μg per egg (Yalcin *et al.*, 2004), but higher amounts can detrimentally affect feed intake and egg production. Iodine has a narrow range between dietary requirements and upper tolerance levels, however, compared with eggs, muscle is more resistant to enrichment but safer in terms of toxicity (De Smet, 2012). It has been shown that increasing dietary concentrations of iodine can increase muscle content, but the source is important as organic sources such as seaweed are cited as more efficacious (De Smet, 2012).

Iron

Iron deficiency anaemia affects around 2 billion people globally (ACC/SCN, 2000) including 24% of children and 21% of all females (WHO/UNICEF/UNU, 2001). Iron enrichment may be able to improve the nutritional status of groups of people at risk of iron deficiency anaemia, especially infants, children and pregnant women. Iron deficiency can detrimentally affect infant development, which is then maintained through childhood (Lozoff *et al.*, 1991), and has been shown to increase the incidence of respiratory infections.

Eggs may be more suitable for fortification than other stable foods as these can have low bioavailability (Revell *et al.*, 2009), although the addition of iron to flour has been mandated since 1953. Iron has been increased by up to 19% (Park *et al.*, 2004) and an iron-enriched egg with 10–20% more iron than a standard egg would provide 10% RDA. The efficacy of the enrichment is affected by a number of factors including source of supplement, mineral–mineral interaction and bird to bird variation, which all need to be taken into consideration (Schiavone, 2011). Chelated iron has been shown to be more effective for egg enrichment than iron sulfate alone, although there is a limit beyond which further supplementation is not efficacious (Park et al., 2004).

With the exception of duck, poultry meat is naturally lower in iron than red meat, with chicken and turkey meat containing between 1.7 and 7.6 μg iron/g (depending on cut). Chicken meat has been enriched with iron, with chelated iron being more efficacious as chelation makes the iron electronically neutral and stable and therefore improves uptake in the small intestine. Seo et al. (2008a) found that organic iron was more effective for enriching broiler meat and that levels in breast and leg meat increased as iron supplementation level and duration increased (Seo et al., 2008b). There is potential for fortification of poultry products with iron, although the bioavailability of any supplement needs to be taken into account to reduce environmental contamination (Leeson, 2003).

Zinc

Zinc is essential for a number of metabolic and physiological functions including reproductive and immune functions. Zinc deficiency will impair metabolic pathways and cell division. Marginal zinc deficiencies are common in both developing and developed countries with serious implications for children, impairing cognitive function and growth. Zinc levels are especially low in elderly populations. Zinc content of eggs has been increased by supplementation (Stahl et al., 1988), although some cooking methods, such as frying, can reduce bioavailability to around 70% (Plaimast et al., 2009). Studies on egg enrichment with zinc have found mixed results, with some authors reporting poor utilization (Skrivan et al., 2005) and others increasing Zn content in eggs by as much as 90% (Stahl et al., 1988). Turkey and chicken are good sources of zinc naturally, with turkey containing 50% RDA per portion. Zinc may also act as an antioxidant and increase levels of α-tocopherol in birds under temperature stress conditions (Powell, 2000). Supplementation of diets for meat birds has not been found to increase content in meat tissues as zinc levels are homoeostatically controlled (Bou et al., 2004, 2005), therefore poultry meat appears to be resistant to enrichment with zinc.

CHOLESTEROL

Meta-analysis has shown that saturated fat is the major diet determinant for blood cholesterol (Clarke et al., 1997) and if cholesterol is consumed there is a compensatory mechanism in the body which reduces production so intake has a limited effect on blood cholesterol (Hu et al., 1999).

Eggs contain around 200 mg cholesterol compared with the recommended daily intake of 300 mg, however the fat in eggs is in the form of emulsified oil and is high in monounsaturated fatty acids (MUFAs), which may minimize the effects of cholesterol. The cholesterol content of eggs has been reduced by 10% (Elkin, 2007). Attempts at using genetic selection to produce a low cholesterol egg have been less effective due to deleterious effects on egg production and hatchability (Hargis, 1988). The additions of 3-hydroxy-3-methylglutaryl coenzyme A reductase inhibitors, red yeast rice and copper have significantly decreased yolk cholesterol (Pesti and Bakalli, 1988; Elkin *et al.*, 1999; Wong *et al.*, 2005).

Misconceptions about eggs and cholesterol abound due to misreading and misuse of research, but consensus from heart and lung health organizations is that one egg a day is not harmful as there is no conclusive evidence of egg consumption being linked to increased cardiovascular risk (BNF, 2009). Several large analyses of data have shown a positive relationship between egg consumption and reductions in cardiovascular mortality (Tunstall-Pedoe *et al.*, 2000).

YOLK PIGMENTATION

Yolk colour is one of the key attributes of the egg in terms of customer perception. Consumer surveys have shown that the priorities for European consumers in terms of their sensory characteristics are shell strength, albumen consistency and yolk colour (Hernandez *et al.*, 2000). A range of raw materials and feed additives is used by the egg industry to achieve the desired yolk score for their particular market. The key carotenoids are broadly split into 'yellow' hued pigments (e.g. lutein, zeaxanthin and beta-apo-8-carotenoic acid) and red hued (e.g. capsanthin, citranaxanthin, canthaxanthin). These carotenoid pigments are present both in raw materials such as maize, maize by-products, lucerne and grass meal but can also be extracted from marigold and red peppers or in the case of citranaxanthin, canthaxanthin and beta-apo-8-carotenoic acid, be synthetically produced. By combining red and yellow carotenoids different yolk colours can be achieved ranging from more yellow colouration to more golden orange. In addition to their visual appearance, yolks can be 'loaded' with yellow carotenoids in order to improve their pigment-carrying properties, for example in the production of pasta.

However, the carotenoids have benefits beyond the visual appearance of the egg. The most common cause of blindness in the developed world is age-related macular degeneration (AMD), which is a chronic degenerative condition. Lutein and zeaxanthin can act as an optical filter to absorb visible light and protect the macular against damage. Studies have shown that a combination of these two dihydroxycarotenoids can be used to reduce the risk of progression to AMD (AREDS2, 2013). Typical European diets contain around 1–3 mg/day and 6 mg is suggested as an appropriate target to decrease the AMD risk (Seddon *et al.*, 1994). On a dry weight basis the lutein content of egg yolk is substantially lower than green vegetables such as kale and parsley (around 10 μg/g cf 80 μg/g), but it is suggested that the lutein and zeaxanthin from egg yolk is more bioavailable due to the increased fat content (Mangels *et al.*, 1993).

The concentrations of lutein and zeaxanthin in chicken egg yolk are 292 ± 117 μg/yolk and 213 ± 85 μg/yolk (average weight of yolk is about 17–19 g), respectively. This is obviously way below the recommended levels, but supplementing the diet with high levels of lutein can increase the yolk content to 1000 μg/yolk. Commercial egg brands are now being marketed as a source of lutein in countries across the globe. Fortification of other staple foods such as wheat with lutein has been tried, but the reduction in bioavailability during the baking process can reduce its usefulness (Abdel-Aal et al., 2013).

There may also be potential for novel ingredients to be used to influence both yolk colour and the nutrient content of eggs. For example, tomato powder has been used in commercial trials to pigment yolks and new xanthophyll sources are being developed through plant breeding.

SUMMARY

Obesity is the biggest challenge to human nutrition and health. Telling consumers to eat more healthily does not work. However, providing tasty, healthy and convenient meal solutions has the potential to contribute to a healthier population and provide opportunities for sustained growth in poultry production.

Poultry meat and eggs already deserve a prominent place in a healthy balanced diet. Rather than encouraging fad diets we need to ensure that our products are safe, tasty and nutritious and provide opportunities for them to become part of the regular rotation of meals that people eat. They could also be specifically positioned to help as part of a weight loss programme.

Beyond the standard meat and egg products there are also opportunities to add value. There are a number of points that need to be taken into account when considering enrichment of poultry meat or eggs:

- efficiency of nutrient transfer to eggs/meat;
- availability of a suitable source of the nutrient for enrichment;
- any risk of toxicity to the bird;
- nutrient potential for enrichment related to RDA per portion;
- established positive benefit of the nutrient and a reported deficiency in human diets;
- potential interactions with other nutrients;
- effect on taste/appearance; and
- potential to add value to the final product through a beneficial claim for human health.

There is consumer demand and interest in altering the composition of poultry meat and eggs to improve their nutritional profile and enhance human health (Wong, 2007). In areas where poultry meat and eggs are lacking nutrients, such as some vitamins and minerals, it may be economically viable to achieve enhancement via diet fortification. However, there is also potential to pursue enrichment post-processing in addition to nutritional strategies at bird level.

The industry needs to take care in how products are developed and marketed. Consumers may want tasty healthy food but they do not want to feel that what they are eating has been manipulated in some unnatural way.

Human nutrition and health is becoming a more important issue. As the population grows, the appetite of that population will also grow disproportionately as will the burden on healthcare systems as a result of increasing levels of obesity. The challenge for the food industry is to provide innovative ways of encouraging people to eat healthily but also meeting the demands for taste and convenience. Poultry products are well placed to meet this demand and fulfil the role of becoming an essential part of a healthy balanced diet.

REFERENCES

Abdel-Aal, E.M., Akhtar, H., Zaheer, H. and Ali, R. (2013) Dietary sources of lutein and zeaxanthin carotenoids and their role in eye health. *Nutrients* 5, 1169–1185.

ACC/SCN: United Nations Sub-Committee on Nutrition (2000) *Fourth Report on the World Nutrition Situation*. United Nations.

Ahn, D.U., Sell, J.L., Jo, C., Chamruspollert, M. and Jeffery, M. (1999) Effect of dietary conjugated linoleic acid on the quality characteristics of chicken eggs during refrigerated storage. *Poultry Science* 78, 922–928.

Al-Fataftah, A.A., Herzallah, S.M., Mabood, F. and Alshawabkeh, K. (2013) Enrichment of vitamin B12 and B6 and lowering cholesterol levels of eggs by lactic acid bacteria. *Journal of Food, Agriculture and the Environment* 11, 674–678.

Antony, A.C. (2003) Vegetarianism and vitamin B12 deficiency. *American Journal of Clinical Nutrition* 78, 3–6.

Age-Related Eye Disease Study 2 Research Group (AREDS2) (2013) Lutein + zeaxanthin and omega-3 fatty acids for age-related macular degeneration: the age-related eye disease study 2 – randomized clinical trial. *The Journal of the American Medical Association* 309, 2005–2015.

Barraj, L.M., Tran, N. and Mink, P. (2009) A comparison of egg consumption with other modifiable coronary heart disease lifestyle risk factors: a relative risk apportionment study. *Risk Analysis* 29, 401–415.

Bernstein, A.M., Pan, A., Rexrode, K.M., Stampfer, M., Hu, F.B., Mozaffarian, D. and Willett, W.C. (2012) Dietary protein sources and the risk of stroke in men and women. *Stroke* 43, 637–644.

Berri, C. (2000) Variability of sensory and processing qualities of poultry meat. *World's Poultry Science Journal* 56, 209–224.

Best, P. (2007) Pig feeds gain the omega factor. *Feed International* April, pp. 10–11.

BNF (2009) *British Nutrition Foundation's Nutrition Bulletin* 34, 66–70.

Bou, R., Guardiola, F., Tres, A., Barroeta, A.C. and Codony, R. (2004) Effect of dietary fish oil, α-tocopheryl acetate and zinc supplementation on the composition and consumer acceptability of chicken meat. *Poultry Science* 83, 282–292.

Bou, R., Guardiola, F., Barroeta, C. and Codony, R. (2005) Effect of dietary fat sources and zinc and selenium supplements on the composition and consumer acceptability of chicken meat. *Poultry Science* 84, 1129–1140.

Cáceres, E., García, M.L. and Selgas, M.D. (2008) Conventional and fat-reduced cooked sausages enriched with folic acid. *Fleischwirtschaft International* 5, 58–60.

Clarke, R., Frost, C., Collins, R., Appleby, P. and Peto, R. (1997) Dietary lipids and blood cholesterol: quantitative meta-analysis of metabolic ward studies. *British Medical Journal* 314, 112–117.

DEFRA (2013) Family Food Datasets. Available at: https://www.gov.uk/government/statistical-data-sets/family-food-datasets (accessed 25 April 2014).

Department of Health (2012) Nutrient Analysis of Eggs. Available at: http://www.gov.uk/government/public/nutrient-analysis-of-eggs (accessed 24 April 2014).

Department of Health (2013) Policy: Reducing Obesity and Improving Diet. Available at: http://www.gov.uk/government/policies/reducing-obesity-and-improving-diet (accessed 24 April 2014).

De Smet, S. (2012) Meat, poultry, and fish composition: strategies for optimizing human intake of essential nutrients. *Animal Frontiers* 2, 10–16.

Doscherholmen, A., McMahon, J. and Ripley, D. (1976) Inhibitory effect of eggs on vitamin B_{12} absorption: description of a simple ovalbumin ^{57}Co-Vitamin B_{12} absorption test. *British Journal of Haematology* 33, 261–272.

Drummond, C. (2014) What can farmers learn from science to improve the nutritional value of our food? Health by stealth. *Nuffield Farming Scholarship Trust, Arden Report.*

Due, A., Toubro, S., Skov, A.R. and Astrup, A. (2004) Effect of normal-fat diets, either medium or high in protein, on body weight in overweight subject: a randomized 1-year trial. *International Journal of Obesity Related Metabolic Disorders* 28, 1283–1290.

Ebrahimi, M., Zare Shahneh, A., Shivazad, M., Ansari Pirsaraei, Z., Tebianian, M., Ruiz-Feria, C.A., Adibmoradi, M., Nourijelyani, K. and Mohamadnejad, F. (2014) The effect of feeding excess arginine on lipogenic gene expression and growth performance in broilers. *British Poultry Science* 55, 81–88.

EC Directive (2008) EC Directive 2008/100/EC of 28 October 2008 amending Council Directive 90/496/EEC on nutrition labelling for foodstuffs as regards recommended daily allowances, energy conversion factors and definitions. Available at: http://eurlex.europa.eu/LexUriServ/LexUriServ.do?uri = OJ:L:2008:285:0009:0012:EN:PDF (accessed 14 April 2014).

Elkin, R.G. (2007) Reducing egg cholesterol content 11: review of approaches utilizing non-nutritive dietary factors or pharmacological agents and an examination of emerging strategies. *World's Poultry Science Journal* 63, 5–32.

Elkin, R.G., Yan, Z., Zhong, Y., Donkin, S.S., Buhman, K.K., Story, J.A., Turek, J.J., Porter Jr, R.E., Anderson, M., Homan, R. and Newton, R.S. (1999) Select 3-hydroxy-3-methylglutaryl-coenzyme A reductase inhibitors vary in their ability to reduce egg yolk cholesterol levels in laying hens through alteration of hepatic cholesterol biosynthesis and plasma VLDL composition. *Journal of Nutrition* 129, 1010–1019.

EU (2006) EU Pledge – Nutrition Criteria White Paper. Available at: http://www.eu-pledge.eu/sites/eu-pledge.eu/files/releases/EU_Pledge_Nutrition_White_Paper_Nov_2012.pdf (accessed 14 April 2014).

Fernandez, M. (2006) Dietary cholesterol provided by eggs and plasma lipoproteins in healthy populations. *Current Opinion in Clinical Nutrition and Metabolic Care* 9, 8–12.

Fife-Shaw, C. and Rowe, G. (1996) Public perceptions of every day food hazard: a psychometric study. *Risk Analysis* 16, 487–500.

Fisinin, V.L., Papazyan, T.T. and Surai, P.F. (2009) Producing selenium enriched eggs and meat to improve the selenium status of the general population. *Critical Reviews in Biotechnology* 29, 18–28.

Fletcher, D.L. (2002) Poultry meat quality. *World's Poultry Science Journal* 58, 131–146.

Givens, D.I. and Gibbs, R.A. (2006) Very long chain n-3 polyunsaturated fatty acids in the food chain in the UK and the potential of animal-derived foods to increase intake. *Nutrition Bulletin* 31, 104–110.

Gonzalez-Esquerra, R. and Leeson, S. (2001) Alternatives for enrichment of eggs and chicken meat with omega 3 fatty acids. *Canadian Journal of Animal Science* 81, 295–305.

Grashorn, M.A. (2005) Enrichment of eggs and poultry meat with biologically active substances by feed modifications and effects on the final quality of the product. *Polish Journal of Food and Nutrition Sciences* 14/55, 15–20.

Grashorn, M.A. (2007) Functionality of poultry meat. *Journal of Applied Poultry Research* 16, 99–106.

Grau, A., Guardiola, F., Grimpa, S., Barroeta, A.C. and Codony, R. (2001) Oxidative stability of dark chicken meat through frozen storage: Influence of dietary fat and a-tocopherol and ascorbic acid supplementation. *Poultry Science* 80, 1630–1642.

Hargis, P.S. (1988) Modifying egg yolk cholesterol in the domestic fowl – a review. *World's Poultry Science Journal* 44, 17–29.

Hasler, C.M. (2000) The changing face of functional foods. *Journal of the American College of Nutrition* 19, 499–506.

Hayat, Z., Cherian, G., Pasha, T.N., Khattak, F.M. and Jabbar, M.A. (2009) Effect of feeding flax and two types of antioxidants on egg production, egg quality and lipid composition of eggs. *Journal of Applied Poultry Research* 18, 541–551.

Health & Social Care Information Centre (HSCIC) (2014) *Statistics on Obesity, Physical Activity and Diet: England 2014.* HSCIC, Leeds, UK.

Health Survey for England (2012) Trend Tables. Available at: http://www.hscic.gov.uk (accessed 14 April 2014).

Hebert, K., House, J.D. and Guenter, W. (2005) Effect of dietary folic acid supplementation on egg folate content and the performance and folate status of two strains of laying hens. *Poultry Science* 84, 1533–1538.

Hernandez, J.M., Blanch, A.J. and Roche, F.H.L. (2000) Perceptions of egg quality in Europe. *International Poultry Production* 8, 7–11.

Hibbeln, J.R., Nieminen, L.R. and Lands, W.E. (2004) Increasing homicide rates and linoleic acid consumption among five Western countries, 1961–2000. *Lipids* 39, 1207–1213.

Hoey, L., McNulty, H., McCann, E.M.E., McCraken, K.J., Scott, J.M., Blaznik Marc, B., Molloy, A.M., Graham, C. and Pentieva, K. (2009) Laying hens can convert high doses of folic acid added to the feed into natural folates in eggs providing a novel source of food folate. *British Journal of Nutrition* 101, 206–212.

Hu, F.B., Stampfer, M.J., Rimm, E.B., Manson, J.E., Ascherio, A., Colditz, G.A., Rosner, B.A., Spiegelman, D., Speizer, F.E., Sacks, F.M., Hennekens, C.H. and Willett, W.C. (1999) A prospective study of egg consumption and risk of cardiovascular disease in men and women. *Journal of the American Medical Association* 281, 1387–1394.

Hvans, N.M., Juul, S., Bech, B. and Nexo, E. (2004) Vitamin B6 level is associated with symptoms of depression. *Pyschothera Psychosoma* 73, 334–343.

Imanari, M., Kadowaki, M. and Fujimura, S. (2008) Regulation of taste-active components of meat by dietary branched-chain amino acids; effects of branched-chain amino acid antagonism. *British Poultry Science* 49, 299–307.

Jayasena, D.D., Ahn, D.U., Nam, K.C. and Jo, C. (2013) Factors affecting cooked chicken meat flavour: a review. *World's Poultry Science Journal* 69, 515–526.

Jiang, Y.H., McGeachin, R.B. and Bailey, C.A. (1994) Alpha-tocopherol, beta-carotene, and retinol enrichment of chicken eggs. *Poultry Science* 73, 1137–1143.

Judd, S. and Tangpricha, V. (2008) Vitamin D deficiency and risk for cardiovascular disease. *Circulation* 117, 503–511.

Kiyohara, R., Yamaguchi, S., Rikimaru, K. and Takahashi, H. (2011) Supplemental arachidonic acid-enriched oil improves the taste of thigh meat of Hinai-jidori chickens. *Poultry Science* 90, 1817–1822.

Klasing, K.C. (2000) *Comparative Avian Nutrition.* CAB International, Wallingford, UK.

Leeson, S. (2003) A new look at trace mineral nutrition of poultry: can we reduce the environmental burden of poultry manure? *Proceedings of Alltech's 19th Annual Symposium*, pp. 125–129.

Leeson, S. and Caston, L.J. (2003) Vitamin enrichment of eggs. *Journal of Applied Poultry Research* 2, 24–26.

Lopez-Ferrer, S., Baucells, M.E., Barroeta, A.C. and Grashorn, M.A. (1999) N-3 enrichment of chicken meat using fish oil: alternative substitution with rapeseed and linseed oils. *Poultry Science* 78, 356–365.

Lozoff, B., Jimenez, E. and Wolf, A.W. (1991) Long term developmental outcome of infants with iron deficiency. *The New England Journal of Medicine* 325, 687–694.

Mangels, A.R., Holden, J.M., Beecher, G.R., Forman, M.R. and Lanza, E. (1993) Carotenoid contents of fruits and vegetables – an evaluation of analytical data. *Journal of the American Dietetic Association* 93, 284–296.

Mendonca, C.X., Almeida, C.R.M., Mori, A.V. and Watanabe, C. (2002) Effect of dietary vitamin A on egg yolk retinol and tocopherol levels. *Journal of Applied Poultry Research* 11, 373–378.

Meynier, A., Leborgne, C., Viau, M., Schuck, P., Guichardant, M., Rannou, C. and Anton, M. (2014) N-3 fatty acid enriched eggs and production of egg yolk powders: an increased risk of lipid oxidation? *Food Chemistry* 153, 94–100.

Miles, S. and Frewer, L. (1999) *Effective Risk Communication about Food Related Hazards: A Review of the Literature.* Ministry of Agriculture, Fisheries and Food, London.

Mutungi, G., Ratliff, J., Puglisi, M., Torres-Gonzalez, M., Vaishnav, U., Leite, J.O., Quann, E., Volek, J.S. and Fernandez, M.L. (2008) Dietary cholesterol from eggs increases plasma HDL cholesterol in overweight men consuming a carbohydrate restricted diet. *Journal of Nutrition* 138, 272–276.

Naber, E.C. (1993) Modifying vitamin composition of eggs: a review. *Journal of Applied Poultry Research* 2, 385–393.

Pal, S. and Ellis, V. (2010) The acute effects of four protein meals on insulin, glucose, appetite and energy intake in lean men. *Circulation* 121, 2271–2283.

Park, S.E., Namkung, H., Ahn, H.J. and Paik, I.K. (2004) Production of iron enriched eggs of laying hens. *Asian Australasian Journal of Animal Science* 17, 1725–1728.

Pesti, G.M. and Bakalli, R.I. (1988) Studies on the effect of feeding cupric sulphate pentahydrate to laying hens on egg cholesterol content. *Poultry Science* 84, 865–874.

Plaimast, H., Sirichakwal, P.P., Puwastien, P., Judprasong, K. and Wasantwsiut, E. (2009) *In vitro* bioaccessibility of intrinsically zinc-enriched egg and effect of cooking. *Journal of Food Composition and Analysis* 22, 627–631.

Powell, S.R. (2000) The antioxidant properties of zinc. *Journal of Nutrition* 310, 1447S–1454S.

Revell, D.K., Zarrinkalam, M.R. and Hughes, R.J. (2009) Iron content of eggs from hens given diets containing organic forms of iron, serine and methyl group donor, or phytoestrogens. *British Poultry Science* 50, 536–552.

Roche, H.M., Noone, E., Nugent, A. and Gibney, M.J. (2001) Conjugated linoleic acid: a novel therapeutic agent? *Nutrition Research Reviews* 14, 173–187.

Ruxton, C.H.S., Derbyshire, E. and Gibson, S. (2010) The nutritional properties and health benefits of eggs. *Nutrition and Food Science* 40, 263–279.

Schiavone, A. (2011) Egg enrichment with vitamins and trace minerals. In: Van Immerseal, F., Nys, Y. and Bain, M. (eds) *Improving the Safety and Quality of Eggs and Egg Products.* Woodhead Publishing, Cambridge, UK.

Schoenthaler, S.J., Bier, I.D., Young, K., Nichols, D. and Jansenns, S. (2000) The effect of vitamin–mineral supplementation on the intelligence of American schoolchildren: a randomized, double blind placebo-controlled trial. *Journal of Alternative and Complementary Medicine* 6, 31–35.

Schubert, A., Holden, J.M. and Wolf, W.R. (1987) Selenium content of a core group of food based on a critical evaluation of published analytical data. *Journal of the American Dietetic Association* 87, 285–299.

Seddon, J.M., Ajani, U.A., Sperduto, R.D., Hiller, R., Blair, N., Burton, T.C., Farber, M.D., Gragoudas, E.S., Haller, J., Miller, D.T., *et al.* (1994) Dietary carotenoids, vitamin A, C and E, and advanced age-related macular degeneration. Eye disease case-control study group. *Journal of the American Medical Association* 272, 1413–1420.

Seo, S.H., Lee, H.K., Ahn, H.J. and Paik, I.K. (2008a) The effect of dietary supplementation of Fe-methionine chelate and FeSO$_4$ on the iron content of broiler meat. *Asian Australasian Journal of Animal Science* 21, 103–106.

Seo, S.H., Lee, H.K., Lee, W.S., Shin, K.S. and Paik, I.K. (2008b) The effect of level and period of Fe-methionine chelate supplementation on the iron content of broiler meat. *Asian Australasian Journal of Animal Science* 21, 1501–1505.

Shin, W., Yan, J., Abratte, M., Vermeylen, F. and Caudill, M.A. (2010) Choline intake exceeding current dietary recommendations preserves markers of cellular methylation in a genetic subgroup of folate compromised men. *Journal of Nutrition* 140, 975–980.

Simopoulos, A.P. (1991) Omega 3 fatty acids in health and disease and in growth and development. *American Journal of Clinical Nutrition* 53, 438–463.

Skov, A.R., Toubro, S., Runn, B., Holn, L. and Astrup, A. (1999) Randomized trial on protein vs carbohydrate in ad libitum fat reduced diet for the treatment of obesity. *International Journal of Obesity* 23, 528–536.

Skrivan, M., Skrivanova, V. and Maurounek, M. (2005) Effects of dietary zinc, iron and copper in layer feed on distribution of these elements in eggs, liver, excreta, soil, and herbage. *Poultry Science* 84, 1570–1575.

Squires, M.W. and Naber, E.C. (1993) Vitamin profiles of eggs as indicators of nutritional status in the laying hen: vitamin A study. *Poultry Science* 72, 154–164.

Stahl, J.L., Cook, M.E. and Greger, J.L. (1988) Zinc, iron and copper contents of eggs from hens fed varying levels of zinc. *Journal of Food Composition Analysis* 1, 309–315.

Surai, P.F. (2006) *Selenium in Nutrition and Health*. Nottingham University Press, Nottingham, UK.

Surai, P.F. and Sparks, N.H.C. (2000) Tissue-specific fatty acid and *a*-tocopherol profiles in male chickens depending on dietary tuna oil and vitamin E provision. *Poultry Science* 79, 1132–1142.

Surai, P.F. and Sparks, N.H.C. (2001) Designer eggs: from improvement of egg composition to functional food. *Trends in Food Science and Technology* 12, 7–16.

Szymczyk, B. and Pisulewski, P.M. (2003) Effects of dietary conjugated linoleic acid on fatty acid composition and cholesterol content of hen egg yolks. *British Journal of Nutrition* 90, 93–99.

Szymczyk, B., Pisulewski, P.M., Szczurek, W. and Hanczakowski, P. (2001) Effects of conjugated linoleic acid on growth performance, feed conversion efficiency and subsequent carcass quality in broiler chickens. *British Journal of Nutrition* 85, 465–473.

Tunstall-Pedoe, H., Vanuzzo, D., Hobbs, M., Mahonen, M., Cepaitis, Z., Kuulasmaa, K. and Keil, U. (2000) Estimation of contribution of changes in coronary care to improving survival, event rates and coronary heart disease mortality across the WHO MONICA project populations. *Lancet* 355, 688–700.

Vander Wal, J.S., Gupta, A., Kholsa, P. and Dhurandhar, N.V. (2008) Egg breakfast enhances weight loss. *International Journal of Obesity* 32, 1545–1551.

Volker, D. and Jade, N.G. (2006) Depression: does nutrition have an adjunctive treatment role? *Nutrition and Dietetics* 63, 213–226.

Wang, Y., Lehane, C., Ghebremeskel, K. and Crawford, M.A. (2009) Modern organic and broiler chickens sold for human consumption provide more energy from fat than protein. *Public Health Nutrition* 13, 400–408.

Wen, H.Y., Davis, R.L., Shi, B., Chen, J.J., Chen, L., Boylan, M. and Spallholz, J.E. (1997) Bioavailability of Selenium from veal, chicken, beef, pork, lamb, flounder, tuna, selenomethionine and sodium selenite assessed in selenium-deficient rats. *Biological Trace Element Research* 58, 543–563.

Wenk, C., Leonhardt, M., Martin, R. and Scheeder, M.R. (2000) Monogastric nutrition and potential for improving muscle quality. In: Decker, E., Faustman, C. and Lopez-Bote, C.J. (eds) *Antioxidants in Muscle Foods*. Wiley, New York, pp. 199–227.

WHO/UNICEF/UNU (2001) *Iron Deficiency Anaemia Assessment, Prevention and Control*. World Health Organization, Geneva.

Wong H.K. (2007) Designer poultry eggs and meat for health enhancement. World Poultry Science Association (WPSA, Malaysia Branch), 5 August 2007, Kuala Lumpur.

Wong, H.K., Engku Azahan, E.A. and Tan, S.L. (2005) Reduction in egg cholesterol content of laying hens through supplementation with red yeast rice. *Malaysian Journal of Animal Science* 10, 15–21.

Yalcin, S., Kahraman, Z., Yalcin, S., Yalcin, S.S. and Dedeoglu, H.E. (2004) Effects of supplementary iodine on the performance and egg traits of laying hens. *British Poultry Science* 45, 499–503.

Yao, L., Wang, T., Persia, M., Horst, R.L. and Higgins, M. (2013) Effects of vitamin D3 enriched diet on egg yolk vitamin D3 content and yolk quality. *Journal of Food Science* 78, 178–183.

Yarashenko, F.A., Dvorska, J.E., Surai, P.F. and Sparks, N.H.C. (2003) Selenium enriched eggs as a source of selenium for human consumption. *Biotechnology, Food Science and Policy* 1, 13–23.

Yarashenko, F.O., Surai, P.F., Yaroshenko, Y.F., Karadas, F. and Sparks, N.H.C. (2004) Theoretical background and commercial application of production of Se-enriched chicken. In: *Proceedings of the XXII World's Poultry Congress*, Istanbul, Turkey, p. 410.

Yeung, R. and Morris, J. (2001) Consumer perception of food risk in chicken meat. *Nutrition and Food Science* 31, 270–279.

Zeisel, S.H. (2013) Nutrition in pregnancy: the argument for including a source of choline. *International Journal of Women's Health* 5, 193–199.

PART VI

The Roles of Genetics and Breeding in Sustainability

CHAPTER 11
Breeding for Sustainability: Maintaining and Enhancing Multi-trait Genetic Improvement

William G. Hill,[1]* Anna Wolc,[2,3] Neil P. O'Sullivan[3] and Santiago Avendaño[4]

[1]Institute of Evolutionary Biology, University of Edinburgh, Edinburgh, UK; [2]Department of Animal Science, Iowa State University, Ames, Iowa, USA; [3]Hy-Line International, Dallas Center, Iowa, USA; [4]Aviagen Limited, Newbridge, Midlothian, UK

INTRODUCTION

Enormous improvements have been effected over the last 60 years or more in the efficiency and product quality of meat and table-egg poultry production. As is well recognized, a large proportion of the improvements has been made by breeding and using genetically improved stocks. However, the need for efficient and sustainable animal protein production has also been growing. The global human population is predicted to grow to 9 billion by 2050 with two-thirds of the population living in cities and increasingly adding animal products to their diets. At the same time, there is a decreasing global availability of resources for agriculture such us land, water and energy (UK Foresight Report, 2011).

To enable this expansion, current and future chickens will have to perform with increasing efficiency and to do so in a wide range of environments, characterized by varying management practices, feed quality (form and density) and gut and immune challenges, among other factors. To ensure the sustainability of poultry breeding, breeding goals will likely continue to broaden, and more traits will be included to simultaneously improve productivity and efficiency, environmental impact, robustness, animal health and welfare, and food quality and safety.

Looking forward, providing there is genetic variation present in the populations, such improvements can be made. It is critical that such variation be present in the commercial breeding nuclei, as these populations far outperform other genotypes available from native, hobbyists' or fanciers' populations for the combination of traits of economic importance.

*Corresponding author: w.g.hill@ed.ac.uk

In this chapter, we shall review theory and evidence from experimental and commercial populations which give some guide on the maintenance of genetic variation. We consider how accuracy of selection and rates of improvement in all traits can be increased best to utilize this variation, notably by incorporation of genomic information. We then discuss what are the many issues associated with other traits of health, welfare, robustness, environmental impact or public perception that are being or may have to be faced and are potential risks to the sustainability of the industry. We consider how breeders can tackle these problems by including new records, new evaluation methods and improved design of their selection programmes.

We focus mainly on chickens at the expense of other poultry species because they dominate both meat and egg markets, there is most information about them, and they are the best developed model for analysis of long-term genetic change; however, most topics transfer across species.

MAINTENANCE OF GENETIC VARIATION IN POPULATIONS UNDER SELECTION

The presence of genetic variation is crucial for the sustainability of poultry breeding and the ability of populations to face new challenges. Evidence for the ability of populations to maintain genetic variation under long-term selection comes from theory, response to selection in experimental conditions, past and current rates of genetic improvement in commercial populations, and more recently from molecular data.

Population size; mutation

First, it is important to note that, even if parents are highly selected and differ little in genotype, half of the useful (the additive) genetic variation is recovered each generation in diploid species as a consequence of Mendelian sampling. Further, there remains residual genetic variation among the selected parents dependent on the accuracy and intensity of selection. Indeed, if the trait is determined by genes of small effect at many loci, in the absence of inbreeding and adverse fitness correlations, variance is retained and selection response is expected to continue indefinitely.

Even so, variation is bound to be lost by fixation, particularly through selection of those genes that do have large effect, but also by chance because nucleus populations are of finite size. The rate of loss of variation is likely to be at least by a proportion $1/(2N_e)$, where N_e is the effective population size. Since the main bottleneck is through males the rate is likely to exceed $1/4N_m$ in a population with N_m males. Clearly, as theory shows (Robertson, 1960) and experiments in *Drosophila*, for example, illustrate (Weber, 2004), maintaining sufficiently large population sizes is an important requirement in retaining variation and avoiding an early and low selection limit.

What should not be ignored is the role of mutation in producing new variation (Hill, 1982). Here we broadly include not just single base changes, but other *de novo* events such as genome rearrangements. Mutants of large effect have been revealed in selection experiments in poultry (Pettersson *et al.*, 2013). In other species, after 50 generations of selection for body weight in a highly inbred mouse line, Keightley (1998) calculated an increment of 0.23–0.57% in the heritability per generation from mutations. Responses have continued over many generations in initially inbred lines of *Drosophila* (Mackay *et al.*, 1994), and increases in fitness assessed by relative population growth have continued over thousands of generations in asexual microbial populations (Wiser *et al.*, 2013). Analyses over many species typically indicate a heritability increment from mutations of about 0.1% per generation, but estimates range from 0.01% to 1% (Houle *et al.*, 1996). The steady-state variance maintained for neutral mutations is proportional to $2N_e$, so a mutational heritability of 0.1% would imply that a population of effective size (N_e) of 150 would be needed to sustain a heritability of 30%, although this is likely to be an underestimate as mutations with unfavourable effects on fitness or other traits would be lost. Effective sizes of breeders' populations may be somewhat below this figure: N_e ~50–200 in broilers (Andrescu *et al.*, 2007) and 50–300 in layers (Qanbari *et al.*, 2010), estimated from linkage disequilibrium, with the lower figures applying to more recent years. Even so, the estimates imply that mutations have been making a substantial contribution to the continuing selection responses over recent decades.

Selection experiments and improvement in production traits of commercial populations

Evidence on continued response comes from selection experiments maintained in small closed populations (Hill and Bünger, 2004; Hill, 2008; Neeteson-van Nieuwenhoven *et al.*, 2013). For example high and low selection for 8-week weight over 40 generations in broilers by Siegel and colleagues (Siegel, 1962; Johansson *et al.*, 2010; Dunnington *et al.*, 2013), for over 90 generations for high body weight in quail by Marks and colleagues (Marks, 1996), and over 30 generations in turkeys (Nestor *et al.*, 1996) each produced over two-fold increases, although these were all experimental populations maintained with limited numbers of parents. Selection in the Illinois maize lines for oil content has continued upwards for 100 generations (years), yet the number of ears ('maternal' families) chosen as parents each generation ranged from only 12 to 24 (Dudley and Lambert, 2004). Although response rates slowed down in some of these lines, these are typically fitness-associated effects such as anorexia in the low weight lines of Siegel (Zelenka *et al.*, 1988) rather than attenuation of variation.

In accordance with the continuing response, for which rates have been fairly consistent, estimates of heritability within broiler populations, at least, seem not to have changed much over the decades. Estimates for three current Aviagen broiler populations of the heritability of juvenile (5 week) weight average 36% (Table 11.1). These differ little from average estimates over many old studies (38% from half sibs, 31% from offspring–parent) for 8-week body weight

Table 11.1. Heritability[a] (diagonal), genetic (above) and phenotypic (below) correlations of juvenile body weight[b] and a number of leg traits[c] (Kapell *et al.*, 2012b).

Trait	BW	LD	CT	TD	HB
BW	0.36	0.18	0.14	0.19	0.3
LD	−0.04	0.05	0.23	0.05	0.02
CT	0.01	−0.02	0.05	−0.04	−0.01
TD	0	0.07	0.08	0.17	0.13
HB	0.11	−0.01	−0.01	0.01	0.08

BW, juvenile body weight; LD, leg defects; CT, crooked toes; TD, tibial dyschondroplasia; HB, hock burn
[a]Average estimates of heritability and correlations over three independent lines
[b]Body weight and leg traits scored at 5 weeks of age
[c]Positive correlations with body weight are unfavourable for all leg traits

summarized by Kinney (1969). In the early generations of Siegel's experiment in a population derived from inbred line crosses, the heritability of 8-week weight was approximately 25% and remained approximately so. Selection response in the high lines has continued almost linearly for 50 generations, though that in low lines has slowed (Dunnington *et al.*, 2013). Unselected control lines derived by relaxed selection in generations 6, 13, 19 and 26 showed little regression, indicative of no more than weak disadvantageous fitness effects, although that is not a uniform finding among all selection experiments.

MAINTENANCE OF GENETIC VARIATION: RATES OF IMPROVEMENT OF PRODUCTION TRAITS IN COMMERCIAL POPULATIONS

Changes over the last six decades in egg and meat yields and in efficiency of production are well documented (e.g. Laughlin, 2007; FAOSTAT, 2015). Separating out the genetic component requires proper design, but fortunately we have this from independent sources up to 2001 for broilers and more limited data for egg layers. These results are well known, but they represent the best documented results of realized selection over many generations in livestock.

Broilers

The Athens Canadian Random Bred Control population was established in 1957 from crosses in 1955 of eight commercial and eight experimental broiler strains, and subsequently maintained without selection or apparent phenotypic trend. Using comparisons with this control, genetic improvement was assessed by Havenstein and colleagues in two trials, the first with 1991 and the second 2001 commercial stock (Table 11.2), in each case using a (then) modern (1991 or 2001) diet and one formulated to 1957 specifications (Havenstein *et al.*, 1994,

2003a, b). Growth rate improved dramatically, of which about 85–90% could be attributed to genetic change, and was seen on both diets. Other traits, including meat yield, also changed greatly, and there was a subsequent reversal of the increases to 1991 in fat content. The differences between 1957 and 1991 and between 1957 and 2001 birds are also compared in Table 11.2, from which rates of change during the 1990s can be seen. Although the D2–D1 comparison is confounded with a change of commercial strain, response still appeared to be continuing apace, i.e. of the order of over 2% of the mean per year in body weight.

Additional evidence can be obtained from industry which has maintained both selected and unselected control populations in the same environment. Comparisons made over the 37 year period 1972–2005 show the continuing changes in live performance and yield while reducing mortality (Table 11.3). Subsequent unpublished data from Aviagen show annual trends fitted by linear regression from 2006 to 2013 of 44 g in live weight at 35 days, with a corresponding reduction of 0.42 days in age and 0.02 in food conversion ratio to 2 kg, and increases of 0.80% in carcass yield and 0.73% in breast yield. These annual changes of approximately 2% are similar to those observed over the previous 40 year period.

Comparative studies have been conducted to assess the changes in broiler phenotype and dietary responses due to intense genetic selection (Mussini, 2012). Four broiler strains, one of heritage stock unselected since the 1950s representing the old meat-type bird and three current populations, were reared. Birds of each genotype were killed weekly from days 7 to 56. Relative to the

Table 11.2. Comparison of growth and body composition of 1957 control and 1991 commercial and of 1957 control and 2001 commercial broilers reared on the diet used in that year. The difference D1 denotes changes between 1957 and 1991, D2 between 1957 and 2001, and D2–D1 the estimated change between 1991 and 2001 (Havenstein et al., 1994, 2003a, b).

Year of trial	1991			2001			
Year of population	1991	1957	D1	2001	1957	D2	D2–D1
Body weight (kg)	3.11	0.79	2.32	3.95	0.81	3.14	0.82
Carcass weight (kg)	2.07	0.50	1.51	2.81	0.48	2.33	0.82
Carcass yield (%)	69.7	61.2	8.5	74.4	60.8	13.6	5.1
Breast yield (%)	15.7	11.8	3.9	21.3	11.4	9.9	6.0
Carcass fat (%)	15.3	9.4	5.9	15.9	10.6	5.3	–0.6

Table 11.3. Comparisons, averaged over sexes, between 2005 broiler and 1972 randomly selected controls for production traits (Fleming et al., 2007).

Genotype	Live weight (g) at 42 days	FCR to 2.0 kg	Carcass yield % at 2.0 kg	Breast yield %	Mortality 0–42 days %
2005 Broiler	2665	1.65	68	17.4	4.15
1972 Control	1210	2.23	65.3	11.05	5

heritage strain, the current strains were shown to have significantly increased body weight and muscle accretion, especially breast meat. Jejunum and ileum segments were longer, but shorter when related to body weight, and relative heart and gizzard relative weights were also reduced. Tibia breaking strength increased. The results confirm the substantial change that has occurred in the broiler due to selection, and show these changes have to be considered in diet formulation.

As we discuss in a subsequent section, notwithstanding the large and continuing changes made in production, substantial improvements have also been made in traits associated with health and welfare.

Egg layers

The primary traits of strains for table-egg production, notably egg number itself, are clearly directly related to fitness and to the 24 h circadian cycle. There has long been concern as to whether continuing progress in such stocks could be realized or limits would soon be reached, exemplified by Dickerson's (1955) analysis of 'genetic slippage' and by attempts to select in ahemeral or continuous light cycles (e.g. Yoo et al., 1984). Nevertheless, they have continued.

Comparisons between unselected control lines derived from commercial stock over three periods from 1950 with the performance of a selected population of 1993 are shown in Table 11.4. All measures of performance, namely age at first egg, egg production and egg weight, improved over the 40 years from 1950, while body weight remained relatively unchanged. The consequent ca. 30% improvement in feed conversion efficiency arose from changes both before and after 1970. Even if selected lines had regressed in performance, this is unlikely to have affected comparisons between the control populations, all long relaxed before 1993. Similar comparative evidence was obtained by McMillan et al. (1990) by comparing production profiles of three strains representing commercial stocks sampled between 1950 and 1970 and subsequently unselected, and by comparing commercial and unselected control flocks during 1970–1980. Production profiles showed that the onset of lay commenced at an earlier age and persistency of egg production after reaching the peak rate of lay increased.

Table 11.4. Progress in egg production: contemporary comparison of egg production traits for unselected control lines (C) established from selected populations in 1950, 1958 and 1973, with a commercial population (S) of 1993 (Jones *et al.*, 2001).

Population	18 week weight (g)	Age at lay (days)	Hen day (%)	Egg weight (g)	Egg mass (g/day)	FCE
1950 C	1440	182.9	56.9	56.5	34.2	0.319
1958 C	1336	172.6	59.7	61.8	37.0	0.345
1972 C	1331	166.3	64.2	61.0	41.2	0.378
1993 S	1429	154.9	73.4	63.6	49.3	0.426

FCE, egg weight (g) per gram feed

Annual improvements in egg production (2.1 eggs better than unselected controls), egg size (0.09 g), body weight (–0.2 g), viability (0.59) and feed conversion ratio (FCR) (–0.026), illustrate that responses can be achieved in these traits despite strong unfavourable correlations among them.

Although random sample tests are not designed to reveal genetic trends they do nevertheless document long-term trends in layer performance. Anderson *et al.* (2013) have documented these for North Carolina tests from 1958 to 2011 (Table 11.5). There have been striking increases in almost all production traits and reduction in mortality, with a consequent 50% increase in feed conversion efficiency (g eggs/g feed). Similarly, data from German random sample tests over the period 1975–1997 show an approximately 30% annual increase in egg mass per hen housed in both white and brown egg strains, i.e. about 1% per year (Hill, 2008).

Traits under selection during this period included age at sexual maturity and peak rate of lay, but with particular pressure on persistency of egg production due to its high genetic variation post-peak. Feed conversion improved from reducing adult body weight and by selection for low residual feed intake. The shape of the egg weight curve during lay has flattened with increases in early egg size and a relative reduction post-peak (Fig. 11.1) to produce more eggs in the desired egg weight classes (Anderson *et al.*, 2013). Other traits under selection during this period in which substantial responses have resulted include shell breaking strength, puncture score and, more recently, dynamic stiffness (de Ketelaere *et al.*, 2004), and responses continued for increased livability, albumen height, dry matter solids, eggshell colour, and freedom from blood or meat spots.

MAINTENANCE OF GENETIC VARIATION – MOLECULAR DATA

The information on continuing amounts of variation discussed so far have all been at the trait level estimated using quantitative theory. Information at the molecular level is rapidly accumulating providing better ways to measure the actual variation at the genome level. Polymorphism at the nucleotide level in

Table 11.5. Changes in mean laying-house performance for the first production cycle from the 1st (1958) North Carolina Random Sample Layer Test and subsequent 37th (2009) North Carolina Layer Performance and Management Test (Anderson *et al.*, 2013).

Test year	Age (days) at 50% production	Hen day production (%)	Hen housed production	FCE (g egg/ g feed)	Mortality (%)	Egg weight (g)	Body weight (g)
Brown egg strains							
1958	166.2	65.9	214	0.326	16.6	60.3	2670
2009	139.4	85.1	281	0.492	5.5	61.4	1896
White egg strains							
1958	173.2	70.1	212	0.345	10.9	59.5	2051
2009	139.1	85.8	276	0.501	6.4	60.2	1682

(a)

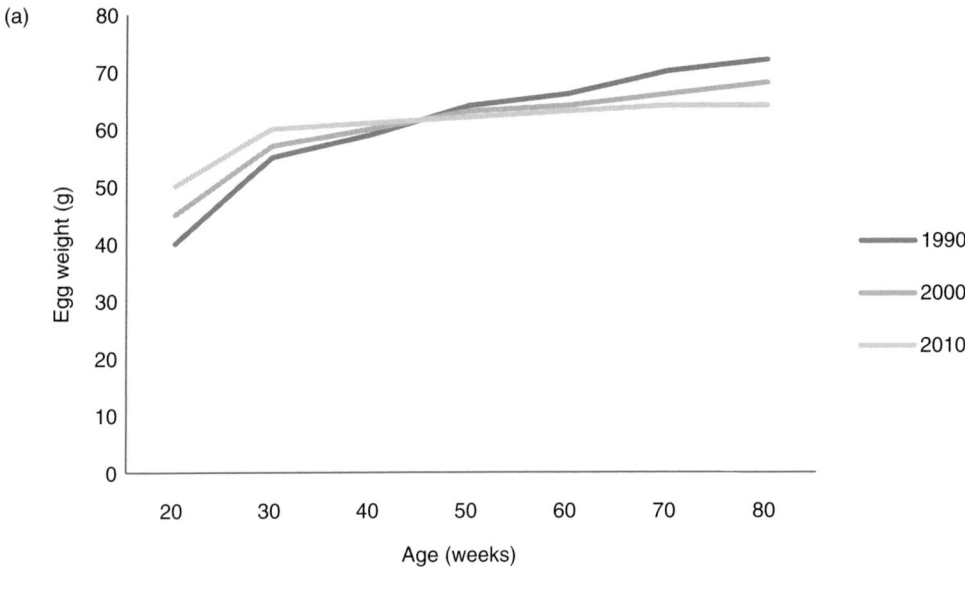

(b)

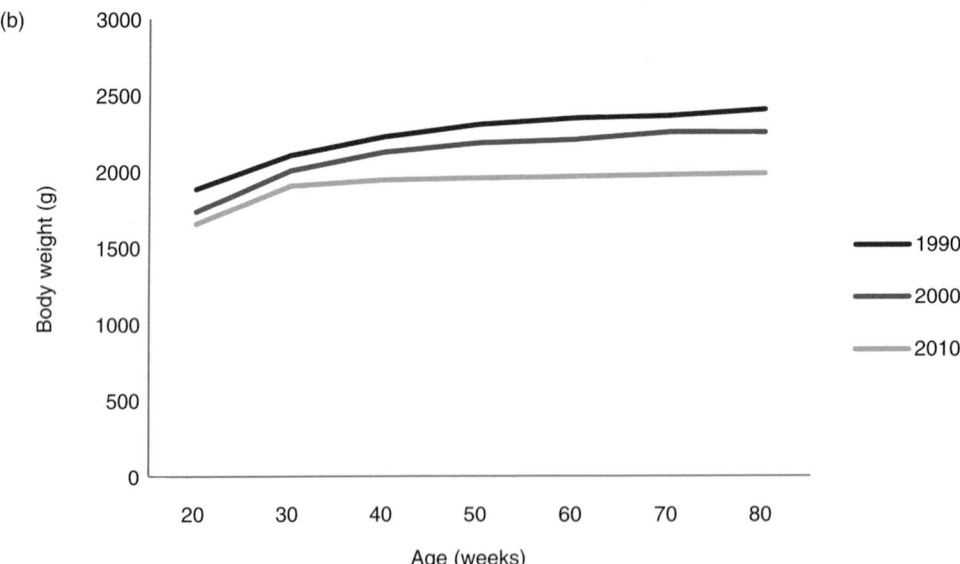

Fig. 11.1. Changes in (a) egg weight and (b) body weight as a function of age in Hy-Line brown birds, 1990–2010 (Hy-Line data).

poultry, estimated at 6.5×10^{-3} on autosomes and 2.0×10^{-3} on the Z chromosome, is substantially higher than that found in humans (Axelsson *et al.*, 2004).

A concern often raised, however, has been the extent to which genetic variation has been lost in breeders' populations. The analysis of Muir *et al.* (2008) showed that modern commercial stocks have lost about half the variation at the molecular level of the foundation native jungle fowl. Most of this loss from the hypothetical ancestral population had occurred, however, before the current populations, including fanciers' strains, were bred, indicating it has not just been a consequence of modern breeding methods and population structure. Further, Rubin *et al.* (2010) showed that levels of heterozygosity within two broiler (average 3.4×10^{-3}) and two layer commercial strains (average 2.4×10^{-3}) were a high proportion of that in a red jungle fowl population (4.1×10^{-3}).

Furthermore, large between and within commercial line differences exist at the genetic level. The extent of between-lines diversity was evaluated in the study by Andrescu *et al.* (2007), shown both by analysis of individual markers and by linkage disequilibrium among multiple markers. Within-line variation was revealed by Kranis *et al.* (2013) by resequencing DNA of 243 birds from 24 lines representing commercial broiler and layer breeds, and several experimental and inbred lines. Approximately 139 million SNPs were found, of which about 78 million were segregating within one or more lines. On average, 8 million segregating SNPs were detected in each commercial line.

Resequencing data provide further evidence of mechanisms contributing to diversity levels in chicken breeds, with the strongest effect being recombination rate and a much lower impact of mutation rate quantified as intergenic divergence and synonymous substitutions (Mugal *et al.*, 2013). A deeper understanding of variability and recombination rates between macro- and micro-chromosomes is expected as sequencing costs fall and more data become available.

Microsatellite markers were widely used with the aim of identifying QTL of large effect by linkage analysis such that these could be exploited individually in breeding programmes using marker-assisted selection. While many QTL were detected (animal genome database http://www.animalgenome.org/), the power and precision of location of QTL of small effect is low. Now, with use of dense molecular markers, much more powerful studies are possible using genome-wide analysis (GWAS). Although there is substantial linkage disequilibrium (LD) within poultry populations which limits discriminating power, many sites where SNP markers are strongly associated with important traits have been revealed (Wolc *et al.*, 2014). Nevertheless, the magnitudes of QTL/gene effects on the traits found in such studies, and indeed in the large data sets on human height, for example, show that only a small proportion of the total variation is accounted for by the loci detected. The highly polygenic structure of quantitative traits revealed by such analysis explains why QTL detection and introgression have had little impact on poultry or other livestock breeding. Further, it implies that the heterozygosity at trait genes is not strongly reduced by selection and therefore that variation at neutral molecular markers is in itself a good indicator of variability in quantitative traits of commercial importance.

The above arguments support the expectation that useful variation can be assumed to remain providing population sizes are not allowed to become too small, and therefore continued response can be expected within populations, providing that selection is practised in the direction of market needs and deleterious side effects are eliminated or at least kept at bay. There are, however, opportunities to increase the rates of progress achieved so far.

INCREASING RATES OF GENETIC IMPROVEMENT OF POULTRY BY INCORPORATING GENOMIC DATA

Principles

Traditionally, genetic evaluation is based on phenotypic records, incorporating the genetic information directly only through the pedigree and estimates of the genetic variances and co-variances among the traits. Methods have followed the principles set out by Lush (1937) on prediction of breeding values and by Henderson (1984) in introducing BLUP to generalize the ideas. Such breeding value prediction is based on data on the animal, its parents, sibs and earlier generations, each combined optimally according to the amount of data, contemporary environmental grouping, numbers of relatives and their pedigree relationships, and the trait's heritability. Simultaneous evaluation of multiple traits is undertaken and weighted according to their genetic correlations and importance in the breeding goal. Critically, however, when assessing individuals that do not have their own or progeny records, notably unproven males for sex-limited traits and for traits recorded post-mortem, ancestral information does not include the effect of Mendelian sampling due to genetic segregation from parent to offspring even though it contributes over half the potentially useful genetic variance each generation.

The availability of very dense SNP markers and indeed whole genome sequencing provide a radically different route to the previous and generally unsuccessful aim of using markers to identify QTL of large effect to introgress into populations. The approach, 'genomic prediction' or 'genomic selection', is based on simultaneous estimation of marker-associated effects across the whole genome in order to provide genome-wide estimates of breeding values (Meuwissen *et al.*, 2001). In a layer breeding programme, for example, the genotypic information on the candidate birds, their aunts and sisters, and indeed more distant relatives can be utilized to increase accuracy, because the relationship or similarity in genotype is assessed for each marker and combined with the individual phenotypic information. Put simply, the best prospective males from a family are those most similar in genotype to their best performing sisters, aunts and other members of the population. Pedigree, molecular and phenotypic data can then be combined within the classical BLUP structure to maximize the accuracy of choosing selection candidates, and the predicted breeding values of each can be compared directly despite being supported by very different amounts and types of data, just as with BLUP based on production data alone.

How best to optimize models and procedures in genomic evaluation programmes is an area of highly active research (e.g. see review by Gianola, 2013 and more specific analyses of broiler data by Morota *et al.*, 2014), but beyond the scope of this review. One major issue is the degree to which different genomic regions should be weighted rather than fitting an overall infinitesimal model. Whilst it is likely that there are real trait differences in the optimum model, however, quite general assumptions seem to be robust in terms of prediction accuracy.

Using simulation studies, genomic selection was shown to have the potential in poultry to provide increased accuracy of selection, reduced generation intervals and better control of inbreeding (Dekkers *et al.*, 2009). Experiences with real data have brought genomics in the poultry industry from a concept to potential deliverables (Avendaño *et al.*, 2010) and subsequently into consolidation stages (Kranis *et al.*, 2013; Wolc *et al*, 2014). We outline current progress in the following sections.

Genomic selection in layers

To quantify gains from genomic selection, a multi-generation selection experiment has been undertaken by Hy-Line International and Iowa State University (Wolc *et al.*, 2014). A brown egg layer line was split into parallel selection lines comprising a pedigree control and a genomic sub-line. The traditional pedigree-based BLUP selection was continued in 13 month cycles using a nested mating structure. The genomic sub-line size was maintained with a generation interval of 7 months and smaller numbers, but without increasing the expected annual rate of inbreeding, and each female was allowed to produce progeny with ten different males. Selection was based on the same index of 16 egg production and quality traits as in the pedigree sub-line but with genomic information included in the breeding value estimation.

In the final generation of the experiment (generations 3 and 5, respectively, of the pedigree and genomic sub-line), the sub-lines were hatched and raised together. For 12 out of 16 traits, the genomic sub-line significantly outperformed the pedigree sub-line. Its greater genetic progress originated from shorter generation intervals and greater accuracy of selection of males (Wolc *et al.*, 2011). The experiment demonstrates the feasibility of genomic prediction and the promise of substantial improvement when using larger populations and more optimal designs.

Genomic selection in broilers

Research was undertaken by Aviagen into developing a sustainable and cost-efficient strategy to implement genomic selection for routine evaluations, not least so as to incorporate accurate information on reproduction traits at the time of selection of juveniles. To avoid the prohibitive cost of high-density genotyping on very many selection candidates, a low-density and imputation strategy for genomic selection offered a viable solution (Habier *et al.*, 2009). The correlation between imputed and real high-density genotype was around 0.97, showing that large-scale imputation is sufficiently robust to implement in genomic selection for routine evaluations in elite broiler lines (Wang *et al.*, 2013). The relative improvement from genomic prediction in terms of selection accuracy, measured as the correlation between phenotype adjusted for fixed effects and pedigree/genomic estimated breeding valve (EBV) when animals had no phenotypic records, ranged between 20% and 70% (Fig. 11.2). Based on these improvements in accuracy, genomics information has been incorporated by Aviagen in routine broiler genetic evaluations since 2012.

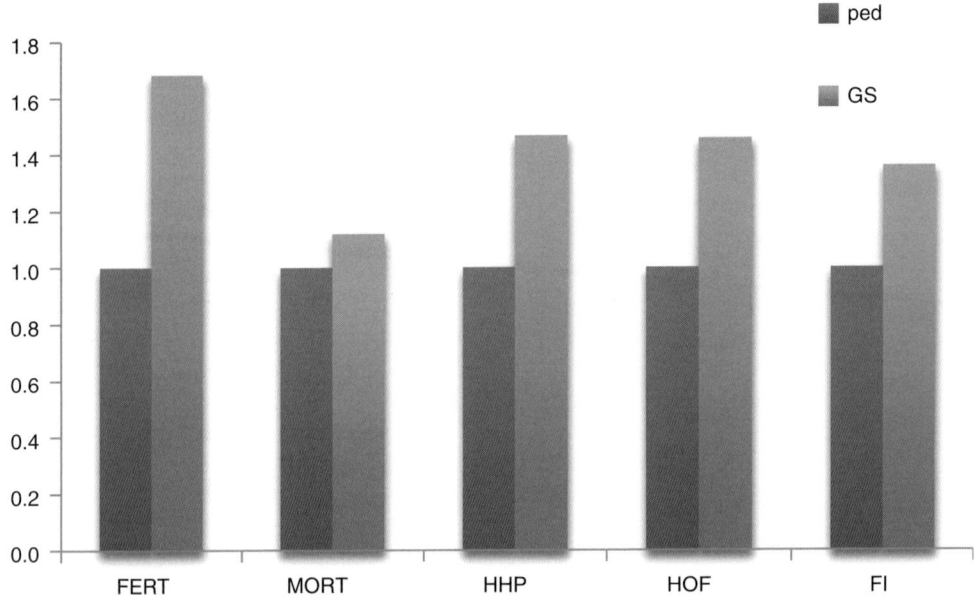

Fig. 11.2. Relative improvement in prediction accuracy of genomic selection (GS) over pedigree-based (ped) EBVs in a broiler population, measured as the correlation of EBVs with adjusted phenotype for five traits: fertility percentage (FERT), laying mortality (MORT), hen-housed egg production (HHP), hatchability percentage (HOF) and feed intake (FI) (Wolc *et al.*, 2014).

Applications of genomics in improvement of disease resistance

As genotyping costs fall and the quality of the sequence assembly and annotation status improves, genomics information has many potential opportunities (Fulton, 2014; Morota *et al.*, 2014). There is the potential to understand the variation in response to disease challenge, clearly offering a novel way to predict breeding values for immune response on selection candidates. Analyses of mortality in response to Marek's disease infection have shown a number of regions of the genome with effects on resistance and show that a genomic selection programme would be effective (Wolc *et al.*, 2013). In addition to work on Marek's disease, there is a growing number of research initiatives on immune response and zoonosis, for instance in relation to coccidiosis, *Salmonella*, *Escherichia coli*, *Campylobacter* and avian influenza (Jie and Liu, 2011), but details are beyond the scope of this review.

BREEDING TO IMPROVE THE COMPLETE BIRD AND SYSTEM: HEALTH, WELFARE, ROBUSTNESS AND ENVIRONMENTAL IMPACT

Breeding goals

We have focused so far on historical rates of improvement, prospects for long-term continued improvement in production, efficiency of production and other traits, and on developments such as genomics aimed at increasing these rates. Although feed availability and environmental load form true constraints to livestock production (e.g. more production means more nitrogen excretion), genetic improvement increases efficiency of use of these resources.

Commercial poultry breeding goals have broadened greatly since the 1970s, with the focus on objectives such as efficiency (productive and environmental), and emphasis on animal robustness and adaptability is increasing relative to that on productivity (Neeteson-van Nieuwenhoven *et al.*, 2013). In the specific case of welfare, for example, broiler breeders have stated that welfare-related traits represent 18% to 33% of the relative weight in the breeding goal (Hiemstra and Ten Napel, 2013). Thus traits subject to genetic selection include skeletal integrity (e.g. leg strength, walking ability, tibial dysplasia), contact dermatitis, heart and lung function and livability (Hiemstra and Ten Napel, 2013).

Robustness

Poultry production occurs globally in very contrasting geographical environments characterized by ambient temperature, humidity, altitude, disease exposure (gut and immune challenge) and feed quality. There is therefore the potential for substantial genotype × environment (G × E) interactions (Neeteson-van Nieuwenhoven *et al.*, 2013). Because of the cost of recording, genotyping and very strict requirements to maintain the highest levels of biosecurity, breeding programmes are typically concentrated in few locations rather than widespread around the world. It is, however, the practice of both broiler and layer breeders to undertake testing of crossbred sibs or progeny in one or more field environments. These field test data can be combined optimally with recording in the pedigree environment so as to breed more robust and resistant birds able to perform well in a wide range of environmental conditions.

Fertility and hatchability in broilers

Notwithstanding the intense selection pressure applied to traits of the growing bird, selection has also been applied to traits of the broiler parent, and there have been significant improvements in reproductive performance. In an analysis of data from the UK, western Europe and south and central America, the performance of broiler breeding stock was regressed against years and so included any management or environmental as well as genetic change. Over the 10 year period to 2002 the mean percentage annual improvements for male line, female line and parent stock, respectively, were 0.4, 1.7 and 1.7 for total egg number and similarly for hatching egg number; 0.6, 1.8 and 1.3 for total chick production; 0.25, 0.47 and 0.09 for hatchability; and –0.69, –0.05 and –0.41 for female mortality (Hocking and McCorquodale, 2008). Results for the different regions were similar even though breeding programmes were international, indicative of little G × E interaction. Recently, Collins *et al.* (2014) have shown that the hatch performance of a modern commercial broiler is very similar to that of the ABRC control population, there being no significant differences in percentage infertile eggs, embryonic mortality, hatchability, or saleable chicks.

Welfare

Aggrey (2010) pointed out that meeting the welfare needs of farm animals has become an integral part of animal agriculture but that, until recently, environmental and management modifications have been the route to meeting these needs. Genetic selection has improved growth, livability and general welfare, but he suggested that aspects of welfare such as metabolic disorders, susceptibility to some diseases, and skeletal problems have increased and that integrated management, genetic and genomic tools should be employed to improve production and welfare traits to address public and consumer concerns.

There have, however, been substantial improvements in recent decades in recording methods for welfare-related traits, and alteration of selection objectives to utilize them in multi-trait indices of breeding objectives. The extent to which that selection can be effective depends on several factors. The first is that records can be taken in a timely manner and, if these are proxies for the trait in question, are sufficiently closely correlated to it. The second is that there is significant genetic variation in that trait to enable its improvement. The third is that it is not so unfavourably correlated with other production, health and welfare traits that selection on it would be ineffective in improving overall productivity, efficiency and welfare. The fourth is that the breeding programme can be appropriately structured to enable such selection to take place. Our knowledge and experience of genetic parameters is such as to be optimistic that all these conditions can be met, at least to some degree. We consider examples, but not a comprehensive review, of issues and successes and remaining problems.

Leg health and contact dermatitis in broilers

Using data from X-ray and visual assessment we showed that while the genetic correlations between body weight and feet and leg health are unfavourable, they are not strongly so for leg traits (Table 11.1, Kapell et al., 2012b) or foot pad dermatitis (Kapell et al., 2012a), and prevalences have been reduced by selection. For example, deformities of the long bones (LD) and crooked toes (CT) were recorded since 1985 and tibial dyschondroplasia (TD) and hock burn (HB) since 1990. The prevalence of CT and HB decreased in the first decade (range among lines 1.2 to 2.3% and 1.3 to 1.5% per year, respectively), then stabilized at low levels, while that of LD decreased by 0.6 to 0.9% and of TD by 0.4 to 1.2% per year (Kapell et al., 2012b). Hence major improvements in leg health have been made along with those in growth and production traits by including them in the breeding goal. Routine assessment of leg health has become standard practice in broiler breeding programmes.

Cardiovascular function and ascites in broilers

Pulse oximetry, which measures the oxygen saturation level of the blood, has a direct relationship with heart and lung function and is an important indicator of susceptibility to developing ascites and sudden death syndromes. Since 1991 selection only of individuals with a family index above the average has been practised by Aviagen, and the incidence of ascites and sudden death at field level has decreased. Oximeter readings (%) in blood of Ross 308 crossbred broilers increased during the 1990s from the low 70s to the low 80s, and to about 88% in 2010–2012. Levels of ascites as measured by the Canadian Food Inspection Agency have fallen correspondingly from about 36% in 1995–1999 to about 8% in 2008 (AAFC, 2015). Understanding of the fundamental biological bases for the pulmonary hypertension/ascites syndrome in broilers and novel indicator traits may come from immunological and genomic investigations (Wideman *et al.*, 2013).

Marek's disease in egg-type chickens

A breeding programme incorporating a challenge test following inoculation with a standard amount of a vv+ strain of Marek's disease (MD) virus in multiple lines and multiple generations of egg-type chickens has been undertaken (Fulton *et al.*, 2013). Mortality of progeny from selected sires has been shown to significantly decrease, at a rate of 1 to 14% per generation, in eight of the nine elite lines studied. This challenge test strategy was very effective in predicting the genetic resistance/susceptibility among families for unchallenged selection candidates in the biosecure pedigree environment.

Competition-related impacts on welfare

A novel approach to dealing with genetic assessment and improvement of traits associated with competition such as feather pecking and mortality in layer strains was initiated by Muir, Bijma and colleagues. The basic idea is to assume the phenotype of an individual reared in, say, a six-bird cage, is dependent both on its own genotype and on the phenotype and genotype of its pen mates; reciprocally its genotype influences the phenotype of the pen mates. Muir (1996) exemplified the utility of this by practising selection in a competitive multi-bird cage environment, each from a single sire family, and selecting the family as a whole. Mortality rates in multiple-bird cages fell dramatically, from 68% to 9% in four generations, the last similar to that in an unselected control kept in single-bird cages; and egg production rose rapidly.

BLUP approaches have been developed for estimating breeding value for both the direct and indirect social effects simultaneously in multi-bird cages with mixed families, on the basis of which selection decisions can be made (Bijma *et al.*, 2007; see also, for example, Ellen *et al.*, 2011 for review). The ideas are being adopted in practice. For example, with the objective of reducing aggression, in their sire progeny field testing Hy-Line house test hens in commercial cages without beak treatment, so that half-sib groups provide data on rates of feather picking and cannibalism to incorporate in selection within the pedigree populations.

Feed intake control of broiler breeders

A stand-out in this area is feed intake control of broiler breeders during puberty. Here the issue from the geneticist's perspective is that the genetic correlation between early and later growth, and similarly between early and late appetite, is very high. Whilst it is possible by selection to bend the growth curve of the chicken (Ricard, 1975), very large sacrifices in early gain have to made in view of its high genetic correlation with late gain if mature weight (and the animal's desire to achieve it) is not to increase.

It has been suggested that broilers have altered food intake control mechanisms and may be constantly hungry due to the effects of selection for fast growth and efficiency. Recent studies have found that feeding behaviour in broilers is governed by non-random bouts consistent with periods of hunger and satiety which are common across selected and unselected lines and that short-term feeding behaviour is similar across poultry species (broilers, turkeys and ducks). Underlying normal controls of feeding behaviour seem to be conserved in broiler birds and short-term feeding behaviour is heritable but independent from selection for FCR (Howie *at al.*, 2009, 2011).

Dawkins and Layton (2012) argue that changing selection goals, sampling other populations and incorporation of appropriate QTL from non-elite populations can all be employed to resolve the problem. Whilst it is the case that multi-objective selection can be effective, as we have seen from changing the distribution of egg weight through the laying year, changing the relative size of egg and body weight in egg-laying stocks, and in improving leg traits in broilers, it does not imply that any trait combination is readily changed.

Hocking and McCorquodale (2008) point out that, while selection to reduce the prevalence of multiple ovulation would lead to long-term welfare benefits, egg production does not reflect ovarian activity and thus the tendency for multiple ovulation. They suggest that modern genomic tools make such an approach more feasible, but we are not aware there have yet been useful developments.

On management and feed control, EFSA (2010) concluded that research in this area is limited and more research is needed to draw firm conclusions about feeding programmes in relation to bird welfare; there is also ongoing research on alternative feeding systems and their impact on feeding amounts, behaviour and stress indicators (e.g. van Emous *et al.*, 2013).

Livability

An ultimate consequence of poor welfare status is increased mortality. In Havenstein's second study, cumulative mortality (averaged over diets) for the 2001 selected broilers and 1957 control to 42 days was 3.6% and 2.1%, respectively, indicating somewhat higher mortality in the modern strain (Havenstein *et al.*, 2003a). At the same live weight, however, mortality was lower in the selected population. Changes can be estimated using industry (Aviagen) comparisons of three (then) modern (M) lines with the lines from which they were derived and maintained as unselected controls (C) since 1972. There were large differences in live weight at 42 days, 2.59 kg for M and 1.22 kg for C, and in FCR to 2 kg live weight, 1.66 and 2.23, respectively, while mortality to 42 days, 4.2% and 3.9%, respectively, differed little (Fleming *et al.*, 2007).

As Table 11.5 shows, laying mortality in egg stocks has fallen substantially in recent decades as a consequence of selecting to reduce it. The testing of cross-bred birds in commercial conditions without beak treatment contributes to improvement of both welfare, as noted above, and to livability.

Environmental impact

Increases in efficiency of production, in terms of feed efficiency, carcass yield, egg weight etc., have a major positive impact of poultry production on the environment in terms of resource use. Calculations by Williams *et al.* (2006) on the effects of genetic selection on the environment give estimated improvements in broiler-related global warming potential (t CO_2) of 23% over 20 years with broilers continuing to be the lowest of meat-producing species, and are expected to show a substantial reduction in dry manure output per broiler processed. Genetic improvements in feed conversion are believed to be largely responsible for many of the positive changes in these environmental impact measures that have and will occur. For layers, the environmental footprint per kilogram of eggs produced in the USA for 2010 was estimated to be lower by 65% to 71% per kilogram eggs produced for each of acidifying emissions, eutrophying emissions and greenhouse gas emissions, and 31% lower in cumulative energy demand compared with 1960 (Pelletier *et al.*, 2014).

CONCLUSIONS

There are well documented large and continuing changes in production traits in both egg- and meat-type poultry over half a century, during which time the efficiency of production in terms of, for example, product output/feed input have increased enormously, and are of the order of 2% per year. This has come from the application of sound genetic modelling and analysis, from estimation of trait heritability and genetic correlation among them, the incorporation of an increasing number of traits in broader breeding objectives and from increasingly sophisticated data recording. A whole new approach through exploiting genomic data is in its early days, so more rapid progress is to be expected. A crucial requirement for continued response for production traits and for others associated with health and welfare is the presence of sufficient genetic variation. We have shown there is little evidence that it is falling in nucleus populations, so one can remain optimistic. New problems will arise but can be met. The critical factor is to maintain programmes with multiple objectives, which maintain fitness while improving production.

REFERENCES

AAFC (2015) 050P Poultry Condemnation Report by Species for Federally Inspected Plants. Available at: http://aimis-simia.agr.gc.ca/rp/index-eng.cfm?action = pR&pdctc = &r = 133 (accessed 14 April 2014).

Aggrey, S.E. (2010) Modification of animals versus modification of the production environment to meet welfare needs. *Poultry Science* 89, 852–854.

Anderson, K.E., Havenstein, G.B., Jenkins P.K. and Osborne, J. (2013) Changes in commercial laying stock performance, 1958–2011: thirty-seven flocks of the North Carolina random sample and subsequent layer performance and management tests. *World's Poultry Science Journal* 69, 489–513.

Andreescu, C., Avendano, S., Brown, S.R., Hassen, A., Lamont, S.J. and Dekkers, J.C.M. (2007) Linkage disequilibrium in related breeding lines of chickens. *Genetics* 177, 2161–2169.

Avendaño, S., Watson, K.A. and Kranis, A. (2010) Genomics in poultry breeding: from Utopia to deliverables. *Proceedings of the 9th World Congress on Genetics Applied to Livestock Production.* Leipzig, Germany, Session 07–01.

Axelsson, E., Smith, N.G., Sundström, H., Berlin, S. and Ellegren, H. (2004) Male-biased mutation rate and divergence in autosomal, Z-linked and W-linked introns of chicken and turkey. *Molecular Biology and Evolution* 21, 1538–1547.

Bijma, P., Muir, W.M., Ellen, E.D., Wolf, J.B. and van Arendonk, J.A.M. (2007) Multilevel selection 2: Estimating the genetic parameters determining inheritance and response to selection. *Genetics* 175, 289–299.

Collins, K.E., McLendon, B.L. and Wilson, J.L. (2014) Egg characteristics and hatch performance of Athens Canadian Random Bred 1955 meat-type chickens and 2013 Cobb 500 broilers. *Poultry Science* 93, 2151–2157.

Dawkins, M.S. and Layton, R. (2012) Breeding for better welfare: genetic goals for broiler chickens and their parents. *Animal Welfare* 21, 147–155.

De Ketelaere, B., Bamelis, F., Kemps, B., Decuypere, E. and De Baerdemaeker, J. (2004) Non-destructive measurements of the egg quality. *World's Poultry Science Journal* 60, 289–302.

Dekkers, J.C.M., Stricker, C., Fernando R.L., Garrick, D.J., Lamont, S.J., O'Sullivan, N.P., Fulton, J.E., Arango, J., Settar, P., Kranis, A., McKay, J., Koerhuis, A. and Preisinger, R. (2009) Implementation of genomic selection in egg layer chickens. *Journal of Animal Science*, 87(E-Suppl. 2), iii.

Dickerson, G.E. (1955) Genetic slippage in response to selection for multiple objectives. *Cold Spring Harbor Symposia on Quantitative Biology* 20, 213–224.

Dudley, J.W. and Lambert, R.J. (2004) 100 generations of selection for oil and protein content in corn. *Plant Breeding Reviews* part 1, 24, 79–110.

Dunnington, E.A., Honaker, C.F., McGilliard, M.L. and Siegel, P.B. (2013) Phenotypic responses of chickens to long-term, bidirectional selection for juvenile body weight – historical perspective. *Poultry Science* 92, 1724–1734.

EFSA Panel on Animal Health and Welfare (AHAW) (2010) Scientific Opinion on welfare aspects of the management and housing of the grand-parent and parent stocks raised and kept for breeding purposes. *EFSA Journal* 8, 1667.

Ellen, E.D., Peeters, K., Visscher, J. and Bijma, P. (2011) Reducing mortality due to cannibalism in layers: taking into account social genetic effects. *World Poultry Science Association, 7th European Symposium on Poultry Genetics*, pp. 12–14.

FAOSTAT (2015) Available at: http://faostat.fao.org/site/339/default.aspx (accessed 15 June 2015).

Fleming, E.C., Fisher, C. and McAdam, J. (2007) Genetic progress in broiler traits – implications for welfare. *Proceedings of the British Society of Animal Science* 2007, 50.

Fulton, J.E. (2014) The value of resequence data for poultry breeding. A primary breeders perspective. *Poultry Science* 93, 494–497.

Fulton, J.E., Arango, J., Arthur, J.A., Settar, P., Kreager K.S. and O'Sullivan, N.P. (2013) Improving the outcome of a Marek's disease challenge in multiple lines of egg type chickens. *Avian Diseases* 57, 519–522.

Gianola, D. (2013) Priors in whole-genome regression: the Bayesian alphabet returns. *Genetics* 194, 573–596.

Habier, D., Fernando, R.L. and Dekkers, J.C.M. (2009) Genomic selection using low-density marker panels. *Genetics* 182, 343–353.

Havenstein, G.B., Ferket, P.R., Scheideler, S.E. and Larson, B.T. (1994) Growth, liveability and feed conversion of 1957 vs. 1991 broilers when fed typical 1957 and 1991 broiler diets. *Poultry Science* 73, 1785–1794.

Havenstein, G.B., Ferket, P.R. and Qureshi, M.A. (2003a) Growth, liveability and feed conversion of 1957 versus 2001 broilers when fed representative 1957 and 2001 broiler diets. *Poultry Science* 82, 1500–1508.

Havenstein, G.B., Ferket, P.R. and Qureshi, M.A. (2003b) Carcass composition and yield of 1957 versus 2001 broilers when fed representative 1957 and 2001 broiler diets. *Poultry Science* 82, 1509–1518.

Henderson, C.R. (1984) *Applications of Linear Models in Animal Breeding*. University of Guelph, Guelph, Ontario.

Hiemstra, S.J. and ten Napel, J. (2013) Study of the impact of genetic selection on welfare of chicken bred and kept for meat production. Final report of a project commissioned by the European Commission (DG SANCO 2011/12254).

Hill, W.G. (1982) Predictions of response to artificial selection from new mutations. *Genetical Research* 40, 255–278.

Hill, W.G. (2008) Estimation, effectiveness and opportunities of long term genetic improvement in animals and maize. *Lohmann Information* 43, 3–20.

Hill, W.G. and Bünger, L. (2004) Inferences on the genetics of quantitative traits from long-term selection in laboratory and farm animals. *Plant Breeding Reviews* 24(2), 169–210.

Hocking, P.M. and McCorquodale, C.C. (2008) Similar improvements in reproductive performance of male line, female line and parent stock broiler breeders genetically selected in the UK or in South America. *British Poultry Science* 49, 282–289.

Houle, D., Morikawa, B. and Lynch, M. (1996) Comparing mutational variabilities. *Genetics* 143, 1467–1483.

Howie, J.A., Tolkamp, B.J., Avendano, S. and Kyriazakis, I. (2009) The structure of feeding behavior in commercial broiler lines selected for different growth rates. *Poultry Science* 88, 1143–1150.

Howie, J.A., Avendano, S., Tolkamp, B.J. and Kyriazakis, I. (2011) Genetic parameters of feeding behavior traits and their relationship with live performance traits in modern broiler lines. *Poultry Science* 90, 1197–1205.

Jie, H. and Liu, Y.P. (2011) Breeding for disease resistance in poultry: opportunities with challenges. *World's Poultry Science Journal* 67, 687–696.

Johansson, A.M., Pettersson, M.E., Siegel, P.B. and Carlborg. Ö. (2010) Genome-wide effects of long-term divergent selection. *PLoS Genetics* 6, e1001188.

Jones, D.R., Anderson, K.E. and Davis, G.S. (2001) The effects of genetic selection on production parameters of single comb White Leghorn hens. *Poultry Science* 80, 1139–1143.

Kapell, D.N.R.G., Hill, W.G., Neeteson, A.-M., McAdam, J., Koerhuis, A.N.M. and Avendaño, S. (2012a) Genetic parameters of foot-pad dermatitis and body weight in purebred broiler lines in 2 contrasting environments. *Poultry Science* 91, 565–574.

Kapell, D.N.R.G., Hill, W.G., Neeteson, A.-M., McAdam, J., Koerhuis, A.N.M. and Avendaño, S. (2012b) Selection for leg health in purebred broilers. Twenty five years of selection for improved leg health in purebred broiler lines and underlying genetic parameters. *Poultry Science* 91, 3032–3043.

Keightley, P.D. (1998) Genetic basis of response to 50 generations of selection on body weight in inbred mice. *Genetics* 148, 1931–1939.

Kinney, T.B. (1969) A summary of reported estimates of heritabilities and of genetic and phenotypic correlations for traits of chickens. *USDA Agriculture Handbook* 363.

Kranis, A., Gheyas, A.A., Boschiero, C., Turner, F., Yu, L., Smith, S., Talbot, R., Pirani, A., Brew, F., Kaiser. P., Hocking, P.M., Fife, M., *et al.* (2013) Development of a high density 600K SNP genotyping array for chicken. *BMC Genomics* 14, 59.

Laughlin, K. (2007) The Evolution of Genetics, Breeding and Production. *Temperton Fellowship Report No.15.* Harper Adams University College, Newport, UK.

Lush, J.L. (1937) *Animal Breeding Plans.* Iowa State College Press, Ames, Iowa.

Mackay, T.F.C., Fry, J.D., Lyman, R.F. and Nuzhdin, S.V. (1994) Polygenic mutation in *Drosophila melanogaster* – estimates from response to selection of inbred strains. *Genetics* 136, 937–951.

Marks, H.L. (1996) Long-term selection for body weight in Japanese quail under different environments. *Poultry Science* 75, 1198–1203.

McMillan, I., Fairfull, R.W., Gowe, R.S. and Gavora, J.S. (1990) Evidence for genetic improvement of layer stocks of chickens during 1950–80. *World's Poultry Science Journal* 46, 235–245.

Meuwissen, T.H.E., Hayes, B.J. and Goddard, M.E. (2001) Prediction of total genetic value using genome-wide dense marker maps. *Genetics* 157, 1819–1829.

Morota, G., Abdollahi-Arpanahi, R., Kranis, A. and Gianola, D. (2014) Genome-enabled prediction of quantitative traits in chickens using genomic annotation. *BMC Genomics* 15, 109.

Mugal, C.F., Nabholz, B. and Ellegren, H. (2013) Genome-wide analysis in chicken reveals that local levels of genetic diversity are mainly governed by the rate of recombination. *BMC Genomics* 14, 86.

Muir, W.M. (1996) Group selection for adaptation to multiple-hen cages: selection program and direct responses. *Poultry Science* 75, 447–458.

Muir, W.M., Wong, G.K.S., Zhang, Y., Wang, J., Groenen, M.A.M., Crooijmans, R.P.M.A., Megens, H.-J., Zhang, H., Okimoto, R., Vereijken, A., Jungerius, A., Albers, G.A.A., *et al.* (2008) Genome-wide assessment of worldwide chicken SNP genetic diversity indicates significant absence of rare alleles in commercial breeds. *Proceedings of the National Academy of Sciences of the United States of America* 105, 17312–17317.

Mussini, J. (2012) Comparative Response of Different Broiler Genotypes to Dietary Nutrient Levels. PhD thesis. University of Arkansas, Fayetteville, Arkansas.

Neeteson-van Nieuwenhoven, A.-M., Knap, P. and Avendaño, S. (2013) The role of sustainable commercial pig and poultry breeding for food security. *Animal Frontiers* 3, 52–57.

Nestor, K.E., Noble, D.O., Zhu, N.J. and Moritsu, Y. (1996) Direct and correlated responses to long-term selection for increased body weight and egg production in turkeys. *Poultry Science* 75, 1180–1191.

Pelletier, N., Ibarburu, M. and Xin, H.W. (2014) Comparison of the environmental footprint of the egg industry in the United States in 1960 and 2010. *Poultry Science* 93, 241–255.

Pettersson, M.E., Johansson, A.M., Siegel, P.B. and Carlborg, Ö. (2013) Dynamics of adaptive alleles in divergently selected body weight lines of chickens. *G3: Genes, Genomes, Genetics* 3, 2305–2312.

Qanbari, S., Hansen, M., Weigend, S., Preisinger, R. and Simianer, H. (2010) Linkage disequilibrium reveals different demographic history in egg laying chickens. *BMC Genetics* 11, 103.

Ricard, F.H. (1975) A trial of selecting chickens on their growth curve pattern experimental design and 1st general results. *Annales de Génétique et de Séléction Animale* 7, 427–444.

Robertson, A. (1960) A theory of limits in artificial selection. *Proceedings of the Royal Society Series B* 153, 234–249.

Rubin, C.-J., Zody, M.C., Eriksson, J., Meadows, J.R.S., Sherwood, S., Webster, M.T., Jiang, L., Ingman, M., Sharpe, E., Ka, S., Hallböök, F., Besnier, F., *et al.* (2010) Whole-genome resequencing reveals loci under selection during chicken domestication. *Nature* 464, 587–593.

Siegel, P.B. (1962) Selection for body weight at 8 weeks of age. I. Short term response and heritabilities. *Poultry Science* 41, 954–962.

UK Foresight Report (2011) Available at: https://www.gov.uk/government/collections/global-food-and-farming-futures (accessed 19 June 2015).

van Emous, R.A., Kwakkel, R.P., van Krimpen, M.M. and Hendriks, W.H. (2013) Effects of growth patterns and dietary crude protein levels during rearing on body composition and performance in broiler breeder females during the rearing and laying period. *Poultry Science* 92, 2091–2100.

Wang, C., Habier, D., Peiris, B.L., Wolc, A., Kranis, A., Watson, K.A., Avendano, S., Garrick, D.J., Fernando, R.L., Lamont, S.J. and Dekkers, J.C.M. (2013) Accuracy of genomic prediction using an evenly spaced, low-density single nucleotide polymorphism panel in broiler chickens. *Poultry Science* 92, 1712–1723.

Weber, K. (2004) Population size and long-term selection. *Plant Breeding Reviews* 24(1), 249–268.

Wideman, R., Rhoads, D., Erf, G. and Anthony, N. (2013) Pulmonary arterial hypertension (ascites syndrome) in broilers: a review. *Poultry Science* 92, 64–83.

Williams, A.G., Audsley, E. and Sandars, D.L. (2006) Determining the Environmental Burdens and Resource Use in the Production of Agricultural and Horticultural Commodities. Main Report Defra Research Project ISO205. Cranfield University and Defra, Bedford, UK.

Wiser, M.J., Ribeck, N. and Lenski, R.E. (2013) Long-term dynamics of adaptation in asexual populations. *Science* 342, 1364–1367.

Wolc, A., Arango, J., Settar, P., Fulton, J.E., O'Sullivan, N.P., Preisinger, R., Habier, D., Fernando, R., Garrick, D.J. and Dekkers, J.C.M. (2011) Persistence of accuracy of genomic estimated breeding values over generations in layer chickens. *Genetics, Selection, Evolution* 43, 23.

Wolc, A., Arango, J., Jankowski, T., Settar, P., Fulton, J.E., O'Sullivan, N.P., Fernando, R., Garrick, D.J. and Dekkers, J.C.M. (2013) Genome-wide association study for Marek's disease mortality in layer chickens. *Avian Diseases* 57, 395–400.

Wolc, A., Kranis, A., Arango, J., Settar, P., Flton, J.E., O'Sullivan, N., Avendaño, S., Watson, K.A., Preisinger, R., Habier, D., Lamont, S.J., Fernando, R., Garrick, D.J. and Dekkers, J.C.M. (2014) Applications of genomic selection in poultry. *Proceedings, 10th World Congress of Genetics Applied to Livestock Production* 080.

Yoo, B.H., Sheldon, B.L. and Podger, R.N. (1984) Cross-breeding performance of White leghorn and Australorp lines selected under continuous light for short interval between eggs: full-year egg production and efficiency of food utilization. *British Poultry Science* 25, 233–243.

Zelenka, D.J., Dunnington, E.A., Cherry, J.A. and Siegel, P.B. (1988) Anorexia and sexual maturity in female white rock chickens. I. Increasing the feed intake. *Behavior Genetics* 18, 383–387.

CHAPTER *12*
Increased Sustainability in Poultry Production: New Tools and Resources for Genetic Management

Jemima Whyte,[1] Elisabeth Blesbois[2] and Michael J. McGrew[1]*

[1] The Roslin Institute and Royal (Dick) School of Veterinary Studies, University of Edinburgh, Easter Bush Campus, Midlothian UK; [2] INRA, UMR-PRC, Nouzilly, France

INTRODUCTION

Each year the global poultry industry produces 99 million t of chicken meat and 69 million t of eggs with 59 billion birds raised and slaughtered to achieve this production level (FAO, 2012). Moreover, in the UK alone, around 100 million broiler chickens, laying hens, turkeys, ducks and game birds are reared annually for food consumption. Poultry meat production is based on a pyramidal structure. At the apex of the pyramid are the modern commercial stocks that have been developed from a few standard breeds of chickens. Poultry companies maintain many select pure lines of chickens, which are crossbred and multiplied to generate this vast number of birds. The efficient and inexpensive production of meat and eggs is largely dependent upon intense genetic selection of desirable traits within these lines such as feed conversion ratio, bird weight and egg number (FAO, 2006).

The drive for animals highly standardized on production criteria has led to an overall reduction in the number and diversity of pure lines (Arthur and Albers, 2003). In addition, it has been estimated that one-half of the genetic diversity present in commercial poultry has been lost from these select pure lines as a consequence of generations of selection and inbreeding (Muir *et al.*, 2008). The global poultry production has expanded vastly over the past century growing by more than 30% in the last decade, presumably because poultry meat production is the most environmentally efficient form of meat production (EAC, 2011). However, even at these high levels of production, global poultry production will need to double in the next 25 years to meet population growth and increasing

*Corresponding author: mike.mcgrew@roslin.ed.ac.uk

demands for animal protein products. To assure sustainable production at these high and ever increasing levels, all areas of biosecurity will need to be improved. A major objective for future biosecurity can be met by preserving and safeguarding all existing genetic diversity in commercial poultry flocks through improved cryopreservation programmes. This objective will retain both disease-resistant traits and potentially climate-specific traits for future exploitation. Our increased ability to manipulate the genome of the chicken through genetic modification will facilitate the transfer of single valuable traits into commercial poultry production lines.

The future growth of the poultry industry will be in developing regions, such as in East and South-east Asia where continued population growth is associated with an increase in meat consumption seen in developing economies (FAO, 2011; Cao and Li, 2013). A major challenge will be adapting commercial flocks to these conditions and potential further climate changes, as poultry performance is reduced in atypical conditions caused by differences in environment, feed quality and disease challenges (Emmerson, 2003; Renaudeau *et al.*, 2012). Development of these markets will require commercial breeds with different production traits than current breeds deliver, which may include resistance to heat stress, higher disease/parasite tolerance or the efficient use of different nutritional sources. The improvement of indigenous local breeds and their potential for integration into commercial breeding programmes may contribute to future production growth in these regions.

A portion of the poultry biodiversity is occupied by local indigenous poultry breeds consisting of locally adapted stocks selected over hundreds of years, currently maintained by hobbyists and small poultry farmers and genetically distinct from most commercial breeds (Leroy *et al.*, 2012; Wilkinson *et al.*, 2012; Pham *et al.*, 2013). As mentioned above, the interbreeding of commercial breeds with these non-commercial native or standard breeds is one way of introducing genetic diversity into the commercial lines and of introducing select genetic traits that may not be present in commercial flocks (Chowdhury *et al.*, 2014). However, many of these breeds are rarely selected for standard production traits and are also facing genetic erosion due to small population sizes and inbreeding (Wilkinson *et al.*, 2012). Moreover, many of these breeds are also considered at risk of extinction in Britain and elsewhere (Rare Breeds Survival Trust, 2014).

This chapter examines the emerging technologies that can be applied in the poultry industry in the coming decade for increased sustainability of genetic resources. Some of these technologies such as semen cryopreservation are currently being used for sustainable production by protecting and maintaining genetic variation in poultry populations and stem-cell biobanks are also under development. New genetic tools to modify the avian genome could potentially be used to introduce novel genetic variation and disease resistance into existing commercial flocks. Future challenges will lie in integration of these emerging technologies into existing production systems and addressing the regulatory policies in both regional and global markets.

SEMEN BIOBANKING IN POULTRY

Semen cryopreservation is a key tool to manage the propagation of the genetic potential of the male. It also constitutes an alternative strategy to limit the erosion of genetic diversity and is of major interest for the preservation of endangered breeds as it greatly facilitates the development of programmes to preserve genetic biodiversity.

Despite the fact that in birds only the male genome can be conserved in semen, as the female is the heterogametic sex, sperm cryopreservation remains the main method for the long-term storage of reproductive cells for the *ex situ* management of genetic diversity in birds. The main reasons are that this method is: (i) non-invasive, conversely to the conservation of diploid germ cells and gonadic tissues; (ii) permissive to the long-term storage of a large number of mature reproductive cells ready for fertilization (millions to billions of spermatozoa/ejaculate); and (iii) easy to reintroduce post-cryopreservation into host females through artificial insemination. Therefore, in backcrossing it would take one to six generations (depending on the genotype and requirement) to return to the desired genotype after semen cryopreservation. By contrast, the conservation of diploid germ cells and gonadal tissues requires the sacrifice of numerous embryos or delicate surgical manipulation. These cell- and tissue-based methods have been improved in the last few years (Nakamura *et al.*, 2010; Macdonald *et al.*, 2012; Silverside *et al.*, 2013; Tajima, 2013) and are also being tested for the production of interspecies chimeras (Wernery *et al.*, 2010; Liu *et al.*, 2012; Van de Lavoir *et al.*, 2012). Diploid germ cells and gonad conservation methods are complementary to but will never replace semen preservation.

Thus, today the ideal avian reproductive cell conservation programme is grounded on the cryopreservation of semen from relevant sires with known reproductive phenotype and complemented with the preservation of other reproductive germ cells such as primordial germ cells (PGCs), gonadal stem cells, or immature oocytes in ovarian tissues.

National programmes of reproductive biobanking

Advances in cryopreservation technology for poultry semen, diploid germ cells and reproductive tissues have resulted in the emergence of cryobanking, which is now being developed in an increasing number of countries as a method to protect avian genetic resources (reviewed by Blackburn, 2006; Woelders *et al.*, 2006; Blesbois, 2007, 2011, 2012). Many countries have developed a germplasm conservation programme for mammalian species such as cattle and some of them, such as the USA (NAGP, 2014), Canada (CAGRP, 2014), France (CNF, 2014), the Netherlands (CGR, 2014) and Japan (NIAS, 2014), have included a poultry collection in their programmes. Most of the national poultry collections contain mainly semen and/or PGCs/ovaries from chickens (*Gallus gallus*). All of them are now storing the reproductive cells from other avian species (turkeys in the USA, ducks in the Netherlands, quail in Canada). In France, the National Cryobank of Domestic Animals contains the semen of four different domestic birds species (Table 12.1). An added dimension is now given in some of these programmes by joining two types of collections: collections of diploid DNA for genomic studies and collections of reproductive cells, thus giving added information on the genomics characteristics and increasing the potential utilization for reproduction of the resources stored in cryobanks (i.e. CRB Anim, 2014).

Semen cryopreservation in birds, efficiency and limits

Efficient methods of bird semen cryopreservation must take into account the general features of spermatozoa cell biology common to all amniotes with internal fertilization, such as the physicochemical parameters compatible with cell survival, cell energetic metabolism and cell membrane plasticity needed for osmotic and temperature challenges (reviewed in Blesbois, 2012). An added challenge in birds is the long-term storage of spermatozoa in the sperm storage tubules (SST) of the female utero-vaginal junction prior to fertilization. Avian spermatozoa previously cryopreserved must still undergo the normal processes of selection and prolonged storage in the SSTs and every small defect occurring during the *in vitro* processing will be exacerbated during the subsequent *in vivo* storage in the SSTs. Thus, cryopreserved spermatozoa reintroduced into the female oviduct may form uncompetitive populations that may be rapidly eliminated from the SSTs. Different methods have been developed to freeze efficiently bird semen, using mainly intracellular cryoprotectants such as glycerol, dimethyl acetamide (DMA) or dimethyl formamide (DMF). These methods were first developed in the chicken and are now being extended to a number of poultry species. With species-specific zootechnical adaptations, semen freezing has been successfully used in chicken, geese, Pekin and muscovy ducks (reviewed by Blesbois, 2011) and more recently guinea fowl (Seigneurin *et al.*, 2013; Váradi *et al.*, 2013). Mean successful fertility rates are 50–60%, but may reach 90% in the most fertile strains/lines (Table 12.2). Of these poultry species, the turkey, in particular, has proven a difficult species as the fertility levels obtained with cryopreserved semen are very low in this species (3–35% fertility, Long *et al.*, 2014). Taking as models the methods developed in domestic birds (diluent, cryoprotectant, freezing and thawing rate), sperm cryopreservation is also now under development in a small number of wild species including wild pheasants, partridges and vultures, although fertility results for these species are still lacking (Saint Jalme *et al.*, 2003; Madeddu *et al.*, 2009, 2010).

Table 12.1. Semen stocks in the French Avian Germplasm Cryobank in 2013.

Cryobank stocks	No. lines or breed	No. donors	No. doses
Gallus gallus	31	659	23,201
Muscovy ducks	9	325	1,001
Pekin duck	5	155	1,164
Lander geese	1	17	367

Semen cryopreservation is thus a very important tool for the management of genetic diffusion and genetic diversity. In other areas such as multiplication of flock size, the extensive use of semen freezing for current reproduction practice remains unrealistic in most bird species, partly due to the high cost of employing frozen compared to fresh semen and partly due to the difficulty of obtaining freeze/thaw avian semen without the significant loss of fertilizing potential. To overcome this problem, impairments in fertility may be partly compensated by increasing the frequency of insemination and the number of spermatozoa deposited at each insemination (Table 12.2). Under optimal conditions, high intra-line and inter-line (Table 12.3) variability still exists, mainly dependent on the initial semen quality and the reproductive status of the females (Blesbois *et al.*, 2007, 2008). However, after taking these limitations into account, we have demonstrated that even breeds with very low reproduction potential can be managed by sperm cryopreservation if appropriate females for the F1 production are used (Table 12.3). The use of semen cryopreservation by poultry primary breeder companies (PPBCs) in order to maintain genetic diversity potential is still being debated by the international community. It has been discarded by some of them

Table 12.2. Parameters of semen cryopreservation, insemination and mean reproductive performances depending on species (from Blesbois and Labbé, 2003; Massip *et al.*, 2004; Seigneurin and Blesbois, 2010; Blesbois, 2012; Seigneurin *et al.*, 2013).

	Chicken	Muscovy duck	Pekin duck	Goose	Guinea fowl
Dilution rate (% semen)	25–40	65	35	70	35
Cryoprotectant	Glycerol (11%)	DMA (6%)	DMA (6%)	DMA (6%)	DMF (6%)
Rate of freezing	7°C/min	60	60	60	30
AI dose ($\times 10^6$ sperm)	400	300	300	40	200
AI frequency (per week)	2	2	2	3	2
Mean fertility (%)[a]	60	50	60	60	50
Highest fertility (%)	90	80	90	90	70
Mean number of chicks hatched per frozen ejaculate	2	1.5	1.5	2	1.5

[a]Fertility: percentage of fertile/incubated eggs

Table 12.3. Examples of fertility obtained with cryopreserved semen from chicken (Blesbois *et al.*, 2007).

Genetic origin	Homologous females	Commercial females
Gauloise dorée breed	21%[b]	39%[a]
R+ experimental line (high heat production line)	14%[c]	25%[b]
B4 experimental line (sub-fertile line with specific histocompatibility haplotype)	7%[d]	43%[a]

[a, b, c, d]represent significant differences (P≤0.05)

due to economic constraints (Fulton, 2006) but is currently used by other companies through private confidential programmes or through combinations of private/national programmes (e.g. CNF, 2014).

In conclusion, semen cryopreservation allows the development of 'reservoirs' of genetic diversity able to be easily reintroduced in living poultry populations when needed. The equilibrium between the costs and benefits of maintaining genetic diversity through such *in vitro* methods is sometimes difficult to maintain by single PPBCs, and may necessitate joint actions between private companies and national programmes of genetic conservation. Finally, the methodologies developed in this area are still in progress and further improvements are expected in the future.

AVIAN STEM CELL BIOBANKS

Poultry breeds can be preserved through the cryopreservation of semen (Petitte, 2006; Blesbois *et al.*, 2007). However, as we have seen, there are limitations with semen cryopreservation for some chicken breeds and the cost for regenerating a genetically diverse pure line of poultry from cryopreserved semen would be expensive and time consuming as generations of backcrosses would be needed to recover the breed genotype from semen alone (Blesbois *et al.*, 2007). Semen cryopreservation is also only applicable for conservation of the male sex chromosome. In birds the female is the heterogametic sex and contains a W sex chromosome. Cryopreservation of early embryos and oocytes would make it possible to retain the W chromosome, but such techniques cannot be used because of the characteristics of the avian egg (Blesbois *et al.*, 2007; Santiago-Moreno *et al.*, 2011). Recently, advances in avian embryo micro-manipulation as well as improved stem cell culture techniques, particularly in the culture of PGCs, offer a novel stem cell platform to preserve the full genetic complement of chicken breeds.

Stem cells for avian conservation

Stem cells are undifferentiated cells with the capacity to self-renew and intrinsically having the ability to differentiate into one or more different cell lineages (Wray *et al.*, 2010). For cryopreservation of the germplasm of an organism to be successful, a ready source of stem cells containing both male and female sex chromosomes is needed, the cells need to expand in culture, withstand cryopreservation and efficiently form functional gametes as measured by effective germ line transmission. The reproductive and developmental biology of birds is less well documented than in mammals. However, the accessibility to the avian embryo within the laid egg and the route of migration of cells of the germ cell lineage do, however, facilitate germ cell isolation and manipulation in birds in comparison to mammalian counterparts.

Blastodermal stem cells

The laid chicken egg comprises a disc of around 50,000 cells lying on the surface of the yolk. This disc is the blastoderm, containing cells that will form the embryo proper and the extra-embryonic tissues. At this time, germ cells are contained within the disc and after harvesting the disc and dispersing the cells in culture, germ cells will begin to become apparent after 1–2 weeks (Petitte, 2006; Nakamura *et al.*, 2013). Blastodermal cells can be cryopreserved although the lipid present at these early stages hinders the efficiency of cryopreservation (Patakiné *et al.*, 2012). When injected back into embryos, these cells will form chimeras and can produce offspring. Studies have shown that it is possible to inject fresh and frozen blastodermal cells back into un-incubated embryos to obtain both somatic and germ-line chimeras sparking interest in the use of blastodermal cells for species reconstitution (Petitte *et al.*, 1990; Thoraval *et al.*, 1994; Kino *et al.*, 1997). The limited cell numbers present at this stage makes the expansion of these cells in culture necessary for practical use in cryopreservation.

Avian embryonic stem cells

It is possible to culture early blastodermal cells from chicken, quail and ducks in a complex medium containing animal sera. In culture, these cells will express the pluripotency factors similar to mouse embryonic stem (ES) cells and can loosely be referred to as avian ES cells (Pain *et al.*, 1996; van de Lavoir *et al.*, 2006a). ES cells can form all tissue types when injected into chicken embryos. However, with prolonged cell culture these 'ES' cells lose the ability to differentiate into germ cells and generate progeny when introduced into chimeras. This result renders avian ES cells unsuitable for breed preservation at this time.

Avian germ cells

The germ cell lineage in birds is different from mammals and is thought to format the earliest stages of embryonic development. This process is thought to have aided the rapid evolution and speciation of birds as genes were found to evolve more rapidly in those species containing a 'pre-formed' germplasm (Evans *et al.*, 2014). The selection of production traits that impact reproduction, primarily egg production and signalling pathways involved in body growth and differentiation, may have benefited from this uncoupling of germ cell formation and later developmental processes. It is believed that this early segregation of the germ cell lineage in the blastoderm is the reason chicken ES cells do not form germline chimeras.

The migration of the avian germ cell lineage begins at gastrulation and the migrating PGCs enter the extra-embryonic vasculature system at day 2 of incubation. It is possible to isolate PGCs directly from the blood of both male and female embryos (~100 PGCs) using a glass needle, cryopreserve the cells and inject them directly back into a same-sex host embryo at a later date to produce

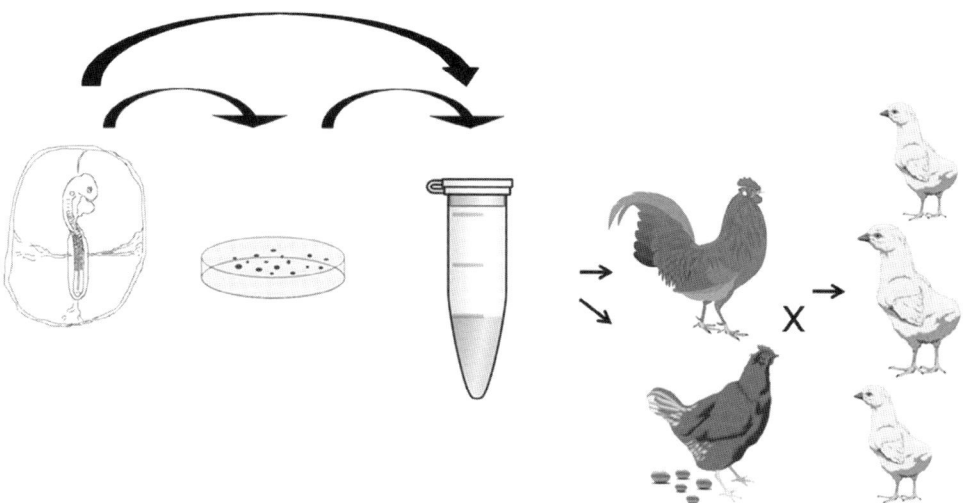

Fig. 12.1. Chicken-breed reconstitution from cryopreserved PGCs. PGCs can be isolated from the blood of early chicken embryos and either cryopreserved or expanded in culture before cryopreservation. The PGCs can subsequently be injected into host embryos. The host embryos are hatched and the adult host birds are crossed. A proportion of the resulting offspring will derive from gametes generated by the introduced germ cells.

chickens containing the donor germ cells. These host chickens can then be bred to produce offspring deriving from the introduced germ cells (Fig. 12.1) (Nakamura *et al.*, 2010). PGCs can also be purified and cryopreserved from the blood of other poultry species such as turkey (Wade *et al.*, 2014). Thus, it is possible to use avian PGCs to create a diploid reproductive cell biobank. Another exciting prospect is the cryopreservation of PGCs from one bird species, followed by xenotransplantation into surrogate host birds of an alternative bird species. The host birds can subsequently be bred to produce offspring derived from the donor species. Recently, it has been shown that male PGCs could be isolated from one bird species and re-injected back into the vasculature system of the embryo of another bird species (Kang *et al.*, 2008; Liu *et al.*, 2012; van de Lavoir *et al.*, 2012). These male chimeric birds can be used to fertilize females from the original donor species and produce purebred offspring of the original species, albeit at low transmission frequencies.

As competition occurs between the reintroduced germ cells and the endogenous germ cells, the use of a host that is depleted of germ cells will produce higher germline transmission rates. Busulphan, a drug commonly used in cancer treatment, and gamma irradiation of the eggs have been shown to significantly reduce the number of endogenous germ cells (Choi *et al.*, 2004; Macdonald *et al.*, 2010). An alternative method is to use transgenic technology to generate a sterile host chicken as has been done in zebrafish (Wong *et al.*, 2011). The generation of a sterile host would be preferential to chemical treatment; however, the use of transgenic technology in germ cell biobanks may not be publically accepted.

Development of a culture medium for PGC propagation

Methods have been developed to purify PGCs from embryonic blood using centrifugation techniques (Zhao and Kuwana, 2003), immuno-magnetic cell separation (Ono and Machida, 1999) and fluorescence activated cell sorting (Mozdziak *et al.*, 2005; Moore *et al.*, 2006). An alternative strategy is to expand the number of germ cells through culturing before cryopreservation and subsequent injection into host embryos (Fig. 12.1). Chicken PGCs can be propagated in a complex growth medium and feeder cells (van de Lavoir *et al.*, 2006b). In this system chicken PGCs proliferate in suspension and can be expanded to several million cells whilst still maintaining germ-line competence (Fig. 12.2). These media conditions support the culture of chicken PGCs, but derivation rates still remain low with efficiencies of 10% to 22% for male lines and generally lower female line derivations (van de Lavoir *et al.*, 2006b; Macdonald *et al.*, 2010). The ability to

(a)

(b)

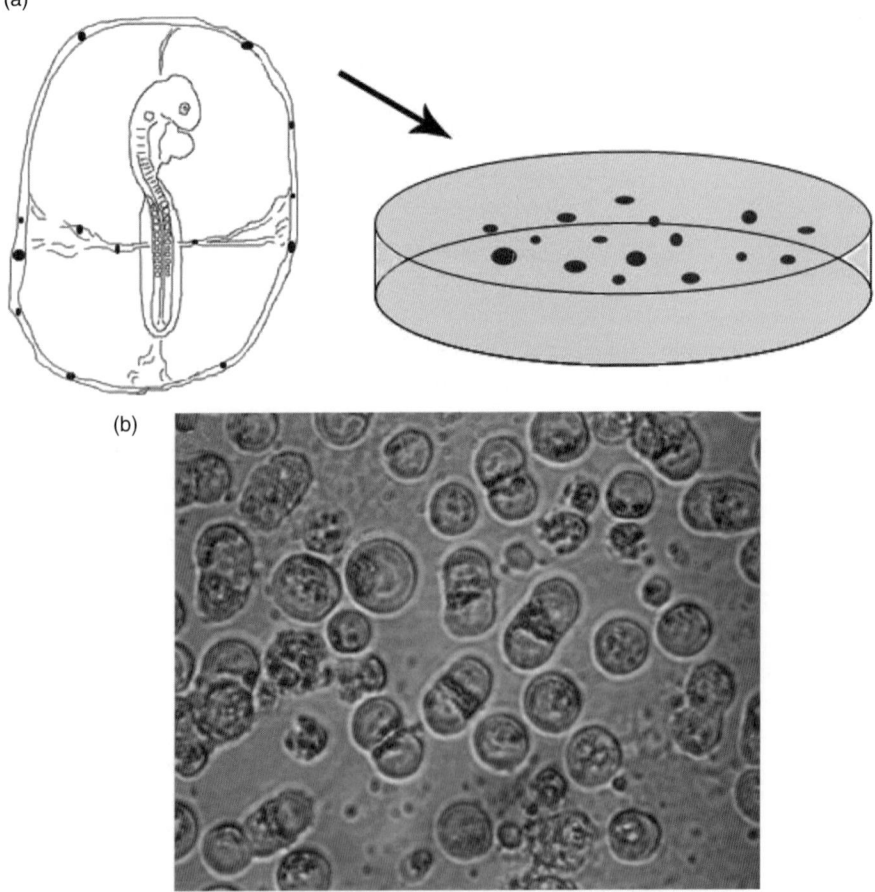

Fig. 12.2. PGCs from chicken embryos. (a) Blood from early embryos can be placed in culture and primordial germ cell cultures can be derived. (b) Micrograph of primordial germ cells in culture (Macdonald *et al.*, 2010).

culture PGCs from other species has also proven to be problematic. The development of new media conditions to culture PGCs with a higher derivation efficiency as well as supporting the growth of PGCs from other species is critical before this strategy can be utilized to create viable frozen stem cell biobanks to restore standard breeds. In comparison to semen cryopreservation, producing an offspring from a PGC is more time-consuming. Frozen semen can produce a hatched chick in just 3 weeks. A minimum of 5 months are needed to produce an offspring derived from cultured PGCs (3 weeks for injected eggs to hatch, 4 months for offspring to reach sexual maturity and 3 weeks to produce offspring from the chicken chimeras). However, the offspring from this cross could potentially contain the entire genome for the cryopreserved breed, which would eliminate the many generations of backcrossing needed to produce a 'pure' breed from frozen semen (Fig. 12.1). PGC culture, expansion, and freezing of cells are also labour intensive in comparison to semen preservation. Nevertheless, PGC biobanks offer the ability to reconstitute the whole genome of an avian species from cryopreserved material and undoubtedly will have an impact on maintaining and safeguarding avian genetic diversity (Glover, 2012).

GENETIC TOOLS TO MODIFY THE CHICKEN GENOME

Transgenic technologies do not currently have a major impact on the poultry industry due to a lack of genetic tools and the perceived public disapproval of the genetic modification of agricultural animals (Arthur and Albers, 2003). However, the recent development of sophisticated molecular tools to modify the animal genome may alter certain aspects of this debate.

Retroviral vectors have been used for the past 20 years to introduce exogenous DNA into the chicken genome (reviewed in McGrew, 2013). The retroviral particles are injected into early-stage chicken embryos and the proviral genome, containing a transgene, is inserted into the chromosomes of cells of the embryo. Breeding from the founder chicken will produce offspring containing an insertion of the provirus in the genome of every cell. The most highly developed vector system for retroviral transgenesis is lentiviral vectors. Replication-defective lentiviral vectors can efficiently deliver a transgene up to 9 kilobases into the genome of chickens, integrate into permissive regions of chromatin for transgene expression, and not result in transgene expression being silenced after germline transmission (McGrew *et al.*, 2004). Many transgenic chicken lines have been produced using this and other retroviral vector systems.

An alternative DNA-based system of transgenesis is the use of transposable elements for the introduction of exogenous DNA into the chicken genome. Transposable elements are mobile genetic elements present in the genome of all animals. These parasitic elements of selfish DNA 'copy' and paste themselves throughout the genome of the host animal. DNA transposons have been engineered to deliver a transgenic cargo carried within the transposon vector. Transposon vectors have been used recently to deliver transgenes into the genomes of chickens, either into PGCs or by direct injection into embryos (Macdonald *et al.*, 2012; Glover *et al.*, 2013; Tyack *et al.*, 2013). DNA transposons offer some advantages over retroviruses in that the construct does not need to be packaged into viral particles. Similarly to retroviruses, the DNA transgene is delivered randomly to the genome and many transposon vectors carry antibiotic resistance genes.

The use of PGCs offers new avenues for genetic modifications of the chicken genome. Classical gene targeting experiments using gene targeting vectors and homologous recombination have recently been used to produce targeted genetic modifications of the chicken genome (Schusser *et al.*, 2013). This 'knockout' technology will be useful to produce genetically targeted chickens as biological models and for biotechnological applications. The recent development of site-specific nucleases for targeting specific genetic loci in the animal genome has opened vistas for specific genetic changes (Wei *et al.*, 2013). Site-specific nucleases have been used to generate defined single nucleotide change in the genome of many mammalian species. It is expected that similar genetic modifications will someday be possible in poultry. Thus, it may be possible to generate defined genetic modifications to the chicken genome, i.e. a single amino acid change in a protein, whilst leaving no other genetic 'footprint'. As stated above, it is difficult to foresee a public acceptance of genetically edited poultry in commercial production systems. Nevertheless, this technology will be useful to investigate the value of select genetic alleles on reproductive and production traits.

A CASE STUDY: THE DEVELOPMENT OF DISEASE RESISTANCE IN POULTRY THROUGH TRANSGENESIS

In the past, poultry disease outbreaks posed a significant threat to national food security and at times devastated the UK and global poultry sectors (Afonso and Miller, 2013; Van Kerkhove, 2013; Zelník *et al.*, 2013). Current vaccination protocols protect against the most virulent chicken pathogens but a recent concern has been the global spread of avian influenza (AI). The UK became officially free from highly pathogenic AI, according to the rules laid down by the World Organisation for Animal Health; however, it still remains one of the main notifiable diseases in the UK and is endemic in many developing countries (World Organisation for Animal Health, 2008). Due to the similarity in structure between the AI virus and human influenza virus, AI can infect humans as well as birds making AI one of the greatest concerns for animal and public health that has emerged in recent times.

Avian influenza

Since the late 1990s, there has been a sharp rise in the number of outbreaks of AI globally (Chmielewski and Swayne, 2011). The origin of AI can be traced back to East/South-east Asia, which is home to an estimated 6 billion domestic poultry (FAO, 2014). Two of the more highly pathogenic strains of the virus (H5N1 and H7N9) were first reported in 1997 and 2013, respectively, and are still prevalent in 2014 (Chan, 2002; Wiwanitkit, 2013). In birds, infection causes a plethora of symptoms ranging from lack of coordination, decreased egg production and sudden death (Liu *et al.*, 2013). AI can be spread from birds to humans as a result of direct contact with infected birds, such as during home slaughter and plucking of infected poultry (FAO, 2014). To date, the transmission of AI from human to human is a very rare event (Wang *et al.*, 2008). Public health concerns centre on the potential for the virus to mutate or combine with other influenza viruses to a form that could easily spread from person to person causing a worldwide pandemic (De Jong *et al.*, 2000).

Genetic engineering of avian influenza resistance in chickens

The most widely used approaches to protecting domesticated flocks of birds against AI is either by culling infected birds (Europe IFAH, 2014) or through vaccination. Large-scale vaccination programmes in countries with endemic AI combined with 'stamping out' have had some success in reducing outbreaks (FAO, 2011). Current vaccinations are incapable of wiping out the virus completely, meaning that low levels can still spread in vaccinated flocks and extensive vaccination programmes such as that seen in Mexico in 2004 were actually found to ignite viral mutation through antigenic drift (Parry, 2005).

A novel and controversial approach to protecting domesticated flocks in countries with endemic AI is to use genetic engineering. The use of genetic modification (GM) is fraught with controversy, however a recent survey in the USA indicated that 40% of individuals surveyed thought that the production of chickens resistant to AI was a very good reason to produce GM animals (Pew Initiative on Food and Biotechnology, 2006). In 2011, the announcement of the world's first AI-resistant chickens was published (Lyall *et al.*, 2011), suggesting such an approach is theoretically possible.

Generation of disease-resistant chickens and challenge studies

The creation of chickens resistant to AI used a 'decoy approach'; that is, by overexpressing an RNA sequence with identical complementarity to the conserved sequence of the viral genome to which the influenza A viral polymerase binds to initiate viral transcription. The 'decoy' RNA has been shown to inhibit viral replication and packaging (Luo *et al.*, 1997). The 'decoy' molecule expression cassette was introduced into a lentiviral vector, packaged, and microinjected into freshly laid chicken eggs (Lyall *et al.*, 2011). The injected eggs were hatched, the hatched chicks raised to sexual maturity and one transgenic founder cockerel was crossed with hens to produce transgenic offspring containing a single integration of the lentiviral construct encoding the 'decoy' RNA. The resulting transgenic and non-transgenic progeny of this cockerel were used in viral challenge studies. In these studies, transgenic and non-transgenic chickens were directly infected with H5N1 virus and then co-housed with uninfected birds (the 'in-contact' group), which were either transgenic or non-transgenic, and the health of the birds was monitored over the next 11 days. All the directly infected birds, both transgenic and non-transgenic, died post-infection. As expected the 'in contact' group housed with the non-transgenic infected birds succumbed to the virus. Surprisingly, all birds of the in-contact group housed with the transgenic infected chicks remained healthy for the duration of the study (Fig. 12.3).

These results show that there were clear differences in viral transmission and/ or susceptibility after exposure to H5N1 virus. Transgenic chickens did not transmit viral infection to co-housed birds, demonstrating for the first time that transgenic modification can have a major impact on susceptibility and propagation of disease infection in a poultry flock. This study in principle supports the concept of using genetic modification for controlling AI infection in poultry.

Prospects for targeted genetic changes

This study remains the single example of disease resistance in poultry through direct genetic modification of the chicken genome. The novel innovation is that a genetic change was introduced that could not be achieved through animal breeding, with the potential to increase the welfare of both poultry and the poultry farmer. This approach could be applied to investigate disease resistance in other domestic species that are affected by influenza such as ducks, quail and pigs. Utilizing a genetic approach to engineer resistance in an agricultural animal

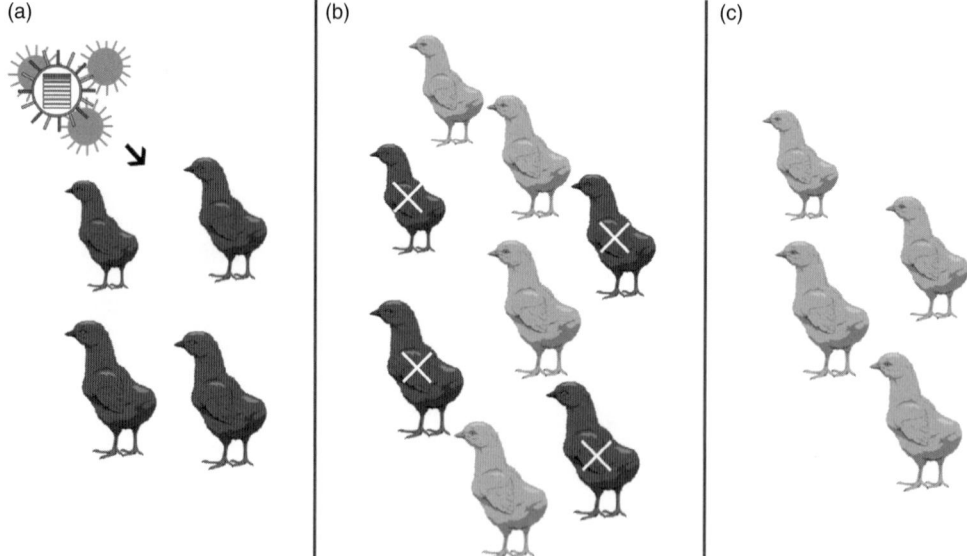

Fig. 12.3. Transgenic chickens with resistance to AI. (a) Transgenic chicks containing a lentiviral transgene encoding a decoy RNA molecule were infected with AI. (b) The chicks were co-housed with brood mates for 10 days. (c) The infected chicks died but the co-housed chicks survived, indicating that the transgenic chicks did not transmit the virus to the co-housed chicks.

obviously has major implications for both consumers and food regulatory policies.

CONCLUSIONS

The development of cryopreservation technology for avian semen and diploid reproductive cells will facilitate efforts to manage avian genetic resources at both local and international levels. National reproductive biobanks containing both diploid and haploid germ cells will help efforts to maintain genetic diversity in poultry flocks and to secure poultry breeds with potential future value to breeding programmes. The development of novel genetic tools for introducing precise genetic changes into the chicken genome is predicted to open new avenues for investigation of gene function during normal growth and development and under disease challenges.

REFERENCES

Afonso, C.L. and Miller, P.J. (2013) Newcastle disease: progress and gaps in the development of vaccines and diagnostic tools. *Developmental Biology (Basel)* 135, 95–106.

Arthur, J.A. and Albers, G.A.A. (2003) Industrial perspective on problems and issues associated with poultry breeding. In: Muir, W.M. and Aggrey, S.E. (eds) *Poultry Genetics, Breeding and Biotechnology*. CAB International, Cambridge, Massachusetts, pp. 1–12.

Blackburn, H.D. (2006) The national animal germplasm program: challenges and opportunities for poultry genetic resources. *Poultry Science* 85, 210–215.

Blesbois, E. (2007) Current status in avian semen cryopreservation. *World Poultry Science Journal* 63, 213–222.

Blesbois, E. (2011) Freezing avian semen. *Avian Biology Research* 4, 44–50.

Blesbois, E. (2012) Biological features of the avian male gamete and their application to biotechnology of conservation. *Journal of Poultry Science* 49, 141–149.

Blesbois, E. and Labbé, C. (2003) Main improvements in semen and embryo cryopreservation for fish and fowl. In: Planchenault, D. (ed.) *Cryopreservation of Animal Genetic Resources in Europe*. BRG, Paris, pp. 55–66.

Blesbois, E., Seigneurin, F., Grasseau, I., Limouzin, C., Besnard, J., Gourichon, D., Coquerelle, G. and Tixier-Boichard, M. (2007) Semen cryopreservation for ex-situ management of genetic diversity in chicken. *Poultry Science* 87, 555–564.

Blesbois, E., Grasseau, I., Seigneurin, F., Mignon-Grasteau, S., Saint Jalme, M. and Mialon-Richard, M.M. (2008) Predictors of semen cryopreservation in chickens. *Theriogenology* 69, 252–261.

CAGRP (2014) Conserving Valuable Canadian Poultry Genetics – You can Help. Available at: http://www.agr.gc.ca/eng/science-and-innovation/science-publications-and-resources/ technical-factsheets/conserving-valuable-canadian-poultry-genetics-you-can-help/?id=1386684173296 (accessed 11 August 2014).

Cao, Y. and Li, D. (2013) Impact of increased demand for animal protein products in Asian countries: implications on global food security. *Animal Frontiers* 3, 48–55.

CGR (2014) Animal genebank collections. Available at: http://www.wageningenur.nl/en/Expertise-Services/Statutory-research-tasks/Centre-for-Genetic-Resources-the-Netherlands-1/Centre-for-Genetic-Resources-the-Netherlands-1/Expertise-areas/Animal-Genetic-Resources/Genebank. htm (accessed 11 August 2014).

Chan, P.K. (2002) Outbreak of avian influenza A(H5N1) virus infection in Hong Kong in 1997. *Clinical Infectious Diseases* 34(Suppl. 2), S58–S64.

Chmielewski, R. and Swayne, D.E. (2011) Avian influenza: public health and food safety concerns. *Annual Review of Food Science and Technology* 2, 37–57.

Choi, Y.J., Ok, D.W., Kwon, D.N., Chung, J.I., Kim, H.C., Yeo, S.M., Kim, T., Seo, H.G. and Kim, J.H. (2004) Murine male germ cell apoptosis induced by busulfan treatment correlates with loss of c-kit-expression in a Fas/FasL- and p53-independent manner. *FEBS Letters* 575, 41–51.

Chowdhury, V.S., Sultana, H. and Furuse, M. (2014) International perspectives on impacts of reproductive technologies for world food production in Asia associated with poultry production. *Advances in Experiment Medicine and Biology* 752, 229–237.

CNF (2014) Available at: http://www.cryobanque.org (accessed 11 August 2014).

CRB Anim (2014) Available at: www.crb-anim.fr (accessed 11 August 2014).

De Jong, J.C., Rimmelzwaan, G.F., Fouchier, R.A. and Osterhaus, A.D. (2000) Influenza virus: a master of metamorphosis. *Journal of Infection* 40, 218–228.

Emmerson, D. (2003) Breeding objectives and selection strategies for broiler production. In: Muir, W.M. and Aggrey, S.E. (eds) *Poultry Genetics, Breeding and Biotechnology*. CAB International, Cambridge, Massachusetts, pp. 113–126.

Environment Audit Committee (2011) *11th Report of Session 2010-2012, Volume II 'Sustainable Food'*; written evidence by the British Poultry Council. The Stationery Office, London.

Europe IFAH (2014) Food producing animals – disease outbreaks. Available at: http://www.ifaheurope.org/food-producing-animals/disease-outbreaks/avianflu.html (accessed 5 May 2014).

Evans, T., Wade, C.M., Chapman, F.A., Johnson, A.D. and Loose, M. (2014) Acquisition of germ plasm accelerates vertebrate evolution. *Science* 344, 200–204.

FAO (2006) *World Agriculture: Towards 2030/2050 Interim Report*. FAO, Rome.

FAO (2011) Approaches to controlling, preventing and eliminating H5N1 highly pathogenic avian influenza in endemic countries. *Animal Production and Health Paper* 171. FAO, Rome.

FAO (2012) FAOStat for 2010. Available at: http://faostat.fao.org (accessed 15 May 2014).

FAO (2014) Avian influenza. Available at: http://www.fao.org/avianflu/en/qanda.html (accessed 1 May 2014).

Fulton, J.E. (2006) Avian genetic stock preservation: an industry perspective. *Poultry Science* 85, 227–231.

Glover, J.D. (2012) Primordial germ cell technologies for avian germplasm cryopreservation and investigating germ cell development. *The Journal of Poultry Science* 49(3), 155–162.

Glover, J.D., Taylor, L., Sherman, A., Zeiger-Poli, C., Sang, H.M. and McGrew, M.J. (2013) A novel piggyBac transposon inducible expression system identifies a role for AKT signalling in primordial germ cell migration. *PloS One* 8, e77222.

Kang, S.J., Choi, J.W., Kim, S.Y., Park, K.J., Kim, T.M., Lee, Y.M., Kim, H., Lim, J.M. and Han, J.Y. (2008) Reproduction of wild birds via interspecies germ cell transplantation. *Biology of Reproduction* 79, 931–937.

Kino, K., Pain, B., Leibo, S.P., Cochran, M., Clark, M.E. and Etches, R.J. (1997) Production of chicken chimeras from injection of frozen-thawed blastodermal cells. *Poultry Science* 76, 753–760.

Leroy, G., Kayang, B.B., Youssao, I.A., Yapi-Gnaoré, C.V., Osei-Amponsah, R., Loukou, N.E., Fotsa, J.C., Benabdeljelil, K., Bed'hom, B., Tixier-Boichard, M. and Rognon, X. (2012) Gene diversity, agroecological structure and introgression patterns among village chicken populations across North, West and Central Africa. *BMC Genetics* 7, 13–34.

Liu, C., Khazanehdari, K.A., Bashar, V., Sleem, S., Kinne, J., Wernery, U. and Chang, I.K. (2012) Production of chicken progeny (*Gallus gallus* domesticus) from interspecies germline chimeric duck (*Anas domesticus*) by primordial germ cell transfer. *Biology of Reproduction* 86, 101–110.

Liu, Q., Liu, D. and Yang, Z. (2013) Characteristics of human infection with avian influenza viruses and development of new antiviral agents. *Acta Pharmacologica Sinica* 34, 1257–1269.

Long, J.A., Purdy, P.H., Zuiberg, K., Hielstra, Q.J., Velleman, S.G. and Woelders, H. (2014) Cryopreservation of turkey semen: effect of breeding line and freezing method on post-thaw sperm quality, fertilization, and hatching. *Cryobiology* 68(3), 371–378.

Luo, G., Danetz, S. and Krystal, M. (1997) Inhibition of influenza viral polymerases by minimal viral RNA decoys. *Journal of General Virology* 78, 2329–2333.

Lyall, J., Irvine, R.M., Sherman, A., McKinley, T.J., Núñez, A., Purdie, A., Outtrim, L., Brown, I.H., Rolleston-Smith, G., Sang, H. and Tiley, L. (2011) Suppression of avian influenza transmission in genetically modified chickens. *Science* 331, 223–226.

Macdonald, J., Glover, J.D., Taylor, L., Sang, H.M. and McGrew, M.J. (2010) Characterisation and germline transmission of cultured avian primordial germ cells. *PloS One* 5, e15518.

Macdonald, J., Taylor, L., Sherman, A., Kawakami, K., Takahashi, Y., Sang, H.M. and McGrew, M.J. (2012) Efficient transmission of primordial germ cells using piggyback and Tol2 transposons. *Proceeding of the National Academy of Sciences USA* 109, 9337–9341.

Madeddu, M., Berlinguer, F., Ledda, M., Leoni, G.G., Satta, V., Succu, S., Rotta, A., Pascui, V., Zinellu, A., Muzzeddu, M., Carru, C. and Naitana, S. (2009) Ejaculate collection efficiency and post-thaw semen quality in wild-caught Griffon vultures from the Sardinian population. *Reproductive Biology and Endocrinology* 7, 18–22.

Madeddu, M., Berlinguer, F., Ledda, M., Pascui, V., Succu, S., Satta, V., Leoni, G.G., Zinellu, A., Muzzeddu, M., Carru, C. and Naitana, S. (2010) Differences in semen freezability and intracellular ATP content between the rooster and the Barbary partridge. *Theriogenology* 74, 1010–1018.

Massip, A., Leibo, S.P. and Blesbois, E. (2004) Cryobiology and the breeding of domestic animals. In: Benson, E., Fuller, B. and Lane, N. (eds) *Life in the Frozen State*. Taylor & Francis, London, pp. 371–392.

McGrew, M. (2013) Avian specific transgenesis. In: Meyers, R. (ed.) *Encyclopedia of Sustainability Science and Technology*. Springer, Berlin.

McGrew, M.J., Sherman, A.S., Ellard, F.M., Lillico, S.G., Gilhooley, H.J., Mitrophanous, K.A., Kingsman, A.J. and Sang, H. (2004) Efficient production of germline transgenic chickens using lentiviral vectors. *EMBO Reports* 5, 728–733.

Moore, D.T., Purdy, P.H. and Blackburn, H.D. (2006) A method for cryopreserving chicken primordial germ cells. *Poultry Science* 85, 1784–1790.

Mozdziak, P.E., Angerman-Stewart, J., Rushton, B., Pardue, S.L. and Petitte, J.N. (2005) Isolation of chicken primordial germ cells using fluorescence-activated cell sorting. *Poultry Science* 84, 594–600.

Muir, W.M., Wong, G.K., Zhang, Y., Wang, J., Groenen, M.A., Crooijmans, R.P., Megens, H.J., Zhang, H., Okimoto, R., Vereijken, A., Jungerius, A., Albers, G.A., Lawley, C.T., Delany, M.E., MacEachern, S. and Cheng, H.H. (2008) Genome-wide assessment of worldwide chicken SNP genetic diversity indicates significant absence of rare alleles in commercial breeds. *Proceedings of the National Academy of Sciences USA* 105, 17312–17317.

NAGP (2014) National animal germplasm program. Available at: http://nrrc.ars.usda.gov/A-GRIN/tax_inv_drilldown_page?record_source = US (accessed 11 August 2014).

Nakamura, Y.F., Usui, D., Miyahara, T., Mori, T., Ono, K., Takeda, K., Nirasawa, H., Kagami, Y. and Tagami, T. (2010) Efficient system for preservation and regeneration of genetic resources in chicken: concurrent storage of primordial germ cells and live animals from early embryos of a rare indigenous fowl (Gifujidori). *Reproduction Fertility and Development* 22, 1237–1246.

Nakamura, Y., Kagami, H. and Tagami, T. (2013) Development, differentiation and manipulation of chicken germ cells. *Development Growth and Differentiation* 55, 20–40.

NIAS (2014) NIAS genebank. Available at: http://www.gene.affrc.go.jp/about-animal_en.php (accessed 11 August 2014).

Ono, T. and Machida, Y. (1999) Immunomagnetic purification of viable primordial germ cells of Japanese quail (*Coturnix japonica*). *Comparative Biochemistry and Physiology – Part A: Molecular & Integrative Physiology* 122, 255–259.

Pain, B., Clark, M.E., Shen, M., Nakazawa, H., Sakurai, M., Samarut, J. and Etches, R.J. (1996) Long-term *in vitro* culture and characterisation of avian embryonic stem cells with multiple morphogenetic potentialities. *Development* 122(8), 2339–2348.

Parry, J. (2005) Vaccinating poultry against avian flu is contributing to spread. *British Medical Journal* 331, 1223.

Patakiné, V.E., Horváth, G., Sztán, N., Váradi, E. and Barna, J. (2012) Vitrification of early avian blastodermal cells with a new type of cryocontainer. *Acta Veterinaria Hungarica* 60, 501–509.

Petitte, J.N. (2006) Avian germplasm preservation: embryonic stem cells or primordial germ cells? *Poultry Science* 85, 237–242.

Petitte, J.N., Clark, M.E., Liu, G., Verrinder Gibbins, A.M. and Etches, R.J. (1990) Production of somatic and germline chimeras in the chicken by transfer of early blastodermal cells. *Development* 108, 185–189.

Pew Initiative on Food and Biotechnology (2006) Available at: http://www.pewtrusts.org/uploadedFiles/wwwpewtrustsorg/Public_Opinion/Food_and_Biotechnology/2006summary.pdf (accessed 7 May 2014).

Pham, M.H., Berthouly-Salazar, C., Tran, X.H., Chang, W.H., Crooijmans, R.P., Lin, D.Y., Hoang, V.T., Lee, Y.P., Tixier-Boichard, M. and Chen, C.F. (2013) Genetic diversity of Vietnamese domestic chicken populations as decision-making support for conservation strategies. *Animal Genetics* 44, 509–521.

Rare Breeds Survival Trust (2014) Available at: http://www.rbst.org.uk (accessed 28 April 2014).

Renaudeau, D., Collin, A., Yahav, S., de Basilio, V., Gourdine, J.L. and Collier, R.J. (2012) Adaptation to hot climate and strategies to alleviate heat stress in livestock production. *Animal* 6, 707–728.

Saint Jalme, M., Lecoq, R., Seigneurin, F., Blesbois, E. and Plouzeau, E. (2003) Cryopreservation of semen from endangered pheasants: the first step towards cryobank for endangered avian species. *Theriogenology* 59, 875–888.

Santiago-Moreno, J., Castaño, C., Toledano-Díaz, A., Coloma, M.A., López-Sebastián, A., Prieto, M.T. and Campo, J.L. (2011) Semen cryopreservation for the creation of a Spanish poultry breeds cryobank: optimization of freezing rate and equilibration time. *Poultry Science* 90, 2047–2053.

Schusser, B., Collarini, E.J., Yi, H., Izquierdo, S.M., Fesler, J., Pedersen, D., Klasing, K.C., Kaspers, B., Harriman, W.D., van de Lavoir, M.C., Etches, R.J. and Leighton, P.A. (2013) Immunoglobulin knockout chickens via efficient homologous recombination in primordial germ cells. *Proceedings of the National Academy of Sciences USA* 110, 20170–20175.

Seigneurin, F. and Blesbois, E. (2010) Update on semen cryopreservation methods in poultry species. Proceedings of the 13th European Poultry Conference (EPC 2010). French Branch of World's Poultry Science Association, Tours (FRA). *World's Poultry Science Journal* 66(Suppl.), 172.

Seigneurin, F., Grasseau, I., Chapuis, F. and Blesbois, E. (2013) An efficient method of guinea fowl sperm cryopreservation. *Poultry Science* 92, 2988–2996.

Silversides, F.G., Robertson, M.C. and Liu, J. (2013) Cryoconservation of avian gonads in Canada. *Poultry Science* 92, 2613–2617.

Tajima, A. (2013) Conservation of avian genetic diversity. *Journal of Poultry Science* 50, 1–8.

Thoraval, P., Lasserre, F., Coudert, F. and Dambrine, G. (1994) Somatic and germline chicken chimeras obtained from brown and white Leghorns by transfer of early blastodermal cells. *Poultry Science* 73, 1897–1905.

Tyack, S.G., Jenkins, K.A., O'Neil, T.E., Wise, T.G., Morris, K.R., Bruce, M.P., McLeod, S., Wade, A.J., McKay, J., Moore, R.J., Schat, K.A., Lowenthal, J.W. and Doran, T.J. (2013) A new method for producing transgenic birds via direct in vivo transfection of primordial germ cells. *Transgenic Research* 22, 1257–1264.

Van de Lavoir, M.C., Mather-Love, C., Leighton, P., Diamond, J.H., Heyer, B.S., Roberts, R., Zhu, L., Winters-Digiacinto, P., Kerchner, A., Gessaro, T., Swanberg, S., Delany, M.E. and Etches, R.J. (2006a) High-grade transgenic somatic chimeras from chicken embryonic stem cells. *Mechanisms of Development* 123, 31–41.

Van de Lavoir, M.C., Diamond, J.H., Leighton, P.A., Mather-Love, C., Heyer, B.S., Bradshaw, R., Kerchner, A., Hooi, L.T., Gessaro, T.M., Swanberg, S.E., Delany, M.E. and Etches, R.J. (2006b) Germline transmission of genetically modified primordial germ cells. *Nature* 441, 766–769.

Van de Lavoir, M.C., Collarini, E.J., Leighton, P.A., Fesler, J., Lu, D.R., Harriman, W.D., Thiyagasundaram, T.S. and Etches, J.R. (2012) Interspecific germline transmission of cultured primordial germ cells. *PloS One* 7, e35664.

Van Kerkhove, M.D. (2013) Brief literature review for the WHO global influenza research agenda – highly pathogenic avian influenza H5N1 risk in humans. *Influenza and Other Respiratory Viruses* 7, 26–33.

Váradi, É., Végi, B., Liptói, K. and Barna, J. (2013) Methods for cryopreservation of guinea fowl sperm. *PloS One* 8, e62759.

Wade, A.J., French, N.A. and Ireland, G.W. (2014) The potential for archiving and reconstituting valuable strains of turkey (*Meleagris gallopavo*) using primordial germ cells. *Poultry Science* 93, 799–809.

Wang, H., Feng, Z., Shu, Y., Yu, H., Zhou, L., Zu, R., Huai, Y., Dong, J., Bao, C., Wen, L., Wang, H., Yang, P., *et al.* (2008) Probable limited person-to-person transmission of highly pathogenic avian influenza A (H5N1) virus in China. *Lancet* 371, 1427–1434.

Wei, C., Liu, J., Yu, Z., Zhang, B., Gao, G. and Jiao, R. (2013) TALEN or Cas9 – rapid, efficient and specific choices for genome modifications. *Journal of Genetics and Genomics* 40, 281–289.

Wernery, U., Liu, C., Guerineche, Z., Khazanehdari, K.A., Saleem, S., Kinne, J., Wernery, R., Griffin, D. and Chang, I. (2010) Primordial germ cell-mediated chimera technology produces viable pure-line Houbara Bustard offspring: potential for repopulating an endangered species. *PloS One* 5, e15824.

Wilkinson, S., Wiener, P., Teverson, D., Haley, C.S. and Hocking, P.M. (2012) Characterization of the genetic diversity, structure and admixture of British chicken breeds. *Animal Genetics* 43, 552–563.

Wiwanitkit, V. (2013) H7N9 influenza: the emerging infectious disease. *North American Journal of Medical Sciences* 5, 395–398.

Woelders, H., Zuidberg, C.A. and Hiemstra, S.J. (2006) Animal genetic resources conservation in the Netherlands and Europe: poultry perspective. *Poultry Science* 85, 216–222.

Wong, T.T., Saito, T., Crodian, J. and Collodi, P. (2011) Zebrafish germline chimeras produced by transplantation of ovarian germ cells into sterile host larvae. *Biology of Reproduction* 84, 1190–1197.

World Organisation for Animal Health (2008) Update on highly pathogenic avian influenza in animals (type h5 and h7). Available at: http://www.oie.int/animal-health-in-the-world/update-on-avian-influenza/2008 (accessed 28 April 2014).

Wray, J., Kalkan, T. and Smith, A.G. (2010) The ground state of pluripotency. *Biochemical Society Transactions* 38, 1027–1032.

Zelník, V., Lapuníková, B. and Kúdelová, M. (2013) Marek's disease: rapid progress in research with unclear biological implementations. *Acta Virology* 57, 265–270.

Zhao, D.F. and Kuwana, T. (2003) Purification of avian circulating primordial germ cells by nycodenz density gradient centrifugation. *British Poultry Science* 44, 30–35.

PART VII
Environmental Sustainability

CHAPTER 13
Reducing the Environmental Impact of Poultry Production

Adrian Williams[1]* and David Speller[2]

[1]Cranfield University, Bedford, UK; [2]Applied Group, Derbyshire, UK

INTRODUCTION

Mainstream poultry production is highly optimized in the context of animal production. Historic analyses have shown substantial improvements over 20 to 50 years (e.g. Jones *et al*., 2008; Pelletier *et al*., 2014). These have included genetic, managerial and nutritional improvements as well as those in the wider feed production system.

Despite this, there are opportunities for improving environmental performance through reduced mortalities and better health, using heat exchangers to save direct energy use, upgrading manure to produce energy, changing diets to reduce the inclusion of high impact feeds, such as soybean meal.

The opportunities vary across the world, given that the starting points are varied. Variations result from different feed availability (and locations and methods of production), electricity and other fuel intensity, genetic stock, health status, building age and management quality.

The main requirements for improving environmental performance relate to the embedded energy and greenhouse gas (GHG) emissions in feeds. Making better use of feeds has a large impact on both these burdens. Reducing ammonia emissions and hence both acidification and eutrophication potentials is closely related to minimizing N excretion and thus requires optimized matching of the amino acid profile in feeds against nutritional needs.

The constraints that prevent immediate progress are disparate. Obtaining low impact feed mixes and optimizing protein needs is limited by the economic availability of alternatives. One illusion to avoid is that of using by-products that are generally considered to have low environmental burdens. If the demand for these increases substantially, their supply must be increased and will almost inevitably increase their production burdens. A clear constraint in balancing protein

*Corresponding author: adrian.williams@cranfield.ac.uk

supply and need is the use of pure amino acids in diets with poorer amino acid balances than can be provided by soybean meal. Pure amino acids are GHG intense to produce and the benefits of having sources with much lower impacts are great. This is a challenge for that industry that could have widespread bene-fits for monogastric animal production. Factors that may prevent rapid uptake of engineering solutions, such as air heat exchangers, anaerobic digestion, litter combustion or belt-drying litter are generally economic rather than environmen-tal, although litter burning needs both careful regulation and control to avoid the production of highly undesirable emissions of compounds such as dioxins.

To date, almost all analyses have been on the main bird in global poultry production, the chicken. The baseline, let alone opportunities for improvement in other species are barely known.

Poultry production is globally very large. World meat production was 124×10^6 t (Mt) in 2012 and egg production was 101 Mt, which equated to 1.9×10^{12} eggs (FAOSTAT, 2014). Chickens dominate world production, with 85% of meat production and about 90% of egg production (Table 13.1). This is equivalent to 5.2×10^9 eggs/day.

Production systems vary widely from backyard poultry, usually primarily laying hens, through to large-scale fully housed systems with highly specialized birds producing either eggs or meat (MacLeod *et al.*, 2013). Almost all produc-tion in the developed world tends towards the latter model. With such a large level of production, some impacts on the environment are inevitable. This paper reviews the nature and extent of these together with methods to reduce impacts.

ENVIRONMENTAL IMPACTS OF POULTRY PRODUCTION

There are two main types of environmental impacts, which relate to emission to the environment and consumption of resources, particularly non-renewable ones. Obvious emissions include ammonia from housing or manure management

Table 13.1. World poultry production (FAOSTAT, 2014).

	$\times 10^6$ t	Proportion by species
Meat, chicken	106	85%
Meat, duck	7.3	6%
Meat, turkey	5.6	5%
Meat, goose and guinea fowl	5.5	4%
Total poultry meat	124	
Eggs, hen, in shell	91	90%
Eggs, other bird, in shell	10	10%
Total eggs	101	
	$\times 10^9$ eggs	
Eggs, hen, in shell	1746	92%
Eggs, other bird, in shell	154	8%
Eggs	1900	

and less obvious ones include nitrous oxide (N_2O) or carbon dioxide (CO_2) from feed-crop production and associated land use change (LUC) or pullet rearing and CO_2 from feed processing or ventilating houses.

Resources are indeed used at every stage in poultry production from feed-crop production and building manufacturing to rearing replacements and the production phases themselves. It is also possible to consider impacts beyond the farm gate and so to include processing, such as egg packing or slaughtering, and further through to retail and consumption in homes or the food service sector. Hence, defining the system boundaries is critical when considering the environmental impacts of poultry production. Different analysts have their own interests and many studies go to the farm gate with a minority going to the consumption stage.

The relevance of considering the whole supply chain both upstream and downstream of the farm is that changes at the farm level may change feed needs and hence upstream activities while changes downstream, e.g. carcass yield or egg fragility, affect the total quantity of farm production that is needed to meet a given consumption demand. Changes in bird performance, whether through genetics or management, can thus impact the whole chain and the quantities of feed of birds needed, so that emissions per unit activity may not change, but the total emissions per unit output do.

A system-based approach is thus helpful, if not essential, and is described below.

LIFE CYCLE ASSESSMENT AND CARBON FOOTPRINTS

Life cycle assessment (LCA) is a systematic approach for analysing production systems; in its ideal form, it covers a whole system from cradle to grave, but cradle to gate analyses are also common. In agricultural production, this is usually to the farm gate. The typical system boundaries of poultry LCAs are given in Fig. 13.1, showing boundaries to the farm gate and final consumption.

The approach requires the output of the system to be defined and this is known as the functional unit (FU). This is defined within each LCA and may have quantitative and qualitative conditions that are specific to that analysis. These may relate to location, time (e.g. every day in the year or seasonal), quality thresholds, such as texture, shape or size, or system definitions or standards (e.g. the UK's Freedom Foods standard).

It is thus only possible to compare LCAs in a general way unless the system boundaries and functional units are the same and if the data sources and timescales are very closely related.

LCA is governed by international standards (BSI, 2006) and a full LCA will usually include analysis of resource use and a variety of environmental emissions. A sub-set of LCA that has become common in recent years is the product carbon footprint (PCF), which is often reduced to the carbon footprint (CF). This is solely concerned with summing all the GHG emissions (GHGE) in the production of a defined functional unit. A Publically Available Specification was first published in 2008 (PAS2050) and gave a method for calculating the PCF using

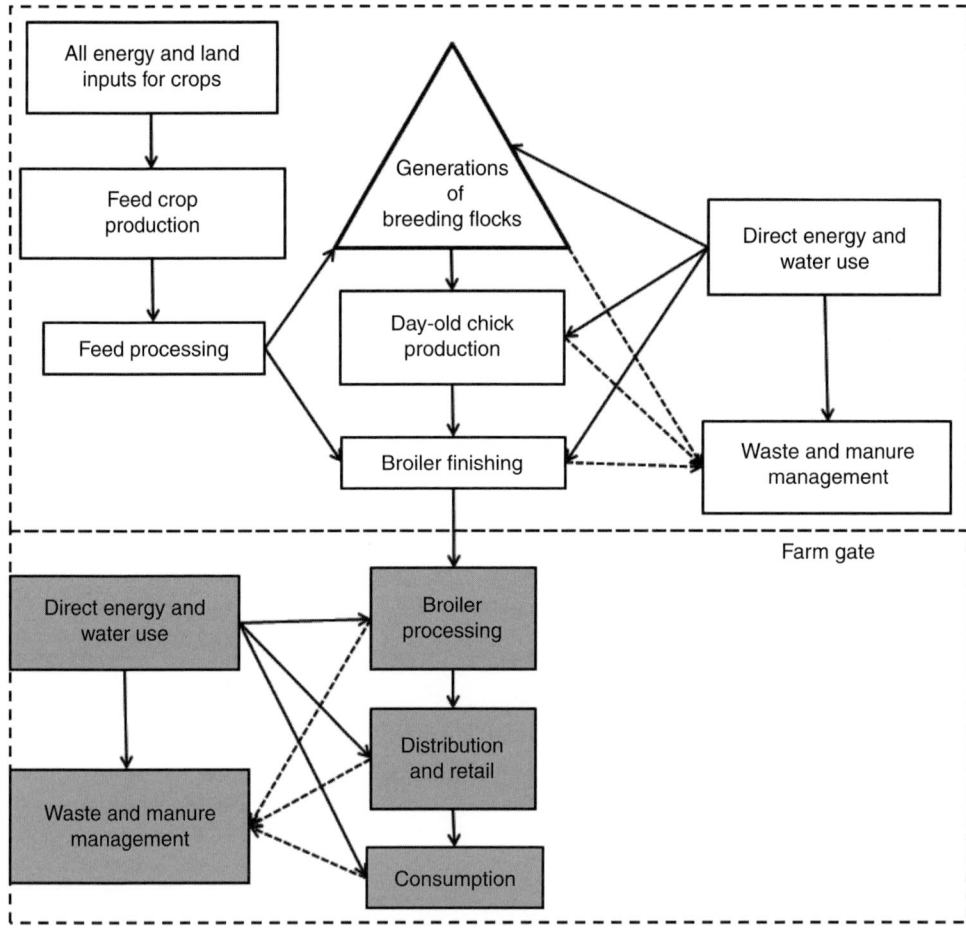

Fig. 13.1. Typical system boundaries in a flow chart for broiler production, including energy flows (solid lines), waste management (dotted lines) and boundaries to the farm gate or consumption (dashed lines). A functional unit must be defined for either case.

LCA, as long as it was based on set thresholds of primary measured data (BSI, 2008). This does not invalidate other approaches and does no more that it says: measuring a PCF. It was revised in 2011 (BSI, 2011) with changes to emissions from land-use change.

IMPACT CATEGORIES

Environmental processes may be measured or calculated on the basis of unit processes, e.g. methane (CH_4) emitted per kilogram manure per day or by area, e.g. nitrate leached per hectare per year. These are summed and aggregated, usually into potentials for causing harm to ecosystems or human health. Those of major interest in agriculture and food are briefly described below.

Resources

These almost always include energy, particularly non-renewable energy, e.g. fossil fuels as well as minerals, like metal ores. All energies are traced back to primary energy resources in the ground, e.g. coal, natural gas and uranium. The overheads of extraction, refining and delivery are always included in deriving the burdens of direct energy use (e.g. as electricity or diesel).

Disparate resources may be aggregated into a common indicator, such as abiotic resource use, which scales the use of a resource to the amounts available globally and relates these to an equivalent weight of antimony (CML, 2014). These can also be quantified with respect to the extra energy that is needed as the resources are incrementally used.

Emissions

Greenhouse gases

GHGE relate to three main sources in poultry agriculture: CO_2 from combustion and LUC, CH_4 from manure management (and enteric emissions from ruminants), N_2O from fertilizer manufacture and N transformations in soil and manure. Secondary N_2O also arises from leached nitrate (NO_3^-) and ammonia (NH_3) volatilization. Further downstream, leaking refrigerants from mobile units used in transport or in retail emit powerful GHGs. The potential impacts are normalized using the global warming potentials (GWP) of each gas (Table 13.2). Note that these are revised about every 5 years by the Intergovernmental Panel on Climate Change (IPCC) and the values do change incrementally, but do not change by orders of magnitude. Most results reviewed in this paper used the 2001 or 2006 revisions of the IPCC (IPCC, 2013). The GWP is related to that of CO_2 on a 100 year time scale as CO_2 equivalents (CO_2-eq.). The GWP is always taken from the IPCC, because of its great interest and the IPPC provides much focused research in this area. Most other impact potentials are derived from a source such as CML (2014).

Acidifying gases

The main contributors to the acidification potential (AP) are sulfur dioxide (SO_2) from fossil fuel consumption and NH_3 from manure and fertilizer. NH_3 is a

Table 13.2. Global warming potentials (GWP) of main greenhouse gases in poultry production on 100 year time scale.

Gas	GWP (2001)	GWP (2006)
CO_2	1	1
CH_4	23	25
N_2O	296	298
R404a (refrigerant)	2780	2780
NH_3 (secondary emission of N_2O)	3	3

counter-intuitive source of acidification, being alkaline, but it is oxidized in soil to nitric acid. AP is quantified as equivalents of SO_2.

Eutrophication

Eutrophication is the over-supply of nutrients to a (usually) natural ecosystem, which can be aquatic or terrestrial. Contributions to eutrophication potential (EP) come mainly from phosphate (PO_4^{3-}), nitrate (NO_3^-) and NH_3. It is usually quantified as equivalents of PO_4^{3-} or NO_3^-, but it can also be quantified with the deoxygenating impact in water as O_2 equivalents. The factors used for AP and EP and are regionally specific and must be viewed with some caution in that the actual impacts on specific receiving ecosystems depend on specific local factors, so they quantify a broad potential for harm.

Ozone depletion

Gases contributing to ozone depletion potential (ODP) include older refrigerants and now banned agricultural chemicals such as methyl bromide. These are all scaled to equivalents of CFC-11.

Other aggregations

Several other aggregations are frequently applied in LCA, such as potentials for causing photo-chemical fog, poly-aromatic hydrocarbons, non-CH_4 volatile organic carbon, radioactivity and other toxins. These and other emissions can be combined into unified potentials for causing harm to ecosystems or human health using sets of weighting factors. There are several approaches to this complex task, such as Eco-indicator 99 or CML (2014). Simpler ones are also applied in some agricultural LCAs, such as pesticide use, which makes the general assumption that use is a good proxy for overall negative impacts (e.g. Williams *et al.*, 2010; Leinonen *et al.*, 2012a, b).

BENEFITS OF LIFE CYCLE ASSESSMENT

An over-riding benefit of a systems approach is that the unexpected should be avoided. For example, a mitigation process for, say, ammonia stripping must be evaluated with regards to its energy use, materials used, emissions of generated gases and disposal of any effluents. This must be included in an assessment of introducing this process. An alternative is a hypothetical change in bird performance that improves daily live-weight gain (DLWG). This may only be achievable by feeding a diet including higher concentrations of specific amino acids (AA) that can only be supplied by high energy production of these AA. It is thus possible, although not conclusive without proper quantification, that a potential improvement may have some negative impacts.

One aspect of LCA to remember is that the results of potentials for causing harm arise at different stages in the production chain. These may be physically remote, e.g. GHGE from LUC in South America, electricity generation in the country where soybean meal is processed and litter in a house in the UK.

SYSTEMS OF POULTRY PRODUCTION

This section focuses on the main production systems found in the developed world. These are mainly fully housed systems.

Species analysed

The species analysed by LCA are, not surprisingly, closely related to the volume of production of each species. This is indeed so extreme that almost all results found in the literature are for chickens and with one early-stage study for turkeys.

Geographical coverage

Most studies relate to Europe and North America with a few in South America, a tropical island in the Indian Ocean and Thailand.

MAIN IMPACTS OF POULTRY PRODUCTION

Broilers

The potential for improved environmental performance cannot be realized until the baseline situation has been analysed. Hence, studies addressing this are considered first. Results from 11 LCA studies of 13 standard and three alternative systems show both systematic variation and similarities (Table 13.3). Please note that these are all individual studies conducted for different purposes in different regions, with different data qualities and access to industry activity data. One feature of these diets is the use of pure amino acids. This is commonplace today, but only the studies of Leinonen *et al.* (2012a) and Leinonen *et al.* (2014) appear to include these.

Cumulative energy demand (CED) ranged from 11 to 21 MJ/kg live weight (LW) and GWP from 0.9 to 3.7 kg CO_2-eq./kg LW. Cumulative energy demand was more consistent with coefficient of variation (standard deviation divided by the mean, CoV) of 18% with GWP at 43%. The larger variation in GWP across these studies relates mainly to differences such as dietary composition and, in particular, how GHGE from land-use change were calculated. Eutrophication potential ranged from 2 to 51 g PO_4^{3-}-eq./kg LW and acidification potential from 16 to 60 g SO_2-eq./kg LW. Reasons for the higher variation in eutrophication potential are not very clear.

These studies convey the magnitude of impacts overall, but the breakdown of contributing sources is also important to understand the totality. Three studies give different levels of disaggregation, which have been simplified to two stages: feed production and broiler production (Table 13.4). Feed production clearly dominated in three categories, with means of 71%, 71% and 64% of CED, GWP and EP, respectively. AP had the lowest impact for feed production (34%) and

Table 13.3. Results of life cycle assessment from recent studies across the world. All results are based on a functional unit of 1 kg live weight ready for slaughter.[a]

Source	System	Country[b]	Cumulative energy demand (MJ)	Global warming potential (kg CO$_2$-eq.)	Eutrophication potential (g PO$_4^{3-}$-eq.)	Acidification potential (g SO$_2$-eq.)	Land occupation (m^2/year)
Spies (2003)	Standard	Br	14.3	1.41	16.4	60.4	
Davis and Sonesson (2008)	Standard	S	20.0	0.89			
Katajajuuri et al. (2008)	Standard	Fin	16.0	2.08	2.1	35.0	5.5
Pelletier (2008)	Standard	USA		1.40	3.9	15.8	
Cederberg et al. (2009)	Standard	S		1.33			
Williams et al. (2006)	Standard	UK	11.2	1.80	14.0	25.9	4.3
Leip et al. (2010)	Standard	EU		3.43			
Leinonen et al. (2012a)	Standard	UK	17.8	3.09	14.2	32.7	3.9
Thévenot et al. (2013)	Standard	F-O	21.4	1.56	51.3	19.3	
MacLeod et al. (2013)	Standard	World		3.71			
da Silva et al. (2014):	Standard	F	19.1	2.22	13.8	28.7	2.7
da Silva et al. (2014): Br-Centre West	Standard	Br	18.0	2.06	14.0	31.4	2.5
da Silva et al. (2014): Br-South	Standard	Br	19.1	1.45	14.4	34.5	2.5
Minimum			11.2	0.9	2.1	15.8	2.5
Maximum			21.4	3.7	51.3	60.4	5.5
Mean			17.4	2.0	16.0	31.5	3.6
CoV (%)			18	43	88	40	34
Williams et al. (2006)	Free range	UK	11.2	2.00	23.5	30.8	6.7
Leinonen et al. (2012a)	Free range	UK	17.2	3.44	16.2	40.0	4.8
da Silva et al. (2014)	Label Rouge	F	29.5	2.70	19.3	47.2	3.9

[a]Functional unit: kg of live weight. Where needed, carcass weights were transformed into live weight assuming a carcass yield of 70% for standard systems and 67% for free-range systems, following da Silva et al. (2014)

[b]UK, United Kingdom; F, France; Fin, Finland; F-O, French Overseas (tropical); S, Sweden; Br, Brazil; USA, United States of America; EU, average of several European Union countries

Table 13.4. Breakdown of sources of environmental burdens (percentages) from four studies of broiler production.

	Standard systems					Alternative systems	
	UK	Fin	F	Br-CW	Br-SO	UK (FR)	F (LR)
CED							
Feed production	65	60	78	74	69	71	78
Broiler production	35	40	22	26	31	29	22
GWP							
Feed production	71	54	74	79	68	72	76
Broiler production	29	46	26	21	32	28	24
EP							
Feed production	52	81	66	75	75	49	57
Broiler production	48	19	34	25	25	51	43
AP							
Feed production	25	29	31	52	56	22	23
Broiler production	75	71	69	48	44	78	77

UK, Leinonen *et al.*, 2012a; Fin (Finland), Katajajuuri *et al.*, 2008; F (France), Br-CW (Brazil Centre West) and Br-SO (Brazil South), da Silva *et al.*, 2014; UK(FR), UK, free range, Leinonen *et al.*, 2012a; F(LR), France: Label Rouge, da Silva *et al.*, 2014

was also the most variable. This results from a systematic difference between Brazilian and European production. Most non-legume feed crops in Brazil are fertilized with urea, which has higher specific ammonia emissions than ammonium nitrate, which is more common in Europe. Hence the mean acidification potential from feed production in the European studies was 26% and 46% in Brazil. There was little evidence of systematic difference between standard and alternative systems.

The key overall message is that feed production tends to dominate climate change and energy impacts in broiler production, but ammonia emissions from the broiler production stage dominate the acidification potential.

BREAKDOWN OF BURDENS IN BROILER PRODUCTION

Burdens of breeding overheads are presented in different ways. The breeding phase (i.e. up to the production of day-old chicks) ranged between 5% and 8% across the burdens, with all sources being aggregated, in Leinonen *et al.* (2012a). Katajajuuri *et al.* (2008) found these to be appreciably higher (8% for AP and EP to 12% for GWP and 17% for CED), although the boundaries were not so clearly defined.

Leinonen *et al.* (2012a) give yet more detail about the sources of burdens during the production phase (Table 13.5). This emphasizes the importance of feed production, accounting for about 70% of CED and GWP, although only about 50% and 24% for EP and AP, respectively. Fuels (electricity and heating fuels) represent 36% and 30% of CED in standard and free range (FR), respectively, hence housing and manure contribute little to CED. It is also essential to realize that about two times more energy enters a broiler house as feed than as fuels. In contrast, fuels represent only about 12% of GWP with housing (and ranging land) contributing about 14%, and manure and bedding 3%. The key difference here is that N_2O from excreta during housing and manure management impacts GWP, but not energy use.

EP is barely influenced by fuel use, but about 7% and 43% are incurred from housing (and ranging land), and manure and bedding, respectively. Most of this is from ammonia, with a smaller contribution from nitrate leaching. AP is influenced by fuel use, with some fuels still emitting SO_2, but direct emissions of ammonia cause most of the 16% contribution to AP from housing and 59% from manure management.

Da Silva *et al.* (2014) did not give this breakdown, but did show these for slaughter and packaging (Table 13.6). The proportional increases in burdens of slaughter and packaging above farm gate production include two major terms: the actual burdens (which differ little per chicken, but are strongly affected by the local energy supplies used) and the actual burdens up to the farm gate. AP and EP and GWP were increased by about 1%, 7% and 2%, respectively. The higher increase in EP is probably mainly accounted for by biological oxygen demand in wastewater. The CED increase was more varied with 12% and 7.5% in France for standard and Label Rouge, respectively, and a mean of 27% in Brazil. The much higher value in Brazil results from a combination of lower CED *per se* in production and the use of firewood as a main energy source. Hence, the increase

Table 13.5. Percentage of burdens of broiler production (excluding breeding overheads) up to the farm gate in the UK (Leinonen *et al.*, 2012a).

	CED		GWP		EP		AP	
	Std.	FR	Std.	FR	Std.	FR	Std.	FR
Feed + water	65	71	71	72	52	49	25	22
Electricity	11	10	4	3	0	0	1	1
Heating fuels	25	20	10	7	0	0	1	1
Housing (and ranging land)	1	1	12	15	5	9	12	19
Manure + bedding	-1	-2	3	3	43	42	61	57

CED, cumulative energy demand; GWP, global warming potential; EP, eutrophication potential; AP, acidification potential; Std., standard; FR, free range

Table 13.6. Increases in environmental burdens (percentages) above those at the farm gate incurred from slaughter and packaging broilers in France and Brazil (da Silva *et al.*, 2014).

	CED	GWP	EP	AP
Fr-ST	12	2.4	8.5	0.8
Fr-LR	7.5	1.8	5.8	0.4
Br-CW	28	1.6	8.3	1.3
Br-SO	26	2.4	8.1	1.1

CED, cumulative energy demand; GWP, global warming potential; EP, eutrophication potential; AP, acidification potential; Fr-ST, France; Fr-LR, Label Rouge; Br-CW, Brazil Centre West; Br-SO, Brazil South

of 12% is likely to be much more typical of mainstream European or North American production.

Primary production is thus clearly the stage that incurs the highest burdens and, for most impacts, feed production dominates. Ammonia emissions in broiler production itself have substantial impacts on both AP and EP, particularly AP. This is important to remember given that the EU and many other countries have internationally agreed binding targets on reducing the emissions of acidifying gases.

IMPROVEMENTS IN BROILER PRODUCTION

Direct

Leinonen *et al.* (2014) included the effects of exhaust-inlet air heat exchangers in broiler houses. These supported lower stocking density than normal in order to be more welfare enhancing (30 instead of $37\,kg/m^2$), so the results will not be quite what may occur in more densely populated houses. The results show that the energy saving effect has the biggest benefit on CED, with a 10% reduction (Table 13.7). GWP was reduced by 5% with AP by 3%, but EP by only 1%. The effects result mainly from 35% less heating fuel use, although 5% more electricity use. Feed use and animal performance was little changed by the use of heat exchangers. It is also evident from Table 13.7 that reducing stocking density by 19% adversely affected environmental performance. This is partly from the increased demand for heating energy for young birds. Note that the effects on heating and ventilation energy needs will vary systematically between climate zones in which the balance of heating and ventilation may change considerably.

Diet modification for broilers

Leinonen *et al.* (2013) examined the environmental impacts of alternative protein crops in broiler diets. This was a modelling study with no farm measurements. Replacing soybean meal with alternatives derived from oilseed rape, peas, beans and sunflower was the focus. Soybean meal has relatively high

Table 13.7. Summary of impacts per kg expected edible carcass weight (live weight × killing out percentage) for standard density broilers and low density broilers with and without heat exchangers.

	Standard density	Low density	Low density + heat exchanger	Reduction through heat exchanger
CED, MJ	24.9	28.0	25.2	10%
GWP, kg CO_2-eq.	4.4	4.4	4.2	5%
EP, g PO_4^{3-}-eq.	20.5	19.2	18.9	1%
AP, g SO_2-eq.	47.0	43.3	42.2	3%

CED, cumulative energy demand; GWP, global warming potential; EP, eutrophication potential; AP, acidification potential impacts owing to processing, long-distance transport and land-use change emissions, which can be accounted for in several ways. The demanding requirements for a good balance of AA meant that the alternative protein sources needed more pure AA supplementation than soybean meal. Pure AA production is very energy intensive and so each has high burdens. This reduced the potential benefits appreciably.

Including alternatives (peas, beans or oilseed rape) at rates of 10 to 30% superficially reduced the GWP per unit output by up to 30%. These were not, however, statistically significant (as determined by Monte Carlo simulations of the whole broiler production system). The magnitude of any substitution was also strongly influenced by the method of calculating land-use change emissions from soybean meal. These could dwarf any other effects and highlights the need for global acceptance of the methods to be used. Da Silva *et al.* (2014) also found large effects of obtaining soybean meal from different parts of Brazil in which land-use change was currently virtually static or dynamic with relatively rapid deforestation. It is worth noting that one of the LUC methods applied by Leinonen *et al.* (2013) would not distinguish between the exact sources of soybean meal, but would apply a value uniformly across the commodity. Meul *et al.* (2012) also showed that method of accounting for LUC emissions has a large effect on the results for feeds that include soybean meal.

Acidification potential could be significantly reduced by 20% using peas and pure amino acids. This is partly a result of the reduced transport burden for soybean meal. Marine fuels used for shipping are not routinely desulfurized, unlike road fuels.

Nguyen *et al.* (2012) investigated the use of environmental constraints to formulate low-impact poultry feeds using crude protein (CP) and metabolizable energy (ME) as dietary constraints. LCA results for feeds were included in least-cost ration formulations for fast- and slow-growing broilers. Only the diets were considered, not the whole production cycle. The overall effects on broiler production would inevitably be lower than benefits seen in feeds alone. The study was conducted in the north-western French region of Britany, with most feeds, apart from soybean meal, sourced in parts of France. The largest effects were seen in fast-growing broilers (with the highest demand for protein and energy). Optimizing diets for the least GWP or EP content reduced these by up to 2 to 14% and 1 to 6%, respectively, but prices were increased by 1–5%. A main factor was the source of maize. EP was chosen because of its regional specificity (e.g. as influenced by rainfall and soil) and GWP with its contrasting global nature. Nguyen *et al.* (2012) did find that much of the environmental benefit (about 70%) could be obtained with appreciably lower cost increases. Prices also changed at different times of year, which affects the cost comparisons. Unlike Leinonen *et al.* (2013), this study did not appear to include the use of pure AA.

Tongpool *et al.* (2012) compared two generic broiler diets in Thailand: 'animal-mixed' and 'vegetarian' feeds. Animal-mixed included fish meal and meat and bone meal (considered waste products with no intrinsic burdens) together with higher inclusion of crop by-products such as distillers' dark grains and solubles (DDGS). In contrast, the vegetarian feed contained only primary crops or crop by-products plus micronutrients. The diets were expected to deliver the same bird performance. The vegetarian diet had higher burdens in 13 of 14 impact categories. These ranged from 4% for AP, through 18% for GWP, 19% for EP and 23% for fossil reserves. It must be stressed that the EU does not permit the use of meat and bone meal or processed animal protein in poultry diets. Furthermore the assumption of zero burdens for these by-products is question-able, certainly for consideration in Europe. These are currently used extensively in the pet food market and so do have a commercial value. If redirected from pet food to poultry feed, the net effect would be to increase demand for crop-derived ingredients in pet foods, which would result in a limited net benefit.

Improving animal performance

Given the importance of feed production, several authors observe that improv-ing feed efficiency (reducing the feed conversion ratio) will help reduce overall impacts. This is self-evidently true in a general way, but there is an important caveat in that the burdens of feed production should not increase (or by less than the reduction in feed use). As noted in previous studies, the composition of feeds has an effect on burdens and attempts to reduce these, e.g. by replacing soybean meal, are highly constrained. If feed conversion ratio (FCR) can only be reduced by increasing feed burdens, the potential gain is reduced or obviated. Improved management such that health is improved and mortalities are reduced will be beneficial, although the achievement is not necessarily easy to quantify or realize.

Increases in DLWG so that the mature weight is reached earlier has some likely benefits. A shorter growing period is likely to decrease the time during which litter is exposed to the atmosphere and act as a source of ammonia emis-sions. These are generally related to the surface area per unit time of occupancy. Hence, AP, EP and ammonia emissions themselves can be reduced per unit out-put. Further, lower heat inputs should be required for younger birds. Again, if this is achieved through a demand for higher protein or energy feeds, much of the potential for improvement will be lost.

Genetic improvement is one area in which environmental gains should be generally applicable. Jones et al. (2008) reviewed the progress made in genetics across the whole livestock sector over the 20 years from 1997 to 2007. Traits include DLWG and FCR. These were examined using the Cranfield systems LCA model and assume no change in the environmental efficiency of feed production or the background environmental efficiency of agriculture.

The historic improvements for broiler chickens are given in Table 13.8.

Table 13.8. Average annual improvement in genetic traits over 20 years for boiler chickens (Jones et al., 2008).

Daily gain (g/day)	FCR (kg/kg)	Mortality (%)	Killing out (%)	Eggs per breeder hen
0.80	−0.015	0.07	0.1	0.9

FCR, feed conversion ratio

The genetic improvement that has been achieved in broilers over the last 20 years has resulted in a substantial reduction in emissions per unit weight of meat produced. The cumulative effects equate improvements of around 1% (of current levels) per year of selection for CH_4 and N_2O, and 0.5% a year for ammonia. Similar annual rates of reduction were also expected over the next 15 years, if current selection practices were to continue. Improvement in each of the traits selected in broilers has contributed to reductions in emissions, but the proportions accounted for by each vary by trait and emission considered. Overall, the greatest reductions have been achieved through the improvements in daily gain and FCR, especially the latter. Improvements in daily gain had the most important effect on reducing ammonia, but had little or no effect on N_2O emissions, which is mostly accounted for by the improvements in FCR.

BROILER LITTER MANAGEMENT

A major change in litter management has occurred in recent years in the UK: centralized combustion for electricity generation. There is also commercial interest in on-farm litter burning to reduce heating energy needs.

Williams *et al.* (2015) compared burning turkey litter to generate electricity centrally with normal manure management as a crop fertilizer. The use of litter as fuel gave positive environmental benefits reductions in CED (14%), EP (55%) and AP (70%), although reductions in GWP were small (3%). Although not exactly comparable, similar results could be expected with broiler litter.

EGGS

Baseline results

Results of nine LCA studies are presented in Table 13.9, most coming from north-west Europe. Production systems are more varied than with broilers. Cages dominate in world commercial production, but alternative free range, barn, aviary and organic systems have become more popular in recent years and so receive some attention. Given the timescales, most if not all of the analyses of caged systems used the older battery cages that are now banned inside the EU. Leinonen *et al.* (2014) explicitly compared production using the replacement colony cage with the now-banned battery cages.

The results show considerable variation, between countries and systems, which reflects different diet formulations, feed delivery distances, energy requirements and heating and ventilation and bird performance (as well as data sources and assumptions made by analysts). The carbon footprints varied from 1.3 to 4.6 for free range in the UK (Taylor *et al.*, 2014) and deep litter with outdoor run in the Netherlands, respectively (Mollenhorst *et al.*, 2006). MacLeod *et al.* (2013) found a world mean of 3.7 kg CO_2-eq./kg eggs for commercial egg production, ranging from 2.3 in Eastern Europe and the Russian Federation to 4.2 in Western Europe (Fig. 13.2). MacLeod *et al.* (2013) used the same methods and

Table 13.9. Summary of studies on egg production. The functional unit is 1 kg eggs at the farm gate.

Source	System[a]	Country[b]	CED (MJ)	GWP (kg CO_2-eq.)	EP (g PO_4^{3-}-eq.)	AP (g SO_2-eq.)	Land occupation (m^2/year)
Dekker *et al.* (2011)	Cage	NL	21	2.2		23	3.3
Leinonen *et al.* (2012b)	Cage	UK	17	2.9	16	53	4.0
Mollenhorst *et al.* (2006)	Cage	NL	13	3.9	25	32	4.5
Pelletier *et al.* (2014)	Cage in 1960	US	18	7.2	70	200	
Pelletier *et al.* (2014)	Cage in 2010	US	12	2.1	20	70	
Wiedemann and McGahan (2011)	Cage	Aus	11	1.3			
Cederberg *et al.* (2009)	Standard[c]	S		1.4			
MacLeod *et al.* (2013)	Standard[d]	World		3.7			
Vergé *et al.* (2009)	Standard	Can		2.5			
Dekker *et al.* (2011)	Barn	NL	23	2.7		64	3.8
Leinonen *et al.* (2012b)	Barn	UK	22	3.5	17	59	4.2
Mollenhorst *et al.* (2006)	Deep litter	NL	13	4.3	31	57	4.8
Mollenhorst *et al.* (2006)	Deep litter with outdoor run	NL	14	4.6	41	65	5.7
Mollenhorst *et al.* (2006)	Aviary with outdoor run	NL	14	4.2	35	42	5.1
Dekker *et al.* (2011)	FR	NL	23	2.8		65	4.1
Leinonen *et al.* (2012b)	FR	UK	19	3.4	19	64	5.1
Taylor *et al.* (2014)	FR	UK	13	1.6			
Dekker *et al.* (2011)	Org	NL	21	2.5		80	6.8
Leinonen *et al.* (2012b)	Org	UK	26	3.4	30	92	16.9
MacLeod *et al.* (2013)	Backyard	World		4.2			

CED, cumulative energy demand; GWP, global warming potential; EP, eutrophication potential; AP, acidification potential
[a]FR, free range; Org, organic
[b]Aus, Australia; Can, Canada; NL, the Netherlands; S, Sweden; UK, United Kingdom; US, United States of America
[c]No LUC emissions
[d]Standard includes cage and other systems, but is dominated by caged birds

assumptions for all their analyses, which show a broad agreement with individual studies reported here. They also reported a world average value of 4.2 kg CO_2-eq./kg eggs for backyard egg production. This is systematically higher than the world average for commercial production owing to lower productivity, higher feed intake, more N_2O emissions from manure and a generally higher proportion of rice in diets, which creates high CH_4 emissions during its production.

In comparisons between systems, the general trend is for standard caged production to have the lowest burdens and alternative systems (free range, barn, aviary or organic) to have higher burdens (Mollenhorst *et al.*, 2006; Dekker *et al.*, 2011; Leinonen *et al.*, 2012b).

Breakdown of burdens in egg production

Leinonen *et al.* (2012b) provides the most detailed breakdown across several burdens (Table 13.10). The laying production phase causes 80 to 84% of the burdens across all systems, with pullet rearing at 15–19% and breeding (including three generations) only 1%. Feed production and water account for nearly 70% of CED and GWP, with most of the remaining energy used directly in housing. The remaining GWP was split almost evenly between fuels and housing and land (14%) with manure and bedding at 3%. In contrast, AP was overwhelmingly dominated by manure + bedding (51%) and housing + land (33%), because of

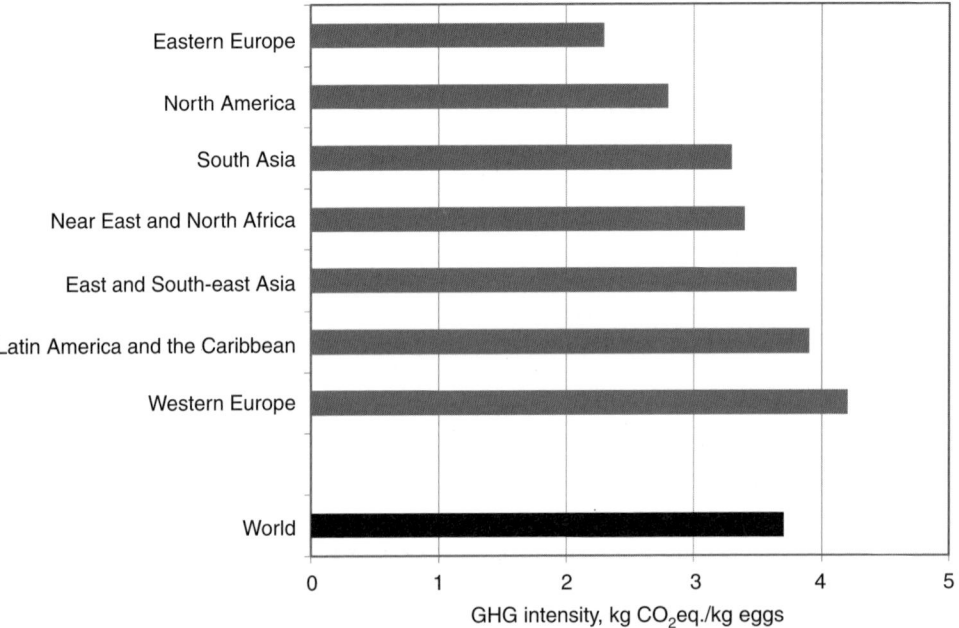

Fig. 13.2. Greenhouse gas intensity of commercial egg production across regions of the world (MacLeod *et al.*, 2013; regions with less than 2% of egg production are omitted).

Table 13.10. Breakdown of environmental burdens for egg production up to the farm gate (Leinonen *et al.*, 2012b). These are the unweighted means (%) of four production systems.

	CED	GWP	EP	AP
Breeders	1	1	1	1
Pullets	19	17	15	17
Layers	80	82	84	82
Feed + water	67	69	45	14
Fuels	34	13	0	2
Housing + land	1	14	17	33
Manure + bedding	−2	3	37	51

CED, cumulative energy demand; GWP, global warming potential; EP, eutrophication potential; AP, acidification potential

ammonia emissions. These and nitrate contributed to there being a roughly equal split between feed production and all the other sources combined.

Mollenhorst et al. (2006) provided a different breakdown, which is not wholly compatible with Leinonen et al. (2012b), but found about 90% of CED was used for feed production and 10% used directly in fuels. Mollenhorst et al. (2006) also found large contributions to AP and EP from manure of similar magnitude to Leinonen et al. (2012b), but GWP from feed production, at 90%, was appreciably higher than found by Leinonen et al. (2012b).

MacLeod et al. (2013) broke down GWP in western European production as 84% feed, 12% manure and 4% fuels, but manure here would include emissions from housing, so giving reasonably close agreement with Leinonen et al. (2012b). MacLeod et al. (2013) broke down GWP in average world production as 78% feed, 17% manure and 5% fuels.

As with broilers, the overall impacts are highest for the main production phase, although the combined breeding and pullet rearing are of considerable importance (about 20% of burdens). Feed production also dominates CED and GWP, but ammonia emissions play a large role in EP and even more so in AP.

IMPROVEMENTS IN EGG PRODUCTION

Direct

Few results exist for practical management changes in LCA studies although there is some speculation. With the high impacts of feed production, most authors observe that better feed utilization is desirable, although it must be stressed that this should be achieved without increasing the emissions intensity of feeds themselves. General observations include reducing mortalities, feed losses (especially in free-range systems) and promoting bird health. The interventions needed to achieve these have not been quantified. Interventions could be from improved management intensity (which implies increased economic rather than environmental burdens), changed or improved buildings, feed improvements or genetic improvements.

The use of heat exchangers in broiler houses to reduce direct energy use, as reported by Leinonen et al. (2014), should be applicable to housed egg production, but this has not been quantified explicitly.

Manure management in caged systems varies. Some, generally older, houses have deep-pit systems, while others have belt-cleaned systems. The belt systems are effective at reducing ammonia emissions from housing and the UK ammonia inventory has emission factors of 35.6% and 14.5% for deep pit and belt cleaned, respectively. This gain is not the total reduction in ammonia emissions for a system in that the subsequent manure will contain more ammoniacal N than for deep-pit collected manure. Hence, there is a greater potential for subsequent loss if the manure is not managed with low-loss techniques.

Taylor et al. (2014) considered potential improvements in FR systems. They suggested that export of manure either raw for fertilization or for use as feedstock for anaerobic digestion could reduce the CF of eggs by 7%. They did, however, assume that the same benefit would accrue from either approach, but this suggests that fertilizer value from layer manure on the farm was not being accounted for fully in their analysis, but only emissions from land application. If their value for gain from FR systems and anaerobic digestion is reasonable, then the potential gain is greater for fully housed systems in that all manure must be managed and is more amenable to export.

Diet modification for egg production

Leinonen et al. (2013) also examined the environmental impacts of alternative protein crops in laying hen diets, as for broiler diets.

Replacing soybean meal with alternatives derived from oilseed rape, peas and beans was examined. As with broiler diets, the need for increasing the inclusion rates of pure AA with alternatives to soybean meal reduced the hoped-for benefits of diet change. Including peas, beans or oilseed rape superficially reduced the GWP per unit output by 4–10%. These were not, however, statistically significant (as determined by Monte Carlo simulations of the whole egg production system). As with broilers, the magnitude of any substitution was also strongly influenced by the method of calculating land-use change emissions from soybean meal.

Acidification potential could be superficially reduced by 3–9% using peas and beans. In contrast, EP superficially increased for all alternative diets by 1–2%. This resulted mainly from the higher nitrate leaching from peas, beans and oilseed rape than from soybean meal.

Nguyen et al. (2012) investigated the use of environmental constraints to formulate low-impact poultry feeds. LCA results for feeds were included in least-cost ration formulations for laying hens. Only the diets were considered, not the whole production cycle. The overall effects on egg production would thus be lower than benefits seen in feeds alone.

Optimizing diets for the least GWP or EP content reduced these by 3% and 5%, respectively, but prices were increased by 0–7%, because prices changed at different times of year. These optimizations also supported reductions in CED and AP of 6 and 2%, respectively.

Taylor et al. (2014) suggested more radical alternative protein sources for poultry including worms from composting organic wastes, algae produced in biological CO_2 absorption systems and processed animal by-products from red meat production. Initial unpublished estimates by Taylor et al. (2014) for replacing soybean meal with these alternative protein sources indicated a potential 60% reduction in the carbon footprint for poultry diets. It was estimated that this change could reduce the egg GHG footprint by 45% and reduce the dependence of the UK on imported soybean meal. This appeared to be a theoretical maximum without consideration of the availability of supply.

Indeed, in all analysis and discussion of alternative feeds, no attention appears to have been paid to whether the land or other resources are available to support mainstream shifts in feed sourcing. Large changes tend to upset markets and if demand increases sharply, prices will increase and potentially make alternatives uneconomic. There is a resource allocation problem to be addressed, which is if demand for say European crops increases to offset imports of soybean meal, what will the opportunity costs be and what other crops might be displaced.

A further consideration that has been addressed by Leinonen *et al.* (2013), Meul *et al.* (2012) and van Middelaar *et al.* (2013) is that the method of calculating emissions from land-use change is a vexed matter and considerably influences the results of analysis of alternative feeds. This is not an arcane academic matter, but one of considerable importance and controversy. Global land-use change emissions are vast and the associated loss of habitat is also vast. How these should be equitably allocated across all human demand for fuel, fibre and food is non-trivial. So fraught is the topic that van Middelaar *et al.* (2013) suggested that GHG emissions from land-use change should not be included in true actual results but reported separately.

Genetic improvement in egg production

Jones *et al.* (2008) reviewed the progress made in genetics across the whole livestock sector over the 20 years from 1997 to 2007. Traits include hen LW, egg weight, yield and FCR (Table 13.11). These were examined using the Cranfield systems LCA model and assume no change in the environmental efficiency of feed production or the background environmental efficiency of agriculture.

They found that the overall reduction in emissions was even higher in layer hens compared to broilers, with emissions per unit of products estimated to be around 30% higher in 1988 compared to 2007 (about 1.5% per year). Similar annual rates of reduction were also expected over the next 15 years. Improvement in all four of the traits considered has contributed to the reductions, but the largest reduction in CH_4 and N_2O is accounted for by changes in FCR and for ammonia by changes in egg weights.

In contrast, Cederberg *et al.* (2009) found no apparent difference in the carbon footprint of Swedish egg production between 1990 and 2005.

Pelletier *et al.* (2014) compared egg production in 1960 with that in 2010 and separated the effects of the foreground (e.g. on-farm hen performance, genetics, feed conversion etc.) with the background (e.g. fertilizer production, feed production, energy carriers and transport). The three primary factors that determined the environmental impacts of US egg production were feed efficiency, feed composition and manure management. The overall annual improvements per kilogram eggs were 0.67%, 1.4%, 1.4% and 1.3% for CED, GWP, EP and AP, respectively. The contributions to the improvements were attributed relatively evenly to improved efficiencies of background systems (27–30%), changes in feed composition (30–44%) and improved bird performance (28–43%).

Table 13.11. Average annual improvement in genetic traits over 20 years for laying hens (Jones *et al.*, 2008).

Mature hen LW, g	FCR (kg/kg)	Egg weight (g)	Eggs per laying hen
−25.20	−0.025	0.11	0.99

LW, live weight; FCR, feed conversion ratio

Table 13.12. Improvements in environmental performance between older, conventional cages and EU-permitted colony cages (Leinonen *et al.*, 2014).

	Conventional cage	Colony cage	Reduction through change
CED, MJ	16.8	15.4	8.3%
GWP, kg CO_2-eq.	2.9	2.8	3.1%
EP, g PO_4^{3-}-eq.	19.0	19.0	0.3%
AP, g SO_2-eq.	2.9	2.8	3.1%

CED, cumulative energy demand; GWP, global warming potential; EP, eutrophication potential; AP, acidification potential

The improvements in bird performance are of the same magnitude as those found by Jones *et al.* (2008), but somewhat smaller. This may result from different average rates of change in the UK and USA, but this is hard to discern. Pelletier *et al.* (2014) addressed change over 50 years compared with 20 years by Jones *et al.* (2008). It is entirely plausible that the rate of change accelerated in the USA with time as agricultural science developed.

Leinonen *et al.* (2014) compared the effect of changing from older cages to the EU-accepted colony cages. In this study, it should be noted that comparison probably includes activity data from older buildings with older cages and newer buildings with colony cages, so that the thermal performance of the latter is likely to be better. However, improvements in bird performance were also realized so that the comparison is a fair one for those considering the change and who are still using older cages in older buildings. The reductions in burdens ranged from 8% in CED through 3% for GWP and AP to 0.3% for EP (Table 13.12).

INTRODUCTION TO CASE STUDY: APPLIED POULTRY

Broiler farming to produce chickens for consumption is a sector that operates on small margins per bird produced but aims to produce a considerable number of birds per farm per year to ensure overall business profitability. The challenge going forward for my business is to see what can be done to reduce environmental impact, whilst also ensuring this is not done at the cost of reducing profits to a point that the units become unviable. The ultimate aim is to find schemes and production methods that offer increased profits at the same time as reducing environmental impacts to truly offer a win for both business and environment.

The aim of this case study is to show what we have already achieved and consider what we are already considering for the future.

CONSTRUCTION OF NEW/REDEVELOPMENT OF OLD AND THE ENVIRONMENTAL BENEFITS

Constructing a new farm or redeveloping an existing unit gives a great opportunity to evaluate the environmental impact of our broiler farm and attempt to mitigate any impact where possible.

Environmental impact of construction

As the company grows we are giving more consideration to the environmental impact of building new sites. We are also finding some planning authorities are demanding such considerations. Initially we were trying to develop sites throughout the year but the potential for weather conditions leading to extra costs and the damage caused to the local environment can be considerable.

An area that, as yet, the sector is not looking at is the environmental sensitivity of the building materials used. Going forward we may need also to consider more the environmental cost of dismantling sites in the future. Asbestos used to be widely used and the potential to cause harm to humans and the difficulties in safely disposing of spent materials was not considered.

ENERGY-SAVING CAPITAL ITEMS

LED lighting

At present we have moved from lighting our poultry buildings using conventional tungsten bulbs to using high-frequency fluorescent tubes. However, we continually review the possibility of using LED lighting to further save on electricity used. Any LED lights used need to offer a good spread of even light and must be fully dimmable. The biggest constraint to date has been cost and justifying the significant capital investment versus the saving in running costs. Initial quotes were in the order of £10,000 per shed versus £2200 for high-frequency tubes. Assuming £0.10/kWh for electricity, the 32 tubes that we currently use cost £3.34/shed/day to operate for 18 h/day. To pay back the investment in LED lighting over 5 years requires a saving of £7800 over 5 years (this being the difference in capital cost excluding interest if the money was borrowed); £7800 over 5 years = £1560/year saving = £1560/7.4 cycles per annum = £210.81 saving per crop required with an average age of chicken 38 days per cycle = £5.55 saving per hour required. When I am only spending £3.34/h currently it is not possible to save £5.55. My estimation is that it would take 10 years to properly save for the extra capital cost benefit, which is too long for current consideration. We need to be able to install LED lighting at no more than twice the cost of the halogen tubes to justify the investment.

Ventilation

Originally we used to ventilate the sheds based on a judgement regarding the environment made by the person working the farm on that given day. This was done by a very simple assessment using their sense of smell and an evaluation of the birds and the condition of the litter/bedding the birds were living on. Any malodour, signs of stressed birds, particularly heat or poor litter quality would lead to a change in the ventilation rates. This was often difficult to assess properly as external weather conditions and temperatures varied throughout a 24 h period. On many occasions the buildings would be over-ventilated at night leading to excessive electricity running fans, excessive gas being burnt to raise shed temperatures and the consumption of food by the birds to maintain body temperatures. All of this had a negative impact on the efficiencies of the unit and increased the use of energy.

Our modern units have fully computer-controlled ventilation systems where we monitor how many kilograms of chicken are in the shed at any given time to work out their physiological demands along with sensors monitoring air temperatures both inside and outside the sheds, relative humidity, air pressures and CO_2 levels. The new computers will also calculate the effects of wind chill on the birds by considering all factors together. This has led to a reduction in heating bills, a reduction in electricity demands per bird and improved overall efficiencies of the birds. Through all of this, litter quality is drier and emissions from the units, particularly ammonia and odour, are down.

COOLING OF THE ENVIRONMENT

I have evaluated many options for keeping my birds cool in the summer. Wherever possible, giving consideration to odour emissions, I like to have tunnel ventilation in place to cool the birds. The principles of this system are very simple in that all we want to do is turn our sheds into wind tunnels and pull the air in one end and extract at the other and by moving that air at speeds of up to 2.3 m/sec we can introduce a chill factor of up to 5°C so on a 30°C day it feels 25°C to the birds. Not only is this process incredibly effective it also operates with relatively little demand for power once the fans are up and running. Alternative measures include high pressure water misting, which requires considerable energy to put the water under high pressure to create the mist.

Underfloor heating

Our birds require a temperature of around 33°C for brooding as chicks, which uses a lot of energy. When the new buildings were built at Lower Farm I was able to install underfloor heating into the concrete floor slab. Whilst initial shed air temperatures are still high, as with traditional buildings, the presence of underfloor heating has allowed me to rapidly drop air temperatures after day 3 whilst maintaining a warm floor for the chicks to continue brooding. The use of this technology has reduced the heating demand of that farm by 30% and further improved bird performance efficiencies.

Renewable energy

Heating and cooling our birds is an essential part of ensuring thermal comfort within our poultry buildings. We have been focusing extensively on the cost of these, particularly the cost of heating the facility. Back in 2004 we joined a UK government scheme that gave a tax relief from the Climate Change Levy imposed on energy bills, which would be granted if we could show year on year reductions in energy used per kilogram of chicken LW produced. Initially some early gains were easily made by basic building maintenance and then for us the redevelopment of the home farm produced huge benefits. One of the significant design changes was the use of underfloor heating. Over subsequent cycles of birds we have been able to compare our sheds to identical sheds without underfloor heating and found the system saves 30% in heat energy.

Initially the air heating at the farm was via direct flame heaters, during the redevelopment this then changed to include hot water underfloor heating heated via LPG gas boilers. In the last 2 years this has again changed to mean that both air heating and the underfloor heating utilize renewable wood fuels via biomass boilers to warm the water and heat the sheds. We achieve the air heating by passing 85°C water through heaters that comprise of a large radiator and a fan to disperse the warm air, and for the underfloor heating the warm water from the biomass simply replaces the gas boilers.

The main differences for our business from doing this have been to reduce the amount of fossil fuels consumed as heating fuel and, as we are not burning gas within the buildings, we have reduced the carbon monoxide and dioxide levels in the shed; this has meant that we do not need to vent the sheds so much to remove the noxious gases and so in turn we need less heat to warm the smaller volumes of incoming air. On a slightly negative note, biomass heating systems need a lot of water to be pumped around a farm and so we have seen our demand for electricity increase slightly. Alternative electricity sources such as wind or solar are always considered for sites, but the original site at Chesterfield is in a National Park and so some of these options are unsuitable.

Rainwater harvesting

At present none of our farms is recycling water. This is an area I greatly want to explore in the coming years. All large intensive broiler units have significant roof areas, which give the potential to catch large volumes of rainwater. All of our modern developments have been designed ready for the implementation of rainwater harvesting with appropriate catchment drains, attenuation ponds, pump stations, etc. The main reasons for not doing so currently are that actually the payback on the investment in setting it up is too long and there are concerns over the consistent quality of harvested water. We will need appropriate treatment of the water to remove contaminants such as bird droppings off the roof and also fail-safe methods of ensuring these cleaning controls remain effective, and clear notification needs to be given should the filtration/cleaning process not meet the required standards.

Digital technology and precision livestock farming

Cameras with image analysis software such as eYeNamic™ cameras help with quick-time decision making and understanding the needs of the birds, hence satisfying retailers and consumers of the welfare standards of these production systems. Improved efficiencies through improved welfare will result in more kilograms of usable chicken produced per reared batch of birds whilst still using the same amount of lighting, heating, labour, bedding and other inputs.

Other camera technology

Initially camera technology was placed at Lower Farm as part of a range of security measures following a break-in during the construction phase. Whilst placing cameras outside the facility, it was minimal extra cost to then place additional cameras inside the sheds with the birds. What I found was that I was able to further improve the environment for the birds and thereby improve the performance and efficiency of the unit. It was very apparent that birds appeared very comfortable in the daytime when management staff were around monitoring them but during the cooler nights the ventilation systems were chilling the birds. On occasions the air temperatures were correct but the airflow rates were causing wind chill.

As faith in the camera technology has grown we find that we are using them instead of going out to the farm during late evenings to continually check on the birds. The average member of staff that would carry out the late-night checks lives 6 miles from the farm, a 12 mile round trip. If one journey can be prevented every cycle between bird ages of 15 days and clear at 38 days that equates to $23 \times 16 = 368$ miles for 7.4 cycles per annum $= 2723$ miles $= 78$ gallons or 354 l of fuel, which is a cost saving of £405 (excluding VAT) and a reduction of emissions.

As the company grows this type of technology is essential, allowing remote monitoring of sites and preventing unnecessary car journeys vising sites.

Feed deliveries

It was never guaranteed in the development phase plans that feed conversion from raw materials to LW would improve as much as it has. Feed is a massive part of a broiler farm's costs and any improvement in feed conversion can easily be justified by its financial implications alone, setting aside the environmental impact of reducing lorry movements.

Lower Farm has gone from an average feed conversion of 1.66:1 to an average 1.60:1. While this increment seems small at 3.6% improvement, when you consider Lower Farm purchases 4750 t/year this 3.6% saving equates to 171 t of feed, which would require seven lorry journeys with an average round-trip journey from the feed mill and back of 85 miles: seven journeys equals 595 miles, with lorries averaging 6 miles to the gallon the amount of fuel saved is 99 gallons or around 450 l. This is a cost saving of £514 (excluding VAT) and a reduction of emissions. We are gaining these efficiencies by focusing on all elements of the bird from optimal environmental controls to ensuring good bird health. Anything that is not deemed perfect by the bird will lead to a degree of stress and the utilization of calories to cope with that stressor, be it coping with temperature variances or the stress of being unduly disturbed. Birds ultimately want a good environment, good health status, good feed, good water and the genetic potential will achieve the rest.

In addition to the improved feed conversion efficiency, as a business we aim to always take full lorry loads of feed; even where a ration change may dictate two part loads, we would order one full load of each. This on average saves an additional 20 lorry journeys per annum.

As an industry that is predominantly an integrated supply chain between the farmer rearing the birds and the processor (where the processor owns or has direct links to the feed mill), farms for bird rearing are often chosen by the processor based on how close they are to the feed mill. The main reason for this is that the feed mill is covering the cost of delivering the feed and so the closer the customer, the cheaper they can deliver. As the processor's customers, particularly the retail customers, demand improved environmental performance, so the processors focus heavily on these types of efficiencies.

Another huge impact on our feed conversion is the quality of feed, both in terms of its composition and also its physical characteristics. The argument for good quality raw ingredients and a well-designed ration are obvious, but physical characteristics are also important. From our own experiences dusty feed or poor quality pellets lead to a decrease in feed intake by the bird, which in turn slows bird growth and negatively impacts on the feed conversion rates of the bird, our estimate is around a 10% decrease in overall feed conversion efficiency.

We have found that feeding a poorly designed ration or restricting feed intake via unscheduled breakdowns etc. can lead the bird to 'scour', which essentially means that the birds increase their water intake whilst decreasing their feed intake and the moisture content of their droppings increases. A bacterial imbalance in the gut can cause the same symptoms. These increased moisture deposits on the floor of the poultry house can, especially in cooler winter temperatures, lead to the bedding material getting wet and 'capping'. We have noticed that when this occurs the bird performance decreases. This may be as a result of the gut imbalance or it may also be as a result of the reduced quality of the immediate environment around the bird, cold wet bedding, dirty feathers, etc. As well as reducing the performance and comfort for the birds a wetter environment can lead to an increase in the levels of ammonia emitted from a facility.

Site disinfection

After thorough cleaning of the facility following the depletion of a batch of birds we use an orchard sprayer that ensures correct application strengths are used and prevents excessive runoff of chemicals. Product choice depends on biosecurity risk of each site, so good biosecurity and low background risks mean we can use less harmful disinfection products, moving away from environmentally damaging products such as formaldehyde.

Emissions reductions

As previously discussed we aim to minimize emissions by feeding healthy birds a good diet and keeping them and their environment in optimal condition. It is very difficult to absolutely remove all emissions from a farm housing a significant number of birds, e.g. over 100,000 birds. As part of our planning process for new sites as well as our evaluation of existing sites we use odour modelling software to predict the potential to cause nuisance of odour from our units. Wherever possible we aim not to emit potentially more than three odour units to any nearby resident or place of work. Three odour units is the level at which the odour is deemed to have the potential to cause offence and become a nuisance.

Over the years we have attempted many things to reduce the odour emitted from our farms. Trials were conducted with various masking agents, ozone was generated on site, different bedding materials were trialled, different ventilation regimes were used and many other things. The best results were gained by dispersing any odorous air around the whole site rather than in concentrated corridors and also we aim to throw the air higher into the air for natural dispersion. To date we have not had any verified odour nuisances since the new dispersion has been occurring.

Going forward, it is our firm belief that ammonia emissions will become more and more important to control, and this is likely to be driven by EU legislation. From our own trials and monitoring, our farms are emitting no more than 8 ppm ammonia and cycle averages are nearer 3 ppm. Some of this low level of ammonia is down to the drier bedding area, which is helped by having under-floor heating.

Feed conversion efficiencies

As already discussed, feed conversion from chicken feed to chicken LW has a massive implication on the profitability of a broiler business as well as the environmental impact of that business. To date my businesses operate around 3% more efficient than the average farms supplying the customers' processing plant. This may not seem a lot, but as already described 3% of a very large tonnage of feed is very significant.

Sustainable intensification

Wherever I travel presenting or listening to conference presentations I hear about the positives of sustainable intensification. The feed sector seems determined to produce more from intensification whilst demanding less by way of inputs and what it takes from the environment. For our part in the intensive broiler sector driven predominantly by the economics of the sector we have become the epitome of this new drive. One clear example of this is when we look at stocking density management within the sector. We are growing birds to a maximum stocking density of 38 kg LW/m^2 floor area.

However, by growing the shed to this stocking density, then taking away for processing 30% of the birds before leaving the remaining 70% of the shed population to grow on up to $38\,kg/m^2$ again, we are actually producing nearer to $44\,kg$ of bird per square metre. This depleting of 30% is called 'thinning' and has been a clear management tool in keeping chicken meat for the consumer a very cost effective option whilst also allowing us to compete with imports that may have been produced at $44\,kg/m^2$ stocking, not our $38\,kg/m^2$.

What has become an issue with this, however, is the results around the presence of *Campylobacter* on chickens sold at the retail level to the consumer. Many years of trials have been unable to reduce significantly the presence of *Campylobacter* and hence there has not been a significant drop in the incidences of food poisoning from *Campylobacter*. Going forward, the cessation of this thinning process may have a massive impact on the sector's ability to continue sustainably intensifying.

CONFLICTS AND SYNERGIES BETWEEN CONSUMER ASPIRATIONS AND ENVIRONMENTAL IMPACT REDUCTION

An area of interest to me as a British poultry farmer has to be the environmental impact of imported meat to the UK. I believe some of this to be due to the disproportionate demand for white breast meat versus the brown thigh and leg meat. As well as issues around balancing the demand for the various parts of the carcass the economics and cost of production also have an impact on levels of imported meat. It is a very real challenge for this business to be able to reduce the environmental impact of the business whilst also competing with other farmers around the world who may or may not have the same level of environmental consideration. As chicken meat is consumed and traded globally and the environment is a global entity, it is important that issues around environmental impact are not simply moved from the UK to other parts of the world.

What is also important is that the consumers' perceptions of which production methods and/or countries are the least damaging to the environment are based on reality. As intensive agricultural practices become more and more focused on reducing environmental impacts, so more traditional practices such as free range are unable to offer the same level of control. For example, fly nuisances from an intensive broiler unit are very rare whilst free-range production models can promote fly populations. No manure leaves a modern intensive broiler unit uncontrolled, from loading out muck on concrete bounded yards to sheeting trailers for dispatch, as compared to free-range birds freely defecating around their roaming areas.

RETAILER AND FARMER CONFLICTS

As a farmer it can sometimes be difficult to reduce every environmental impact when at the same time there are commercial pressures from retailers and outside pressures from non-governmental organizations. An example would be the requirement to give birds their dark period for rest during the dark of the night. It would make better sense to allow birds to have the lights out during the day when there is more demand for electricity and keep them awake with lights on overnight whilst the human population are asleep, thus balancing the overall demand for electricity locally.

As has been shown through previous text within this chapter there are technologies that are improving bird efficiencies, reducing the generation of GHG (such as camera technology allowing remote monitoring of birds), that we can use but these have to be accepted by our customer and we must show that bird welfare is not compromised during the process.

TECHNOLOGY AND ITS ROLES IN IMPROVING EFFICIENCY

Technology has helped get the broiler sector to its current position, year on year increasing output per area of shed whilst also year on year reducing the environmental impact of that business. Going forward there appear to be very few barriers to the possibilities that technology can offer, be it relating to energy harvesting and generating, waste management and recycling, increased outputs or negative environmental impact prevention. As a business we are ready to embrace all that technology can offer to remain competitive and sensitive to our impact on the environment.

POLLUTION CONTROL MEASURES

Some of the recent (past 10 year) reductions in environmental impacts for the intensive broiler sector have been led by European legislation; one in particular, the Integrated Pollution Prevention & Control Regulations, meant all broiler units over 40,000 birds in size had to meet minimum demanding standards. In many cases, including ours at Lower Farm, it led to the complete redevelopment of a farm that was having an impact on its local environment. Whilst no one having to meet the demands of regulations welcomes them with open arms, looking back it has transformed many parts of the sector and definitely helped to reduce the occurrence of pollution incidences.

Manure as fertilizer

All manure generated from our poultry units is sold to local farmers to be used as a fertilizer on their land. This keeps movement of the manure to a minimum, which helps reduce our environmental impact but also reduces the local farmer's dependence on artificial fertilizer sources, which helps reduce his business' impact on the environment. We do ensure they use our manure in accordance with recommended codes of good practice and this does include the safe storage of the manure prior to applying it to their land.

CONSIDERATION TO THE VIABILITY OF OLDER UNITS

On a final note, as a business we are moving towards becoming a contract management company that manages other people's units as well as our own. This
means that going forward we will be introducing our production methods and
environmental considerations to many more businesses; this will have a greater
impact on what we can influence regarding minimizing environmental impact of
those farms as our business grows. What this business model does rely on is the
embracing of modern efficient broiler farms, and when a contract farming model
is considered on older more traditional broiler units it very often does not work
due to the inefficiency inherently built in to outdated sites. To truly continue
reducing the environmental impact of our broiler industry it is essential to continually challenge tradition and embrace modern technologies and methods,
and to do this we as an industry may need to evaluate the true life expectancy of
our broiler units and not leave inefficient units in production for any longer than
is needed.

CONCLUSION

Mainstream poultry production is highly optimized in the context of animal production. Historic analyses have shown substantial improvements over 20 to 50
years (e.g. Jones *et al*., 2008; Pelletier *et al*., 2014). These have included genetic,
managerial and nutritional improvements as well as those in the wider feed production system.

   Despite this, there are opportunities for improving environmental performance through reduced mortalities and better health, using heat exchangers to
save direct energy use, upgrading manure to produce energy, changing diets to
reduce the inclusion of high impact feeds, such as soybean meal.

   The opportunities vary across the world, given that the starting points are
varied. Variations result from different feed availability (and locations and methods of production), electricity and other fuel intensity, genetic stock, health status, building age and management quality.

   The main requirements for improving environmental performance relate to
the embedded energy and GHG emissions in feeds. Making better use of feeds
has a large impact on both these burdens. Reducing ammonia emissions and
hence both acidification and eutrophication potentials is closely related to minimizing N excretion and thus requires optimized matching of the amino acid profile in feeds against nutritional needs.

   The constraints that prevent immediate progress are disparate. Obtaining
low-impact feed mixes and optimizing protein needs is limited by the economic
availability of alternatives. One illusion to avoid is that of using by-products that
are generally considered to have low environmental burdens. If the demand for
these increases substantially, their supply must be increased and will almost inevitably increase their production burdens. A clear constraint in balancing protein
supply and need is the use of pure amino acids in diets with poorer amino acid
balances than can be provided by soybean meal. Pure amino acids are GHG

intensive to produce and the benefits of having much lower impact sources are great. This is a challenge for the industry that could have widespread benefits to monogastric animal production. Factors that may prevent rapid uptake of engineering solutions, such as air heat exchangers, anaerobic digestion, litter combustion or belt-drying litter, are generally economic rather than environmental, although litter burning needs both careful regulation and control to avoid the production of highly undesirable emissions of compounds such as dioxins.

To date, almost all analyses have been on the main bird in global poultry production, the chicken. The baseline, let alone opportunities for improvement in other species, is barely known.

REFERENCES

BSI (2006) *Environmental Management – Life Cycle Assessment – Principles and Framework*, BS EN ISO 14040:2006, Incorporating Corrigendum No. 1. British Standards Institution, London. Available at: http://www.bsi-global.com (accessed 10 July 2015).

BSI (2008) *Specification for the Assessment of the Life Cycle Greenhouse Gas Emissions of Goods and Services*. PAS 2050:2008.

BSI (2011) *Specification for the Assessment of the Life Cycle Greenhouse Gas Emissions of Goods and Services*. PAS 2050:2011. Available at: http://www.bsigroup.com/Standards-and-Publications/How-we-can-help-you/Professional-Standards-Service/PAS-2050 (accessed 10 July 2015).

Cederberg, C., Sonesson, U., Henriksson, M., Sund, V. and Davis, J. (2009) *Greenhouse Gas Emissions from Swedish Production of Meat, Milk and Eggs 1990 and 2005*. SIK report 793, Swedish Institute for Food and Biotechnology, Gothenberg, Sweden.

CML (2014) CML-IA Characterisation Factors. Available at: http://cml.leiden.edu/software/data-cmlia.html (accessed 10 July 2015).

da Silva, V.P., van der Werf, H.M.G., Soares, S.R. and Corson, M.S. (2014) Environmental impacts of French and Brazilian broiler chicken production scenarios: an LCA approach. *Journal of Environmental Management* 133, 222–231.

Davis, J. and Sonesson, U. (2008) Life cycle assessment of integrated food chains – a Swedish case study of two chicken meals. *The International Journal of Life Cycle Assessment* 13, 574–584.

Dekker, S.E.M., de Boer, I.J.M., Vermeij, I., Aarnink, A.J.A. and Groot Koerkamp, P.W.G. (2011) Ecological and economic evaluation of Dutch egg production systems. *Livestock Science* 139, 109–121.

FAOSTAT (2014) Food and Agriculture Organization of the United Nations, Statistics Division. Available at: http://faostat3.fao.org/download/Q/*/E (accessed 10 July 2015).

IPCC (2013) Anthropogenic and Natural Radiative Forcing. Available at: http://www.climatechange2013.org/images/report/WG1AR5_Chapter08_FINAL.pdf (accessed 14 June 2014).

Jones, H., Warkup, C., Chop, A., Coffey, M., Wall, E., *et al.* (2008) A Study of the Scope for the Application of Research in Animal Genomics and Breeding to Reduce Nitrogen and Methane Emissions from Livestock Based Food Chains. Final report to Defra on project AC0204. Available at: http://tinyurl.com/DEFRA-AC0204 (accessed 14 June 2014).

Katajajuuri, J.M., Grönroos, J. and Usva, K. (2008) Environmental impacts and related options for improving the chicken meat supply chain. In: *Proceedings of the 6th LCA-Food Conference*, Zurich.

Leinonen, I., Williams, A.G., Wiseman, J., Guy, J. and Kyriazakis, I. (2012a) Predicting the environmental impacts of chicken systems in the United Kingdom through a life cycle assessment: broiler production systems. *Poultry Science* 91, 8–25.

Leinonen, I., Williams, A.G., Wiseman, J., Guy, J. and Kyriazakis, I. (2012b) Predicting the environmental impacts of chicken systems in the United Kingdom through a life cycle assessment: egg production systems. *Poultry Science* 91, 26–40.

Leinonen, I., Williams, A.G., Waller, A.H. and Kyriazakis, I. (2013) Comparing the environmental impacts of alternative protein crops in poultry diets: the consequences of uncertainty. *Agricultural Systems* 121, 33–42.

Leinonen, I., Williams, A.G. and Kyriazakis, I. (2014) The effects of welfare-enhancing system changes on the environmental impacts of broiler and egg production. *Poultry Science* 93, 256–266.

Leip, A., Weiss, F., Wassenaar, T., Perez, I., Fellmann, T., Loudjani, P., Tubiello, F., Grandgirard, D., Monni, S. and Biala, K. (2010) Evaluation of the Livestock Sector's Contribution to the EU Greenhouse Gas Emissions (GGELS). Final Report, European Commission, Joint Research Centre. Available at: http://ec.europa.eu/agriculture/analysis/external/livestock-gas/full_text_en.pdf (accessed 15 July 2014).

MacLeod, M., Gerber, P., Mottet, A., Tempio, G., Falcucci, A., Opio, C., Vellinga, T., Henderson, B. and Steinfeld, H. (2013) *Greenhouse Gas Emissions from Pig and Chicken Supply Chains – A Global Life Cycle Assessment*. Food and Agriculture Organization of the United Nations (FAO), Rome.

Meul, M., Ginneberge, C., Van Middelaar, C.E., de Boer, I.J.M., Fremaut, D. and Haesaert, G. (2012) Carbon footprint of five pig diets using three land use change accounting methods. *Livestock Science* 149, 215–223.

Mollenhorst, H., Berentsen, P. and de Boer, I. (2006) On-farm quantification of sustainability indicators: an application to egg production systems. *British Poultry Science* 47, 405–417.

Nguyen, T.T.H., Bouvarel, I., Ponchant, P. and van der Werf, H.M.G. (2012) Using environmental constraints to formulate low-impact poultry feeds. *Journal of Cleaner Production* 28, 215–224.

Pelletier, N. (2008) Environmental performance in the US broiler poultry sector: life cycle energy use and greenhouse gas, ozone depleting, acidifying and eutrophying emissions. *Agricultural Systems* 98, 67–73.

Pelletier, N., Ibarburu, M. and Xin, H. (2014) Comparison of the environmental footprint of the egg industry in the United States in 1960 and 2010. *Poultry Science* 93, 241–255.

Spies, A. (2003) The Sustainability of the Pig and Poultry Industries in Santa Catarina, Brazil: a framework for change. Thesis, University of Queensland, School of Natural and Rural Systems Management, Brisbane, Australia, p. 408.

Taylor, R.C., Omed, H. and Edwards-Jones, G. (2014) The greenhouse emissions footprint of free-range eggs. *Poultry Science* 93, 231–237.

Thévenot, A., Aubin, J., Tillard, E. and Vayssières, J. (2013) Accounting for farm diversity in life cycle assessment studies – the case of poultry production in a tropical island. *Journal of Cleaner Production* 57, 280–292.

Tongpool, R., Phanichavalit, N., Yuvaniyama, C. and Mungcharoen, T. (2012) Improvement of the environmental performance of broiler feeds: a study via life cycle assessment. *Journal of Cleaner Production* 35, 16–24.

Van Middelaar, C.E., Cederberg, C., Vellinga, T.V., van der Werf, H.M.G. and de Boer, I.J.M. (2013) Exploring variability in methods and data sensitivity in carbon footprints of feed ingredients. *The International Journal of Life Cycle Assessment* 18, 768–782.

Vergé, X., Dyer, J., Desjardins, R. and Worth, D. (2009) Long-term trends in greenhouse gas emissions from the Canadian poultry industry. *Journal of Applied Poultry Research* 18, 210–222.

Wiedemann, S. and McGahan, E. (2011) *Environmental Assessment of an Egg Production Supply Chain Using Life Cycle Assessment*. Australian Egg Corporation Limited, North Sydney, New South Wales, Australia.

Williams, A.G., Audsley, E. and Sandars, D.L. (2006) Final Report to Defra on Project IS0205: Determining the Environmental Burdens and Resource Use in the Production of Agricultural and Horticultural Commodities. Defra, London. Available at: https://tinyurl.com/defra-IS0205 (accessed 10 July 2015).

Williams, A.G., Audsley, E. and Sandars, D.L. (2010) Environmental burdens of producing bread wheat, oilseed rape and potatoes in England and Wales using simulation and system modelling. *The International Journal of Life Cycle Assessment* 15, 855–868.

Williams, A.G., Leinonen, I. and Kyriazakis, I. (2015) Environmental benefits of using turkey litter as a fuel instead of a fertiliser. *Journal of Cleaner Production*, doi: 10.1016/j.jclepro.2015.11.044.

Part VIII
Horizon 2050

Chapter 14
Horizons and Prospects – a Role for WPSA?

Colin Fisher*

EFG Software, Edinburgh, UK

INTRODUCTION

Can the poultry industry of Europe sustain itself to 2050? Can the poultry industry of Europe be sustained until 2050? Is the European poultry industry sustainable up to 2050? Asking this question in different ways emphasizes the wide range of issues that are inescapably raised and also the contradictions and conflicts that any answer must lead to. The first viewpoint will reflect the development of consumer demand and the ability of a profitable industry to make the necessary investment at the right time to meet this demand. The second viewpoint concerns responsibility for ensuring a food supply into the future and this presumably rests mainly with government and other public bodies. The third way of posing the question raises questions of resources, competing demands and some global technical issues such as biosecurity.

The speakers in this conference and the chapters in this volume have addressed different aspects of this topic and have analysed the various and many factors that are raised. In trying to draw some of the many threads together it seems appropriate to look especially at the implications for poultry science and for the World's Poultry Science Association who organized this meeting.

SCANNING THE HORIZON

Horizon scanning is defined on the UK Government website, which promotes the idea as 'exploring what the future might look like to understand uncertainties better'. An alternative morphological analysis used to predict possible futures for the French poultry industry (Jez *et al.*, 2011), uses scenario building to explore future options. For this conference the organizers limited the geographic horizon

*Corresponding author: cfisher345@gmail.com

to Europe and the time horizon to 2050. These limitations help to focus the discussion but raise the possibility that other horizons may be more important. The future of the European poultry industry may well be determined by events occurring outside the EU and many of the likely outcomes may be resolved during the next decade and will be resolved by 2050 (Mulder, Chapter 3, this volume). It is important to remember that one thing that is always over the horizon is another horizon!

The existential horizons for the future of the poultry industry are easily recognized. Future societies may decide that it is immoral to eat animals, alternative synthetic products may become available with less ethical and resource demands, limiting resources used for crop production may be directed to human consumption and all of these will be played out against the vagaries of climate change. Elements of all of these can be seen already but discussion of them does not really lead anywhere either for poultry science or for the poultry industry.

A more likely scenario is outlined above (Neeteson-van Nieuwenhoven *et al.*, Chapter 1, this volume; Mulder, Chapter 3, this volume). This is of a growing population creating increased demand for poultry products but with many constraints and uncertainties, especially about feed resources. A similar picture is painted by Herrero (2013) for animal agriculture as a whole.

Sustainability of agricultural systems currently receives a lot of attention. A sustainable industry can be defined as one that can persist for many generations and is far-seeing enough, flexible enough and wise enough not to undermine its physical or social systems of support. Amongst many others, the Food and Agriculture Organization of the United Nations (FAO, 2014) and the National Research Council (2010) have produced reports and there are more than a dozen scientific journals dedicated to the topic (Table 14.1). These are listed here, not because of their direct relevance to the sustainability of the poultry industry, but because by their very number they illustrate the complexity of the background against which this must be viewed. Alternative views of this complexity can be found in the multi-author paper by Pretty *et al.* (2010), which considers the 100 most important issues requiring research and consideration in the quest for sustainability, and in Garnett *et al.* (2013).

Table 14.1. Journals specializing in agricultural sustainability.

Journal of Sustainable Agriculture
Renewable Agriculture and Food Systems
International Journal of Agricultural Sustainability
International Journal of Sustainable Agriculture
Sustainable Agricultural Research
International Journal of Sustainable Agricultural Research
Journal of Developments in Sustainable Agriculture
Journal of Agriculture and Sustainability
Agriculture for Sustainable Development
International Journal of Sustainable Agricultural Management and Informatics
Journal of Agricultural Biotechnology and Sustainable Development
Agronomy for Sustainable Development
Asia Pacific Journal of Sustainable Agriculture, Food and Energy

In the UK, the government document 'The Strategy for Sustainable Farming and Food' (Government Office for Science, 2011) outlines the physical, economic and scientific framework within which the future of the poultry industry will have to develop, but gives no indication at all about what shape that future might take. It is obvious that if the poultry industry wishes its potential contribution to future food supplies to be fully understood and recognized, then continuing analysis and presentation of this argument will be essential.

LOOKING BACKWARDS

Opinions about this topic are inevitably both political and personal. The writer is at the end of a career in poultry research and looking back at previous experiences might give some useful pointers about how technical issues that arise in a discussion of sustainability can best be approached. How is the need for technical innovation identified, how is the research organized and how and at what stage do institutions respond? Four past issues might be suggested for further study in this light.

The BSE crisis

These catastrophic events emphasized the importance of pre-emptive guarding against zoonoses and also revealed widespread public opposition to the practice of feeding animals waste products derived from the same species. This has led to a significant loss of nutrients from the poultry production cycle and a loss of system efficiency, which will be of increasing importance in the future.

The removal of antibiotics

This remains a huge issue both for poultry science and for industry and is still a problem for a sustainable future (Walker and Garland, Chapter 9, this volume). The testing of alternative substances to attempt to replace antibiotics has come to dominate applied poultry nutrition research in recent years. When the industry started to grow and intensify it was not possible to keep flocks healthy without the use of antibiotics. One of the major challenges for future expansion is to make sure that such a need is avoided in the future through high level biosecurity and careful infrastructure planning.

The introduction of phytase

This is a classic sustainability story from which a great deal can surely be learned. The original stimulus was to solve an environmental issue in the Netherlands, but phytase use has now grown to be a major technical component of a sustainable future. Production efficiency is enhanced, environmental losses are reduced and a major limiting agricultural resource, phosphate, is saved on a huge scale. This is an outstanding story of technical development depending mainly on genetic engineering and provides an excellent example of how the technical aspects of sustainability can be approached.

Reorganization of the European egg industry

Here we have a significant response to a public concern about welfare translated into change by both legislation and by the supermarkets. The horizon for the whole industry changed and very high investment by producers was required to respond to this change (Ellis and Kempsey, Chapter 4, this volume). Several environmental indices of efficiency are in fact made worse by these changes, but this had little weight in the argument. Some would argue that even welfare itself has been harmed in some respects and undoubtedly the biosecurity of the egg production industry has been reduced by adoption of new methods of production. This is one aspect of sustainability that has not been met by sustainable intensification. In the Netherlands similar changes seem to be occurring in the broiler market with the introduction of the slower growing Kip van Morgen chicken.

Study of past events of this sort using the techniques of historical and sociological analysis might well indicate the pitfalls that arise, illustrate the dangers that following one path can limit future options and show how the timing of events, interventions and investment can influence the success or otherwise of developments to improve sustainability.

SUSTAINABILITY AND POULTRY SCIENCE

Many of the papers in this symposium have discussed the way in which poultry science can contribute to resolving questions raised by sustainability. A further question is whether the issue of sustainability should influence the choice of research priorities in the future. Several groups have addressed this question (Animal Task Force, 2013; Dumont et al., 2014; National Research Council, 2015), but these papers consider animal production as a whole and are largely concerned with large animal systems and agroecology. Jez et al. (2011) consider the future of the French poultry industry using a scenario-setting technique, but only extend their considerations to 2025. They also conclude that 'the scenarios also confirm the relevance of the current research orientations, and reinforce those that improve sustainability, influence product quality, and organise and regulate the sector'. It seems that a more critical analysis will be required before the role of research in helping the poultry sector be sustainable can be understood. The short time period considered of course tends to reinforce the view that the current situation is satisfactory.

Three outstanding technical issues seem likely to arise during the growth of a sustainable poultry industry: breeding and genetics, feed supply and biosecurity.

Breeding and genetics

From the reviews in this volume it is apparent that the breeding industry has the tools to respond to future selection needs (Hill *et al.*, Chapter 11, this volume) and that steps are being taken to conserve future genetic resources (Whyte *et al.*, Chapter 12, this volume). Much genetic research will be carried out in areas other than poultry science and the industry will have a continuing interest in maintaining the expertise required to draw on these developments. This appears to be the case at the present time.

Nutrition and feed supply

The poultry industry probably has sufficient nutritional tools for future developments (Roosendaal and Wahlstrom, Chapter 7, this volume), but training for new recruits in this area of applied science is of the highest importance. The scientific and technical basis of micronutrient and supplement use is mainly driven by companies active in this area but appears to be well founded. The research required to ensure the continuing supply of commodity ingredients (carbohydrates, fats and proteins) cannot be influenced directly by the interests of the poultry industry to any great extent (van der Aar *et al.*, Chapter 6, this volume). Applied poultry science could contribute to the development of new nutrient sources, e.g. insects, and more innovative work on the use of a wider range of nutrient substrates (wastes and fibre for example) will probably need to be encouraged.

Biosecurity

If poultry production is to increase at the high levels foreseen by some commentators then maintaining a high level of biosecurity is going to be a key issue. If large, but sustainable, production units are created the infrastructure and operation of these will have to meet very high standards of biosecurity. If sustainability is to involve a reduction in intensification to satisfy perceived welfare considerations, then biosecurity will be more difficult but an even more important consideration. Globalization of the poultry breeding industry can only succeed with the highest levels of biosecurity. In all cases the research, technical development and training to achieve these objectives will be critical.

SUSTAINABILITY AND THE WORLD'S POULTRY SCIENCE ASSOCIATION

The World's Poultry Science Association (WPSA) is concerned with research, education and organization. The association seeks to encourage dissemination of technical information and collaboration and contacts between poultry scientists, educators, advisors and all those involved in the poultry sector. Its

membership is widely drawn from all these sectors. It does this mainly by organizing various sorts of conference and also through its journal *The World's Poultry Science Journal* and its web page (http://www.wpsa.com). On a small number
of occasions the association has taken a formal position on some technical issues,
e.g. defining standard methods or procedures. The WPSA maintains good rela-
tions with other poultry science organizations and with some public bodies such
as FAO. It will also generally have consultative relationships on a national basis
with government bodies that are responsible for creating the legal technical
framework within which the industry operates (Pym, 2012). The WPSA should
therefore be ideally placed to consider the question of sustainability.

In a sense this symposium and WPSA involvement in the topic has its gen-
esis in a paper presented by Hodges (2009) at the opening ceremony of the 23rd
World's Poultry Congress held in Brisbane, Australia. In his paper entitled
'Emerging boundaries for poultry production; challenges, opportunities and
dangers', Hodges summarized his argument by concluding; 'poultry production
has achieved outstanding biological and economic performance in the last 50
years as outstanding leadership harnessed science and business. But unacceptable negative effects are now evident contributing to the threatened collapse in
Western society due to an unacceptable culture of materialistic consumption'.
This was undoubtedly a political speech and this rather apocalyptic conclusion
makes it easy to reject the substance of the paper. This would be a mistake
because the issues raised by Hodges and discussed in his paper are the key
issues that arise in discussing the future of the poultry industry.

To progress the topic Hodges proposed that a Poultry Industry Leadership
Think-Tank should be established to study and evaluate the many options for
legislative, market based and scientific changes in the poultry industry worldwide. This led WPSA and FAO joining with Dr Hodges to hold a think-tank
meeting together with representatives of the poultry breeding industry (WPSA,
2009). Among several further actions that were proposed, a special session entitled 'Guidance for the poultry sector – issues and options' was planned for the
13th European Poultry Conference to be held in France in 2010 and a further
think-tank meeting was scheduled for April 2010. No report of either of these
events has been found and the initiative started by Dr Hodges seems to have
petered out although Pym (2012) reports that a global summit for business
leaders in the worldwide poultry industry is being discussed with the Global
Challenges Forum (http://www.globalchallengesforum.org).

Support for small scale and village poultry farming in developing countries,
which is widely seen as part of the sustainability story, has blossomed in WPSA
mainly under the guidance of former president Dr Bob Pym. Relationships with
FAO and with the International Network for Family Poultry Development
(INFPD) have been formalized, Mediterranean and African Poultry Networks
have been started and the Asian-Pacific Federation of WPSA Branches has
Working Group No 1: Small scale family poultry farming.

It appears that some institutional reorganization will be required if the discussion of sustainability is to be continued within WPSA. The obvious way to do
this is to create a new working group, although within the European Federation
this would have to be reconciled with the remit of WG No. 1 Economics and

Marketing and in the Asian Pacific Federation with WG No. 1 Small scale family poultry farming. If a biannual symposium on the sustainability of the poultry industry were held, perhaps alternating between the two federations, the topic could be kept under review and new insights into technical and industry problems could be considered in the context of future developments. This might achieve much of what Hodges had in mind for a Leadership Think-Tank providing that industry participation in the activities could be accomplished. Alternative, more modest, proposals could include adding sustainability to the list of topics at European Poultry Conferences and World Poultry Congresses or for Branch, such as the UK, to institute and fund an annual lecture on the topic. Overall it seems desirable that the discussion initiated by the present symposium should be continued.

CONCLUSIONS

The most important conclusion from this symposium is that the discussion of sustainability of the poultry industry, within Europe and beyond, is an important topic that should continue to be analysed, discussed and debated. The World's Poultry Science Association appears to provide an ideal vehicle for such discussions. The most likely scenario for both poultry meat and egg production seems to be one of rapid growth under sustainable intensive conditions to meet the demands of both population growth and demand growth. But this scenario will be accompanied with some serious opposition from pressure groups and with a lot of technical complexity. The global supply of feed resources and biosecurity seem likely to provide the greatest technical challenges. The maintenance of technical expertise and the training of new generations of poultry scientists will be essential.

REFERENCES

Animal Task Force (2013) Research and Innovation for a Sustainable Livestock Sector in Europe. Available at: http://www.animaltaskforce.eu/Portals/0/ATF/documents%20for%20scare/ ATF%20white%20paper%20Research%20priorities%20for%20a%20sustainable%20 livestock%20sector%20in%20Europe.pdf (accessed 2 June 2015).

Dumont, B., González-Garcia, E., Thomas, M., Fortun-Lamothe, L., Ducrot, C., Dourmad, J.Y. and Tichit, M. (2014) Forty research issues for the redesign of animal production systems in the 21st century. *Animal* 8, 1382–1393.

FAO (2014) *Building a Common Vision for Sustainable Food and Agriculture; Principles and Approaches*. Food and Agriculture Organzation of the United Nations, Rome. Available at: http://www.fao.org/3/919235b7-4553-4a4a-bf38-a76797dc5b23/i3940e.pdf (accessed 2 June 2015).

Garnett, T., Appleby, M.C., Balmford, A., Bateman, I.G., Benton, T.G., Bloomer, P., Burlingame, B., Dawkins, M., Dolan, L., Fraser, D., Herrero, M., Hoffmann, I., *et al.* (2013) Sustainable intensification in agriculture: premises and policies. *Science* 341, 33–34.

Government Office for Science (2011) *Foresight. The Future of Food and Farming*. Final Project Report. The Government Office for Science, London. Available at: http://www.uk/government/

uploads/system/uploads/attachment_data/file/288329/11-546-future-of-food-and-farming-report.pdf (accessed 2 June 2015).

Herrero, M. (2013) Feeding the planet: key challenges. In: Oltjen, J.W., Kebreab, E. and Lapierre, H. (eds) *Energy and Protein Metabolism and Nutrition in Sustainable Animal Production.* EAAP Publication No. 134, Wageningen Academic Publishers, the Netherlands.

Hodges, J. (2009) Emerging boundaries for poultry production; challenges, dangers and opportunities. *World's Poultry Science Journal* 65, 5–22.

Jez, C., Beaumont, C. and Magdelaine, P. (2011) Poultry production in 2025: learning from future scenarios. *World's Poultry Science Journal* 67, 105–113.

National Research Council (2010) *Toward Sustainable Agricultural Systems in the 21st Century.* The National Academies Press, Washington, DC.

National Research Council (2015) *Science Research in Food Security and Sustainability.* The National Academies Press, Washington, DC.

Pretty, J., Sutherland, W.J., Ashby, J., Auburn, J., Baulcombe, D., Bell, M., Bentley, J., Bickersteth, S., Brown, K., Burke, J., Campbell, H., Chen, K., *et al.* (2010) The top 100 questions of importance to the future of global agriculture. *International Journal of Agricultural Sustainability* 8, 219–236.

Pym, R.A.E. (2012) WPSA: 100 years of service to the worldwide poultry industry. *World's Poultry Science Journal* 68, 543–549.

WPSA (2009) Report on think-tank meeting on poultry production systems. *Newsletter*, December 2009. Available at: http://www.wpsa.com/newsletter/2009_december.html (accessed 29 May 2015).

PART IX
Poster Abstracts

POSTER 1

The Effect of Protected Aromatic Compounds on the Performance of Broilers Challenged with Salmonella Enteritidis and on Ileal Lactic Acid Microflora

F. Goodarzi Boroojeni,[1]* M. Shahbaz Yousaf,[2]
S. Keller,[3] H.M. Hafez,[4] K. Männer,[1] W. Vahjen,[1]
H. Ur-Rehman[2] and J. Zentek[1]

[1]Institute of Animal Nutrition, Department of Veterinary Medicine, Freie Universität Berlin; [2]University of Veterinary & Animal Sciences, Lahore, Pakistan; [3]Novus Deutschland GmbH, Gudensberg, Germany; [4]Institute of Poultry Diseases, Department of Veterinary Medicine, Freie Universität Berlin

INTRODUCTION

Understanding animal nutritional requirements, together with a proper farm management and adequate feeding programme, is vital to efficient and sustainable poultry meat production. The gastrointestinal tract with its complex microflora plays a key role in growth performance and profitability of modern poultry operations and can significantly be influenced by the diet composition. An innovative, scientifically tested nutritional solution, based on a blend of protected aromatic compounds (BPAC) including benzoic acid (AVIMATRIX®, Novus Int.), was introduced in the poultry industry. Its slow release formulation is specifically developed to act optimally in a poultry intestinal tract environment, improving the status of microbial ecology in the chicken gut.

*Corresponding author: farshad.goodarzi@fu-berlin.de

AIM AND OBJECTIVE

This research assessed the benefit of a BPAC in pelleted feed on broiler performance under a *Salmonella*-challenged condition (trial 1) and on ileal microbiota in non-challenged birds (trial 2).

MATERIAL AND METHODS

The first trial included three different treatment groups: (i) a diet without benzoic acid (BA); (ii) a diet supplemented with 2 g/kg of BPAC; and (iii) a diet supplemented with 4 g/kg of BPAC. Each treatment involved 20 birds, ten replicates per treatment. All birds were challenged with 1 ml of broth culture containing 1×10^6 cfu/ml *Salmonella enteritidis*. The second trial involved two groups of 40 birds, one receiving only a control diet without BA and another fed with a diet supplemented with 2 g/kg of BPAC, eight replicates per treatment. The effect of the BPAC on the ileum microbiota, including various lactobacilli, *E. coli*, and enterococci in the gut, was measured with quantitative real-time PCR.

RESULTS

The birds in the BPAC group demonstrated a significant higher bodyweight gain (BWG; $P < 0.05$) and reduced feed conversion ratio (FCR) compared to the non-supplemented birds, particularly 7 days after challenge. The total *Lactobacillus* spp. count and ratio of lactobacilli/*E. coli* in the gastrointestinal tract of BPAC-fed birds was significantly increased versus the control birds.

CONCLUSION

These data suggest that balancing the gut microflora with a BPAC can play a key role to achieve a more sustainable poultry production. The findings suggest that a BPAC positively influences the microflora balance. This can explain the better performance of *Salmonella*-challenged birds as seen in increased weight and feed efficiency compared to unsupplemented birds. Other research in broilers has demonstrated a more stable gut flora will also result in higher litter quality, which supports healthier feet and thus a better animal welfare status.

POSTER 2

A New More Welfare-Friendly and Feed-Efficient Way to Grow Intensive Broilers

David Filmer*

FLOCKMAN, UK

INTRODUCTION

A growing, more affluent world population demands more meat from ever scarcer, more expensive resources. Broilers are the best candidates to achieve this, but ad-lib feeding does not exploit their digestive system fully. Forbes (2003) showed feeding broilers wet feed improved feed efficiency whilst Stacey *et al.* (2004) showed daily feedback information was essential to optimize growth (see Filmer, 2001 for references). This experiment compared mealtime feeding and feedback information to ad-lib feeding, and no feedback information.

MATERIAL AND METHODS

Nine pairs of broiler houses were used containing a minimum of 25,000 birds fed standard commercial feeds using the 'paired comparison' technique. Each house within a pair was the same regarding structure, equipment, management, breed, sex, parent stock, stocking density and date of housing. The two treatments were allocated at random within each pair.

RESULTS

Birds on the new system had 1.01 percentage units less mortality ($p = 0.0053$), 55 g more live weight ($p = 0.0010$), better feed conversion ratio (FCR) of 0.053 ($p = 0.00028$) and 20.5 units better European Performance Efficiency Factor (EPEF; $p = 0.00006$). Lower 95% confidence limits were 0.50%, 30 g, 0.033 and 14.3, respectively. Average extra margin over feed costs was £0.0379 per bird

*Corresponding author: david@flockman.com

housed and £1083 per house (p = 0.002). Lower 95% confidence limits were £0.0259 per bird housed and £798 per house, respectively. Birds on the new system were visually cleaner and more active but no statistics are available.

CONCLUSIONS

Results show that the new system significantly improved bird performance and profitability. Confidence limits show a 97.5% probability that the new system improved mortality by a minimum of 0.50% and FCR by a minimum of 0.033 and produced an extra margin over feed cost by a minimum of £798 per house. With six to seven crops of broilers per year, this represents a satisfactory return on capital. Less feed intake also lowers scarce water and energy use and the carbon footprint of the meat produced.

SUBSEQUENT WORK

A SPARK Award enabled joint work with Bristol University comparing intermittent dark periods with a single dark period at night. Now nine field trials, each with minimum four replicates, have taken place on integrator's units worldwide, including Thailand, South Africa, Brazil and Turkey, totalling 2.2 million birds. Average benefits were 1.01 percentage units less mortality (p = 0.0121), 40 g more live weight (p = 0.00022), better FCR of 0.062 (p = 0.00001), 17.6 units better European Performance Efficiency Factor (EPEF; p = 0.00020), £0.0432 per bird housed more margin (p = 0.00002) and £1463 per house (p = 0.00002). The lower 95% confidence limits were 0.28%, 10 g, 0.046, 11.0, £0.0314 per bird housed and £1053 per house, respectively.

REFERENCES

Filmer, D. (2001) Nutritional management of meat poultry. In: *Integrated Management Systems for Livestock*. Occasional Publication No. 28. British Society of Animal Science, Edinburgh, pp.133–146.

Forbes, J.M. (2003) Wet foods for poultry. *Avian and Poultry Biology Reviews* 14(4), 175–193.

Stacey, K.F., Parsons, D.J., Frost, A.R., Fisher, C., Filmer, D. and Fothergill, A. (2004) An automatic growth and nutrition control system for broiler production. *Biosystems Engineering* 89, 363–371.

Smothering and Predation Reduce Sustainability of Egg Production

C.A. Weeks,* P.E. Baker, A. Scrase and J. Walton

University of Bristol, School of Veterinary Sciences, Langford, UK

INTRODUCTION

Since the ban in 2012 on conventional cage production for layers within the EU, the proportion of hens kept in loose housed systems has increased. Levels of mortality are higher in such systems compared with all cage systems (e.g. Weeks *et al.*, 2012). The principal causes of mortality are disease, injurious pecking, smothering and injury. In many free-range (FR) flocks predation also causes substantial losses that often are poorly documented. High losses of birds reduce the sustainability of the enterprise. To find solutions there needs to be more accurate recording of both the causes and the levels of mortality. As smothering and predation are causes that are more easily identified by producers, we focus on these here.

MATERIALS AND METHODS

For study 1, data from farmers' records of cumulative mortality (CM) for ten commercial FR flocks of brown genotypes geographically spread over the UK were used. Flock size ranged from three to 10,000 (mean 5,950) and birds were depleted at a mean age of 71.9 weeks (range 65 to 76) during 2013/14. Farmers were asked to record the cause of mortality.

In study 2, which was a short 8-week trial in 2014 of a mobile phone app, eight other UK farmers recorded levels and causes of mortality in 15 FR flocks (size 3,000–14,000) ranging from 16 to 65 weeks at the start. The app enabled up to six causes of mortality (including 'other') to be ascribed to cull or found dead categories.

*Corresponding author: claire.weeks@bristol.ac.uk

RESULTS

The combined smothering and predation data for study 1 were 37.7% of total CM, which averaged 9.3%. Study 2 recorded smothers of up to 54 birds per case, mainly on litter, with 36% of recorded deaths due to smothers and a further 3% to predation. Farmers recorded more information into the app (e.g. numbers and location of a smother) than were found in paper records.

DISCUSSION

These and other published results indicate smothering (Bright and Johnson, 2011) and predation to be substantial causes of reduced production efficiency as well as being a welfare concern for hens in loose housing systems. Many losses to predators are unrecorded as birds are removed from the site. Moreover, losses from both causes are likely to occur early during lay when hens are at peak production, exacerbating the impact on production and sustainability, whereas the rate of loss from ill health tends to peak towards the end of lay with reduced consequence for sustainability.

REFERENCES

Bright, J. and Johnson, E.A. (2011) Smothering in commercial free-range laying hens: A preliminary investigation. *Veterinary Record* 168, 512–513.
Weeks, C.A., Brown, S.N., Richards, G.J., Wilkins, L.J. and Knowles, T.G. (2012) Levels of mortality in hens by end of lay on farm and in transit to slaughter in Great Britain. *Veterinary Record* 179, 647–650.

POSTER 4
Effect of Drying Process on Amino Acid Digestibility of Yeast Protein Separated From Potable Alcohol Distilleries

D. Scholey,[1]* P. Williams[2] and E. Burton[1]

[1]Nottingham Trent University, Brackenhurst, UK; [2]AG-Bio, Brixworth, UK

INTRODUCTION

There is much interest in the use of alternative protein sources in poultry feed and industrial co-products can provide high quality protein with the additional advantage of reducing potential waste. The bioethanol industry has been rapidly expanding and as such is a prime candidate for co-product optimization. Yeast protein concentrate (YPC) has been separated from both bioethanol and potable alcohol distilleries and been fed successfully to broiler chicks at inclusion levels up to 17.5%. However, the drying process of this material is the most costly and energy-intensive part of the process and optimization of this procedure is vital for the viability of the product.

METHOD

In this study, YPC from both bioethanol and potable alcohol distilleries was separated using a continuous centrifugation process and dried by one of three methods: freeze drying (FD), spray drying (SD) and ring drying (RD). These six products were then assessed against soya as a standard protein source for amino acid digestibility in broiler chicks by linear extrapolation from three inclusion levels. Birds were fed the test diets for 3 days and ileal digesta pooled from four birds. Digestibility and total content for each amino acid were compared across YPC sources by one-way ANOVA (SPSS v19).

*Corresponding author: dawn.scholey@ntu.ac.uk

RESULTS

Spray-dried bioethanol sourced YPC was comparable with soya for total amino acid content within this study. Bioethanol-sourced YPC had higher coefficients of digestibility (COD: average across all amino acids measured) than potable YPC within each drying method, with the COD for SD bioethanol being 0.73 compared with 0.58 for potable, and FD bioethanol YPC being 0.63 compared with 0.52 for the equivalent potable YPC. This is potentially due to the use of exogenous xylanase enzymes in bioethanol production (prohibited in potable alcohol production) as these have been shown to lead to reduced viscosity in co-products (Scholey *et al.*, 2011). SD YPC had higher COD than FD (0.73 to 0.63 for bioethanol and 0.58 to 0.52 for potable), which may be due to poor separation of the material and fibre contamination from the stillage. FD and SD both had significantly higher digestibility than RD YPC (0.39 for potable and 0.62 for bioethanol), which may be due to Maillard reactions from overheating the material. Lysine digestibility was particularly affected, with COD ranging from 0.42 to 0.67 compared with 0.81 recorded for soya.

CONCLUSION

Appropriate drying can be seen to be crucial for the viability of YPC as a protein source as the material can easily be overheated resulting in reduced amino acid availability.

REFERENCE

Scholey, D.V., Williams, P. and Burton, E.J. (2011) Potential for alcohol co-products from potable and bioethanol sources as protein source in poultry diets. *British Poultry Abstracts* 7(1), 23–24.

POSTER 5
Improving the Methodology for Assessing Bone Ash in Broilers

C. Sanni* and E. Burton
Nottingham Trent University, Brackenhurst, UK

INTRODUCTION

Poorly utilized phytate phosphorus in broilers is often reflected in poor growth performance and inadequate bone mineralization with the occurrence of fragile and brittle bones, necessitating the inclusion of dietary inorganic phosphorus. There are, however, concerns of the limited global phosphate reserves and the environmental effects of land application of high phosphorus poultry litter. It is thus important to routinely evaluate dietary phosphorus inclusions to ensure requirements are not wastefully exceeded. Routine evaluation of bone ash is hampered by inconsistencies in methodology and a lack of commercial derived reference values for comparison.

AIMS AND OBJECTIVE

Bone ash is a widely accepted measure of phosphorus supply in broilers because of its simplicity. The methodologies cited in available literature for determining bone ash are varied and are sometimes not concise. A study was conducted to compare two common methods of determining the degree of bone mineralization in broilers: one containing a fat extraction step prior to ashing and a second method omitting fat extraction.

MATERIAL AND METHODS

A total of 288 Ross 308 male broilers were fed one of six graded levels of inorganic phosphorus. Each diet had eight replicate pens and six birds per replicate.

*Corresponding author: colin.sanni@ntu.ac.uk

At day 35, right and left tibiae were collected from two birds per replicate. All adhering tissue was removed from bones prior to laboratory analysis. Fat was extracted from right tibiae before ashing, while fat was not extracted in left tibiae. Analysis of variance was conducted for each method to determine method sensitivity in identifying differences between diets at the $P < 0.05$ level. Following this, all bones were ashed at 650°C for 24 h. The strength of relationship between two methods was examined using SPSS to perform Pearson Correlation tests based on the individual bird bone ash values derived from each method.

RESULTS

The fat extraction method was more sensitive in elucidating differences in both ash weight and ash percentage for the two methods compared. Correlation for bone ash weight comparing the two methods ($r = 0.67$) was stronger than the strength of correlation for bone ash percentage ($r = 0.42$).

CONCLUSION

Despite environmental concerns on the use of organic solvents and the laborious nature of fat extraction, results from this study suggest the fat extraction method is a more reliable method for ash determination in finisher broilers, particularly when comparing results from different studies and differences in lipid metabolism.

POSTER 6
The Use Of β-Mannanase in Sustainable Commercial Poultry Production

Karl Poulsen[1] and Marco Martínez-Cummer[2]*

[1]*Elanco Animal Health, Greenfield, Indiana;* [2]*Elanco Animal Health, Basingstoke, UK*

INTRODUCTION AND AIMS

Feed induced immune response (FIIR) is an inflammatory response to β-mannans in feed and a potential threat to broiler production as the leading molecules, β-mannans are highly prevalent in a wide variety of feed ingredients including soybean, sunflower, palm kernel and sesame meals. Since soybean meal is a globally sourced protein ingredient, β-galactomannan is present in most feeds (Hsiao et al., 2006). β-galactomannans are highly viscous, water soluble, heat-resistant compounds that survive the drying/toasting phase of soybean processing (Dale, 1997). Experiments using ingredients rich in β-galactomannan have demonstrated that these molecules are intensely anti-nutritional in monogastric species. A feed-grade enzyme, β-mannanase, hydrolyses non-starch polysaccharide strands (β-galactomannan) found in leguminous feedstuffs and prevents said feed-induced immune responses (Anderson and Hsiao, 2006).

AIMS AND OBJECTIVES

Flock uniformity is a criterion affecting the bottom line of broiler processing plants and an improvement in live weight uniformity in grow-out barns will translate to an improvement in the consistency of processed poultry products. The objective is to show that β-mannanase can have a positive impact on broiler productivity and live-weight uniformity.

*Corresponding author: martinez_marco_antonio@elanco.com

MATERIAL AND METHODS

A broiler trial was done to evaluate the effects of β-mannanase on live performance and uniformity at market age. Treatments included a positive control and
a reduced energy negative control treatment with β-mannanase added to the
negative control at 225 or 400 g/t feed.

RESULTS

β-mannanase added at 225 and 400 g/t significantly increased 43 day weight by
4.6 and 8.7% with decreases in the percentage coefficient of variation (% CV) by
1.4 to 2.7 units. Adding β-mannanase at 225 and 400 g/t also improved feed
conversion ratios (FCR) by 0.8 points and 15.9 points, respectively. This study
demonstrates that FIIR is a reality and can impact broiler performance.

CONCLUSIONS

By using β-mannanase to degrade β-galactomannans in feed we are able to
prevent unnecessary FIIRs that lead to costly inflammatory responses. This has
been shown to conserve valuable nutrients that would otherwise be used to
mount unnecessary innate immune responses and rather can be used to improve
FCR and body-weight uniformity (Anderson *et al.*, 2001). β-mannanase can also
reduce the % CV by approximately 2% by reducing the proportion of birds in
lightweight category and this improvement will most likely be realized at the
processing plant.

REFERENCES

Anderson, D. and Hsiao, H. (2006) Effect of β-mannan (Hemicell® feed enzyme) on acute phase
    protein levels in chickens and turkeys. Abstract 231, *Poultry Science Association, 95th Annual
    Meeting*, 16–19 July, University of Alberta, Edmonton, Alberta, Canada.
Anderson, D.M., McNaughton, J.L., Hsiao, H.-Y. and Fodge, D.W. (2001) Improvement of body
    weight uniformity in broilers, turkeys, ducks and pigs by use of the *Bacillus lentus* mannanase
    (Hemicell®). *International Poultry Scientific Forum* January 2001 Paper 36 (Abstract).
Dale, N. (1997) Current status of feed enzymes for swine. Hemi-cell, Poultry and Swine Feed
    Enzyme. ChemGen Crop, Gaithersburg, Maryland.
Hsiao, H.-Y., Anderson, D.M. and Dale, N.M. (2006) Levels of β-mannan in soybean meal. *Poultry
    Science* 85(8), 1430–1432.

POSTER 7
Efficacy of Butyric Acid and Monolaurate to Combat Bacterial Enteritis Problems in Broilers

F.E. Dias,[1]* A. Schwarz,[2] T. Rogge,[3] J. De Gussem,[4] H. Van Meirhaeghe[2] and M. De Gussem[2]

[1]Vetworks, Cambridge, UK; [2]Vetworks, Poeke, Flanders; [3]Proviron, Oostende, Flanders; [4]Poulpharm, Izegem, Flanders, Belgium

INTRODUCTION

Several trials have shown that butyric acid plays an important role in the development of the intestinal wall. It serves as an important energy source for enterocyte proliferation, speeds up gut repair, strengthens gut barrier functions, has anti-inflammatory and anti-oxidant activity and regulates intestinal water intake. This study reports on a series comprising two floor-pen trials and a field trial involving multiple commercial houses.

OBJECTIVE

To evaluate the efficacy of a specific formulation of butyric acid and monolaurate for the control of bacterial enteritis (BE) in broilers.

MATERIAL AND METHODS

BE induction

To provoke bacterial enteritis (BE) in experimental conditions, a feed rich in non-starch polysaccharides (NSP), wheat/rye and highly methylated citrus pectin (HMC) was given. In trial 1 a high performance feed with high energy and protein level was used to compare BE reduction strategies to a positive control. In

*Corresponding author: francisco.dias@vetworks.eu

trial 2 a lower performance feed, with lower energy and protein content was used to compare BE reduction strategies to a negative control.

Floor pen trials

Trial 1

- Study groups: negative control: feed with moderate protein and energy level; positive control: feed with high protein and energy level (BE feed); positive control + antibiotic: tylvalosin treated group (5 mg/kg LW, 0.04 mg/ml drinking water) on BE feed;
- Study groups: BE feed supplemented with Provifeed™ Optigut, esterified C4/C12 specific blend in different dosages.
- Animals: 2000 commercial broilers (Ross 308): 20 treatments with 5 replicates per group and 20 birds per pen trial.
- Statistical analysis: the statistical test was 2-sided with a $\alpha = 0.10$. Analysis of variance (ANOVA).
- Coccidiosis control and feed strategy: coccidiosis vaccination, without anticoccidials in feed.
- Feed: high performance feed, high energy and protein level.

Trial 2

- Study groups: negative control, positive control, positive control + antibiotic;
- Study groups: salinomycin group: BE feed supplemented with Provifeed™ Optigut and 60 ppm salinomycin in feed, as coccidiosis control product.
- Animals: 1700 commercial broilers (Ross 308): 20 treatments with 5 replicates per group and 17 birds per pen trial.
- Statistical analysis: identical to trial 1.
- Coccidiosis control and feed strategy: coccidiosis vaccination except in group with salinomycin.
- Feed: lower performance feed, lower energy and protein.

Field trial

From the results of the floor pen trials a specific blend of butyric acid and monolaurate was selected (Provifeed™ Optigut), because of its beneficial effect on average end weight (AEW) and feed conversion ratio (FCR) to perform field trials.

- Trial group: 176,500 birds, 5 houses: Provifeed™ Optigut (4 kg/t) day 0–10, (2 kg/t) day 11–31 and (0 kg/t) day 32–42.
- Control group: 207,700 birds, 6 houses: commercial feed, including traditional organic acids.
- Evaluation parameters: AEW and FCR.

RESULTS

Floor pen trials

Improvement in AEW 8.5% (197 g), FCR 6 points.

Study 1

The AEW and the p-values (P) of the comparison with the reference group were 2.839 kg for the Provifeed™ Optigut group (P = 0.051), 2.402 kg for the negative control (P = 0.025), 2.786 kg for the positive control + antibiotic group (P = 0.148) and 2.635 kg for the positive control group (reference).

Study 2

The AEW and the p-values (P) of the comparison with the reference group were 2.210 kg for the Provifeed™ Optigut group (P = 0.011), 2.161 kg for the positive control + antibiotic group (P = 0.098), 2.318 kg for the salinomycin group (P = 0.001) and 2.020 kg for the positive control (reference).

Field trials

Improvement in AEW 4.5% (113 g), FCR 5 points.

CONCLUSION

The product Provifeed™ Optigut, a specific combination of butyric acid and monolaurate, has achieved the best improvement in the weight gain and FCR in the bacterial enteritis model used in the floor pen trials. This specific combination clearly and largely outperformed the separate ingredients. In the field trials, this result was confirmed. Our trials demonstrated that Provifeed™ Optigut can be considered as a potential alternative to antibiotic treatment of BE in broilers.

POSTER 8

Gut Antibacterial Effects of Even- and Odd-Numbered Medium Chain Fatty Acids in the Diet of Broilers

S. De Smet,[1]* J. Michiels,[2] A. Ovyn,[1] N. Dierick,[1] M. Laget[3] and A. Lauwaerts[3]

[1]Laboratory for Animal Nutrition and Animal Product Quality, Department of Animal Production, Ghent University, Melle, Belgium; [2]Department of Applied Biosciences, Ghent University, Ghent, Belgium; [3]Taminco BVBA, Ghent, Belgium

INTRODUCTION AND AIMS

Reducing the use of antibiotics and improving gut health are key issues in making broiler production systems more sustainable. Feed additives based on even-numbered medium-chain length fatty acids have been shown to exert antibacterial effects in the gut of monogastric animals, and concomitantly to improve their performances. Less is known about the effects of odd-numbered fatty acid derivatives and the type of carboxylate compound. New carboxylate formulations that result in a slower release of fatty acids in the gut might be interesting.

MATERIAL AND METHODS

One-day-old male Ross 308 chicks were fed a wheat–maize–soybean based meal diet in two feeding trials, that each consisted of a control diet and six experimental diets with two replicate pens per treatment and 20 birds per pen. In trial 1, Na-carboxylates of heptanoic (Na-C7), octanoic (Na-C8) and nonanoic acid (Na-C9) were added at 0.25% and 0.50%. In trial 2, Na-octanoic acid, methyl-octanoic acid (Me-C8) and isopropyl-octanoic acid (IP-C8) were tested at 0.25% and 0.50%. At day 27/28, three birds per pen were randomly selected, killed and digesta contents were collected from the proventriculus + gizzard, duodenum,

*Corresponding author: Stefaan.DeSmet@UGent.be

jejunum + ileum and caeca. Aliquots were taken for counting of total anaerobic bacteria, coliform bacteria, streptococci and lactobacilli using selective media (Michiels *et al.*, 2009). Tissue samples from the mid-duodenum were taken for measuring the histo-morphological parameters villus length and crypt depth as described by Van Nevel *et al.* (2003). Results were analysed by general linear model procedures with bird as the experimental unit (n = 6 per treatment).

RESULTS

Bird performances are not presented because of the limited number of pen observations. In trial 1, the supplemented diets did not result in lower counts for any of the bacterial groups in any of the digesta compared to the control diet (P > 0.05). However, across digesta and concentrations, lower lactobacilli and total anaerobes counts were found on the Na-C9 and Na-C8 diets compared to the Na-C7 diet (P < 0.05). In trial 2, lower counts of all bacterial groups were observed on the supplemented diets compared to the control diet in the gizzard samples (1–2 $\log_{10}$ cfu difference for coliforms, streptococci and total anaerobes on the 0.5% Na-C8 diet compared to the control diet; P < 0.05). Na-C8 appeared to be more effective than the ester compounds. No effects in the other digesta samples were observed. In trial 1, no effect of diet on the histo-morphological measurements in the duodenal mucosa was found (P > 0.05), whereas in trial 2 villi height was numerically greater for all treatments compared to control (P < 0.05 for 0.5% Me-C8 and 0.25% for IP-C8).

CONCLUSIONS

Medium chain fatty acids tended to have a selective chain length-dependent antibacterial effect. No benefit of ester compounds above Na-salts was observed.

REFERENCES

Michiels, J., Missotten, J., Fremaut, D., De Smet, S. and Dierick, N. (2009) *In vitro* characterisation of the antimicrobial activity of selected essential oil components and binary combinations against the pig gut flora. *Animal Feed Science and Technology* 151, 111–127.

Van Nevel, C., Decuypere, J.A., Dierick, N. and Molly, K. (2003) The influence of *Lentinus edodes* (Shiitake mushroom) preparations on bacteriological and morphological aspects of the small intestine in piglets. *Archives of Animal Nutrition* 57, 399–412.

Improved Utilization of Peanut Meal-Based Diets Supplemented with Enzymes by Laying Hens

P.A. Onimisi,* J.J. Omage and C. Kahuwai

Department of Animal Science, Ahmadu Bello University, Zaria, Nigeria

INTRODUCTION

Peanut meal, though cheaper and more available in Nigeria than soybean meal, is a less nutritive protein source in poultry diet due to its deficiencies in methionine and lysine and its relatively high crude fibre content. Certain measures can improve its nutrient availability and utilization.

AIMS AND OBJECTIVES

Evaluation of the improvement in nutrient utilization through enzyme supplementation of corn–peanut diets and the effect on egg-laying performance of hens.

MATERIAL AND METHODS

A total of 300, 25-week-old laying hens were used. Five treatment diets were formulated to meet standard nutrient requirements of layers: corn–soybean meal (C-SM) diet; corn–peanut meal (C-PM) diet; C-PM diet plus phytase; C-PM diet plus protease; C-PM diet plus G2G (an enzyme complex containing carbohydrases) and C-PM diet plus a combination of protease and G2G. Birds were distributed in a completely randomized design into the five dietary treatments, each having three replicates with 20 birds/replicate housed in deep-litter pens. Birds were supplied appropriate diet and water ad libitum for 12 weeks. All data collected were subjected to ANOVA and treatment means were separated using t-test.

*Corresponding author: onimisiphil@gmail.com

RESULTS

Body weights of birds were better sustained by C-PM diets supplemented with enzymes and particularly improved by the C-PM diet supplemented with G2G compared with the C-SM diet. Feed consumption was not significantly ($P > 0.05$) different for all the treatment groups. All C-PM diets supplemented with enzymes had significantly ($P < 0.05$) lower feed cost than the C-SM diet. Best results were obtained for the C-PM diet supplemented with a combination of protease and G2G in terms of total egg number (54.5 eggs/hen), average egg weight (56.9 g/egg), egg mass (3101.1 g/hen), income above feed cost (756.4 Naira/hen), hen-day production (67.9%) and hen-housed production (67.9%) compared with the performance of birds fed the most expensive C-SM diet with values of 50.4 eggs/hen, 56.6 g/egg, 2854.5 g/hen, 686.6 Naira/hen, 63.2% and 63.2%, respectively. Cost of producing a dozen eggs was highest for C-SM diet (170.4 Naira/12 eggs) than for C-PM diets; supplemented with protease (153.2 Naira/12 eggs), G2G (162.1 Naira/12 eggs) and with a combination of protease and G2G (163.4 Naira/12 eggs). Zero mortality was recorded for C-SM, C-PM plus protease and C-PM plus protease and G2G diets, while the highest value of 7.9% was recorded for the C-PM diet not supplemented with any enzyme.

CONCLUSIONS

The utilization of the peanut meal can be optimally enhanced by the use of enzymes, particularly a combination of protease and G2G supplementation in layer diets. Production cost is lowered while income is increased by using C-PM diets with enzyme supplementation rather than using the C-SM diet.

Poster *10*

Quality and Oxidative Stability of Broiler Meat as Affected by Dietary Supplementation of the Bioflavonoids Naringin and Hesperidin

M. Goliomytis,* P. Simitzis, N. Kartsonas, M. Charismiadou, A. Kominakis and S. Deligeorgis

Department of Animal Science and Aquaculture, Agricultural University of Athens, Athens, Greece

INTRODUCTION AND AIMS

Naringin and hesperidin are naturally occurring flavonoids well known for their antioxidant properties. They are abundant in citrus fruits, especially in pulp, a by-product of the citrus processing industry, which often is treated as waste. Hesperidin concentration in orange peel is between 13 and 24 g/kg, whereas naringin concentration in grapefruit peel is between 0.7 and 17 g/kg. In the present study we evaluated the effect of dietary supplementation with naringin and hesperidin on broiler meat quality parameters and oxidative stability.

MATERIALS AND METHODS

As hatched, 240 day-old broiler chickens were randomly assigned into six treatment groups. The control group C, without any flavonoid dietary supplementation, the N1 and N2 groups dietary supplemented with 0.75 and 1.5 g naringin/kg feed, respectively, the E1 and E2 groups supplemented with 0.75 and 1.5 g hesperidin/kg feed, respectively, and finally the VE group that was supplemented with 0.2 g α-tocopheryl acetate/kg feed. Dietary supplementation with bioflavonoids and α-tocopheryl acetate lasted from 11 days to 42 days of age when ten broilers per treatment group were slaughtered for pectoralis major meat quality assessment (colour-CIE $L^*a^*b^*$, pH24, cooking loss and shear force). Oxidative stability, expressed as nanograms malondialdehyde (MDA) per gram tissue, was

*Corresponding author: mgolio@aua.gr

assessed, on six out of the ten slaughtered birds per treatment, in the pectoralis major and biceps femoris after 3 days and 6 days of storage at 4°C and 120 days of storage at −18°C. MDA is a secondary lipid oxidation product formed by the hydrolysis of lipid hydroperoxides during lipid oxidation. High levels of MDA indicate high rates of lipid oxidation. MDA data were subjected to ANOVA with dietary treatment, muscle and their interaction as fixed effects. No muscle effect or interaction of muscle by added substances was detected and therefore data for MDA determinations were pooled. Finally, all data were analysed with dietary treatment as the fixed effect. The linear dose response (P-linear) to hesperidin or naringin was determined with contrasts among C and E or N group means.

RESULTS

Dietary supplementation with the bioflavonoids did not affect meat quality traits measured except or red colour that was lower in VE group compared with the E1 group (P < 0.05). Oxidative stability improved with increasing levels of both dietary naringin and hesperidin at 6 and 120 days (P-linear < 0.05) but not at 3 days of storage (P-linear > 0.05) compared to C group. Statistical differences (P < 0.05) were obtained, at 6 days of storage, between the C group (12.2 ng/g tissue) and E2, N1 and N2 groups, 8.1, 9.0 and 8.8 ng/g tissue, respectively. At 120 days of storage MDA differences (P < 0.05) were obtained between the C group (16.6 ng/g tissue) and N1 and N2 groups, 12.1 and 13.1 ng/g tissue, respectively. At 6 days of storage the MDA value for the VE group (6.6 ng/g tissue) was lower (P < 0.05) than that of the E1 group (10.1 ng/g tissue) but not lower than any of the naringin group (P > 0.05). At 120 days of storage the VE group (10.3 ng/g tissue) had lower MDA values than the E1, E2 and N2 groups, 14.2, 13.7 and 13.1 ng/g tissue, respectively.

CONCLUSION

Present results indicate that dietary supplementation with the flavonoids naringin and hesperidin to broiler chickens' diet improved meat oxidative stability without affecting its quality parameters. Hesperidin supplementation improved meat oxidative stability to a lesser extent compared to vitamin E whereas naringin's improvement was comparable to that of vitamin E. Further research is warranted in evaluating the efficiency of the citrus pulp as a dietary agent that may increase broilers' meat oxidative stability and quality since it is the main source of the naturally occurring antioxidants hesperidin and naringin.

ACKNOWLEDGEMENT

This research project was implemented within the framework of the Project 'Thalis – The effects of antioxidant's dietary supplementation on animal product quality', MIS 380231, Funding Body: Hellenic State and European Union.

POSTER 11
The Use of Wheat Dried Distillers Grain in Diets for Broilers

Elizabeth Ball*

Agri-Food and Biosciences Institute, Belfast, UK

INTRODUCTION AND AIM

Due to increased interest in biofuel production in the UK and Europe, more wheat dried distillers grain (DDGS) will be available for animal diets. However, there is a lack of information in the literature regarding the use of wheat DDGS in diets for broilers and research is needed to establish the effect of inclusion of wheat DDGS on both broiler performance and nutrient digestibility. The aim of this project was to ascertain the effect of wheat DDGS inclusion in diets for broilers.

MATERIALS AND METHODS

Starter, grower and finisher broiler diets were formulated to contain 0%, 5%, 10%, 15% or 20% wheat DDGS. DDGS replaced a proportion of the wheat and soybean meal within the diet formulation. A performance trial was conducted using 400 broilers with eight pen replications per treatment (10 birds/pen) from 0 to 35 days to determine dry matter intake (DMI), live-weight gain (LWG) and feed conversion ratio (FCR). A digestibility trial was also conducted using 40 broilers (eight replications/treatment) from 7 to 35 days to determine DM retention and ileal DM digestibility. Results were analysed by ANOVA. The effect of dietary treatment, linear and quadratic effects of DDGS inclusion were tested for.

*Corresponding author: elizabeth.ball@afbini.gov.uk

RESULTS

LWG and feed efficiency in the starter period (0–14 days) were significantly ($P < 0.001$) reduced when DDGS was included at 20% (e.g. for LWG the average of the first four treatments was 360 g/day versus 259 g/day). Conversely, LWG and feed efficiency were significantly ($P < 0.001$) improved in the grower period (14–21 days) when DDGS was included at 20% (e.g. FCR was 1.245 for birds offered the 20% inclusion of DDGS versus 1.791 for the average of the other inclusion rates). For the finisher period (21–35 days), there was a linear ($P < 0.01$) reduction in feed efficiency as DDGS inclusion increased (1.199 to 1.431). Overall (0–35 days), DDGS did not significantly affect DMI or LWG but there was a linear (< 0.05) reduction in feed efficiency as DDGS inclusion increased. There was also a linear (< 0.05) reduction in ileal digestibility as DDGS increased (0.702 to 0.611).

CONCLUSION

Inclusion of wheat DDGS to diets for broilers resulted in a linear reduction in feed efficiency and ileal DM digestibility and suggests that wheat DDGS may not be a suitable feed ingredient for broilers. However, depending on cereal cost and availability of DDGS it may be of some economic use at low levels of inclusion.

The Use of Rapeseed Meal in Diets for Broilers

Elizabeth Ball*

Agri-Food and Biosciences Institute, Belfast, UK

INTRODUCTION AND AIM

Soybean meal is the main source of protein in diets for broilers due to its high crude protein level and ratio of amino acids. However, as soybean meal must be imported, home-grown protein sources such as rapeseed meal (RSM) have been considered as an alternative. The optimum inclusion level of RSM in broiler diets has not yet been established and there are conflicting recommendations for inclusion in the literature. Therefore, the aim of this study was to investigate the response, in terms of broiler performance, to different dietary inclusion rates of RSM.

MATERIALS AND METHODS

The starter diets contained 0%, 2.5%, 5%, 7.5% and 10% RSM, the grower diets contained 0%, 5%, 10%, 15% and 20% RSM and the finisher diets contained 0%, 7.5%, 15%, 22.5% and 30% RSM. Crude protein and apparent metabolizable energy were formulated to be 230 g/kg and 12.55 MJ/kg, 210 g/kg and 12.87 MJ/kg and 190 g/kg and 13.3 MJ/kg for the starter, grower and finisher diets, respectively. A performance trial was conducted using 400 broilers with eight pen replications per treatment (10 birds/pen) from 0 to 35 days to determine dry matter intake (DMI), live-weight gain (LWG) and feed conversion ratio (FCR). Results were analysed by ANOVA. Unfortunately, the results for the 15% RSM treatment in the finisher stage had to be removed from the analysis due to extremely poor bird performance. LWG was 30% lower than the average of the other treatments although there was no increase in mortality. Analysis of the diet

*Corresponding author: elizabeth.ball@afbini.gov.uk

and determination of apparent digestible energy may explain why performance was so poor on this treatment, but these results are not available at this time. Performance results are presented for all the treatments in the starter and the grower phase and without the 15% RSM inclusion in the finisher period. For overall lifetime performance, results are presented without the middle inclusion of RSM (i.e. 5% RSM starter, 10% RSM grower and 15% RSM finisher).

RESULTS

There was a significant linear (< 0.05) and quadratic (< 0.001) effect on starter DMI and starter LWG (e.g. for DMI reduced from 347 g to 297 g, SEM = 9.82) when RSM inclusion increased from 0% to 10%. Feed efficiency in the starter period (0–14 days) was linearly (< 0.001) reduced when RSM inclusion increased (i.e. FCR ranged from 1.39 to 1.61, SEM = 0.046). Similar effects were observed in the grower phase (14–21 days) with DMI and LWG being linearly and quadratically (< 0.01) affected by increasing levels of RSM. For example, LWG reduced from 410 g to 345 g (SEM = 10.4) for birds offered 0% or 20% RSM, respectively. There was no effect of RSM inclusion on DMI in the finisher stage (21–35 days). However, LWG and feed efficiency were linearly (< 0.01) reduced as RSM inclusion increased. LWG ranged from 965 g to 843 g (SEM = 31.6) and FCR ranged from 1.35 to 1.58 (SEM = 0.054) for birds offered 0% and 30% RSM inclusion in the finishing stage. During the overall period (0–35 days), there was a quadratic ($P < 0.01$) effect on DMI and a linear (< 0.001) effect on LWG and FCR. LWG reduced from 1626 g to 1374 g (SEM = 39.9) and FCR ranged from 1.31 to 1.65 (SEM = 0.050) for birds offered 0% RSM and birds offered the dietary treatments containing the highest level of RSM inclusion.

CONCLUSION

Inclusion of RSM to diets for broilers resulted in reductions in DMI, LWG and feed efficiency at higher levels of inclusion.

Effects of Dietary Protease on Environmental Impacts of Broiler Production: a Holistic Comparison Using Life Cycle Assessment

I. Leinonen[1]* and A.G. Williams[2]

[1]School of Agriculture, Food and Rural Development, Newcastle University, Newcastle upon Tyne, UK; [2]School of Applied Sciences, Cranfield University, Bedford, UK

INTRODUCTION AND AIMS

Life cycle assessment (LCA) is a useful tool to analyse the overall environmental impacts of agricultural production, as it evaluates the production chain systematically to account for all inputs and outputs that cross a specified system boundary and relates these to the useful outputs. The aim of this study was to apply LCA modelling to quantify the effects of the use of a protease (RONOZYME ProAct – DSM Nutritional Products, Switzerland) in broiler diets on the environmental impacts of standard indoor broiler production.

MATERIAL AND METHODS

All experimental and other primary data (e.g. feed use) were provided by the industry. The bird performance data came from seven separate trials where the effects of low-protein diets including protease on broiler performance were evaluated. Additional background data were obtained from earlier studies on UK broiler production. A structural model for the broiler production system automatically took into account the effect of input variables, such as bird performance, and calculated all of the inputs required to produce the final product, which in this study was a mass unit of expected broiler carcass weight at the farm gate. Separate sub-models were used to quantify the environmental impacts of production of the main feed ingredients and the impacts of manure management.

*Corresponding author: ilkka.leinonen@ncl.ac.uk

The model is detailed in Leinonen *et al.* (2012). As an output of the LCA model, the emissions were aggregated into environmentally functional groups: global warming potential (GWP), eutrophication potential (EP) and acidification potential (AP). The analysis was carried out for two alternative system boundaries: either the feed production chain only, or the whole broiler production chain up to the farm gate, i.e. to the point before transport to slaughterhouse. The slaughter weight of the birds varied between 2 and 3.4 kg, depending on the experiment.

RESULTS

The results for the feed production chain showed that there was a reduction in all environmental impact categories and especially in GWP (on average by 5%) per mass unit of feed, when protease was used in the diets. The main reason for this is the reduction of the amount of soya used in the diets. When the whole broiler production chain was considered, there were relatively bigger reductions in EP and especially in AP (on average by 5%) compared to the feed production chain alone. The reason for this is that a major part of EP and AP arises from the emissions of ammonia from housing and manure management and in the model are directly dependent on the amount of nitrogen excreted by the birds. In this study all manure was assumed to be used entirely as fertilizer in crop production. When protease was used in the diets, the total crude protein content of the feed was reduced, which reduced the amount of nitrogen in manure and therefore also the emissions of ammonia (affecting AP and EP) and leaching of nitrate (affecting EP). A substantial benefit is the reduction of ammonia emissions *per se*, as these are problematic in poultry production. They are the subject of regulation from large units and we have binding international targets to reduce them.

CONCLUSIONS

The use of protease in the broiler diets and the resulting reduction of the use of soya reduced the environmental impacts of both the feed production (mainly GWP) and broiler production (mainly AP) chains. The latter is mainly through reduced ammonia emissions, which has a range of environmental benefits.

REFERENCE

Leinonen, I., Williams, A.G., Wiseman, J., Guy, J. and Kyriazakis, I. (2012) Predicting the environmental impacts of chicken systems in the United Kingdom through a life cycle assessment: Broiler production systems. *Poultry Science* 91, 8–25.

POSTER *14*
Susceptible Phytic Acid Content of Common Feed Ingredients Fed to Poultry

N.K. Morgan,[1]* C.L. Walk,[2] M.R. Bedford[2] and E.J. Burton[1]

[1]*Nottingham Trent University, Brackenhurst, UK;* [2]*ABVista, Marlborough, UK*

INTRODUCTION

The antinutritive effects of phytate and response of phytase are dictated by phytate 'susceptibility', defined as the percentage of total phytate that is receptive to hydrolysis by phytase under gastrointestinal tract (GIT) conditions. The aim of this study was to determine how much variation exists in susceptible phytate content between batches of feed ingredients fed to poultry in the UK.

METHOD

A minimum of ten batches of wheat, soybean meal (SBM), rapeseed meal (RSM), barley, maize and wheatfeed were collected from various sources throughout the UK. Susceptible phytate content of each sample was analysed by exposing the samples to conditions that mimicked the GIT (using warmed acetate buffer), and measuring the free and total phosphate release from the sample using a modified colorimetric assay.

RESULTS

Results showed that susceptible phytate content varies considerably between batches of ingredients (see Table P14.1). There was no correlation between phytate content and susceptibility in any ingredient.

*Corresponding author: nat.morgan@ntu.ac.uk

Table P14.1. Susceptible phytate content of feed ingredients fed to poultry.

Ingredient	Free phosphorus (g/100 g)	Phytic acid phosphorus (g/100 g)	Total phosphorus (g/100 g)	Phytic acid (g/100 g)	Susceptible phytic acid (%)	Range in susceptible phytate (%)
Wheat	0.04	0.07	0.11	0.24	57.44	49.51–63.45
Soybean meal	0.03	0.10	0.12	0.34	48.55	43.62–53.34
Rapeseed meal	0.09	0.24	0.33	0.83	50.81	46.84–56.16
Barley	0.09	0.12	0.21	0.43	55.82	52.68–61.45
Maize	0.05	0.13	0.18	0.47	53.65	51.82–60.78
Wheatfeed	0.10	0.13	0.24	0.47	70.55	66.34–73.36

CONCLUSION

Results suggest that the organic P component of feed ingredients exists in enzyme-susceptible and enzyme-resistant forms, and that binding of divalent cations to phytic acid may render a portion of dietary phytic acid resistant to hydrolysis by phytase. Consequently, it is imperative that diets are formulated and phytase matrix values are developed based on the reactive phytate content of the individual ingredient being fed as opposed to an accepted total phytate content value for the ingredient.

Poster 15
Cost Analysis of Alternative Feed Meals

R.A. Holser*

Russell Research Center, USDA, Athens, Georgia

INTRODUCTION AND AIMS

Poultry feed continues to be a significant expense in poultry production as the cost of maize and soybean meals remains elevated. Alternative meals are under investigation to reduce production costs while maintaining high feed conversion rates and body weight gain. Two promising alternative feed components include dried distillers grain with solubles (DDGS), a co-product from the maize dry milling process, and field pennycress (*Thlaspi arvense*), an emerging energy crop.

AIMS AND OBJECTIVES

Economic analyses were performed to evaluate the use of field pennycress meal as an alternative to soybean meal as a protein source in a formulation with DDGS that could reduce poultry production costs.

MATERIALS AND METHODS

Feeding trials were conducted with Cobb 500 male broiler chicks (0–18 day) fed a modified diet that included pennycress meal at 0, 5, 10 and 15% by weight levels with 3.5% by weight levels DDGS. The costs of feed formulated with pennycress meal were compared to a standard soybean meal formulation. The evaluation was based on a feed with 26.7% crude protein and 3.025 Kcal metabolizable energy. Amino acid supplements, e.g. 0.47% methionine, 1.27% lysine and 0.83% threonine, were used to obtain the recommended nutrient

*Corresponding author: ronald.holser@ars.usda.gov

specifications (Aviagen). A cost function was created and tested with varying inputs to determine the sensitivity of feed costs on components and potential savings.

RESULTS AND DISCUSSION

Birds consumed the modified feed formulations with no significant differences in consumption or weight gain. Associated feed costs were calculated for the control feed and those containing pennycress meal (pcm): 0% pcm, US$1.17642/kg feed; 5% pcm, US$1.1680/kg feed; 10% pcm, US$1.1596/kg feed; 15% pcm, US$1.1512/kg feed. These results show potential savings of 0.7%, 1.43% and 2.14% per kilogram feed with increased levels of pennycress meal. The analysis also considered the impact of transportation costs since poultry production in the USA is predominately in the South-eastern region while maize and soybean crops are grown extensively in the Mid-western region. Pennycress is planted in the winter and harvested in the spring, which complements the maize and soybean crop rotation practiced in the Mid-west. Transportation costs for pennycress and soybean meal produced in the Mid-west and shipped to the South-east are equivalent. However, pennycress cultivated in the South-east and distributed in the South-east reduced shipping charges by 85%.

CONCLUSIONS

Poultry feed formulated with pennycress meal and DDGS showed cost savings when used to replace soybean meal. Pennycress is cultivated as an industrial oilseed crop to supply the biodiesel industry. These results indicate the defatted meal would be a valuable co-product to replace soybean meal in poultry feed at reduced cost.

Identification of Endemic Avian Viruses: Poultry Production Problems and Quantification Using Novel Diagnostic Tests

R. Devaney,[1]* V.J. Smyth,[2] H. Jewhurst[2] and J. Trudgett[2]

[1]*Queen's University, Belfast;* [2]*AFBI, Belfast, UK*

INTRODUCTION AND AIMS

Runting-stunting syndrome (RSS) is a performance disease affecting broiler chickens and is potentially caused by multiple viruses, such as astroviruses and other endemic, enteric viruses, some of which are known and others unknown. The symptoms associated with RSS can be wide and varied and range from decreased body weight, a poor feed conversion ratio (FCR), enteric problems such as diarrhoea, lesions within the intestinal tract of the birds, and in severe cases mortality. This project aims to diagnose the viruses associated with RSS and subsequently aim to control it as a disease using a range of classical virology, molecular techniques, and novel methodologies such as next generation sequencing (NGS). This disease has been poorly researched to date due to the lack of convenient diagnostic tests and the complexity of the community of viruses responsible. High-throughput shotgun sequencing by NGS facilitates the *de novo* discovery of novel viruses and new strains of known viruses, which would be otherwise very difficult to detect.

AIMS AND OBJECTIVES

The project objectives include establishing a sampling regime of affected flocks at different time points in order to study the disease within the flocks as they progress. Affected flocks will also be compared to control birds unaffected by the disease to exclude virus strains that are non-pathogenic. Once sampled, the

*Corresponding author: rdevaney03@qub.ac.uk

Roche 454 NGS platform will be used to identify viruses in samples from growth-affected flocks and compare the viral profiles to those from birds affected at different time points. From these results quantitative polymerase chain reaction (PCR) assays will be developed in order to establish novel diagnostic tests that can be used to diagnose the disease in future flocks and determine virus load in a time- and cost-efficient manner. This will facilitate identification of the most important viruses (highest load) at specific time points. When the disease-causing viruses have been thus identified experimental infections of specific-pathogen-free chicks will be established in order to determine whether the viruses identified are actually responsible for the disease. These studies could lead to the possibility of control measures such as vaccines against the disease, which currently do not exist.

MATERIALS AND METHODS

Methods include a wide range of techniques such as a refined sample preparation using centrifugation, filtration and enzymatic treatment, use of viral extraction protocols in order to gain genomic material for study, quantification techniques in order to establish concentration of genomic material, NGS protocols in order to prepare libraries efficiently for runs on the Roche GS Junior platform (including appropriate bioinformatic analysis), use of PCR techniques and bioinformatics to develop novel diagnostic tests, and classical methods of virus isolation to produce material for use in experimental infection of SPF chicks.

RESULTS

Current results gained include data pertaining to the integrity and quantification of genomic material, revised sample preparation procedures in order to make the process more efficient, purification of viruses from a number of samples from commercial broiler flocks with RSS problems samples and preliminary sequencing trials performed on the NGS platform.

CONCLUSION

Work performed to date has shed some insight into the RSS disease, however future work to be performed will allow for deeper investigation into the disease and the factors controlling it. Through the use of current and novel molecular techniques, this is a prime opportunity to research the causes of RSS with the potential to discover interesting and novel results that will benefit the poultry industry as a whole.

POSTER 17

Impact of a T-2 Contaminated Feed and a Mycotoxin Eliminator (Elitox®) on Immunological, Liver and Intestinal Parameters in Broilers

V. Van Hamme,[1]* P.C. Machado Júnior,[2] S. Dhaouadi[1] and L.F. Caron[3]

[1]Impextraco nv, Heist-op-den-berg, Belgium; [2]Impextraco Latin America, Curitiba, Brazil; [3]Department of Basic Pathology, Universidade Federal do Paraná, Curitiba, Brazil

INTRODUCTION AND AIM

T-2 toxin is regarded as an acutely toxic mycotoxin, known to affect tissues with a high cell division rate, inducing cell apoptosis, and causing severe oral lesions, immunological dysfunctions and impairments in both liver function and intestinal integrity. The aim of the trial was to evaluate the impact of a mycotoxin eliminator (Elitox®) on immunological, liver and intestinal impairments caused by an early T-2 toxicosis.

MATERIALS AND METHODS

A trial was carried out during 28 days at the Federal University of Paraná, Brazil, with 96 1-day-old male Cobb® broilers housed in three 2 m^2 isolators with negative pressure ventilation system and divided into three treatments (n = 32): control (C), T-2 contaminated feed (using purified T-2 toxin at 800 ppb) (T-2) and T-2 contaminated (800 ppb) feed supplemented with a mycotoxin eliminator (Elitox®) at 0.2% (ET-2). On day 14, blood samples of eight animals per treatment were collected and used to determine the profile of circulating immune cells. On day 28, blood samples (n = 8/treatment) were taken for the evaluation of serum markers of liver function and samples of jejunum (n = 6/treatment) were collected. Early toxicological impacts in immune cells were evaluated

*Corresponding author: valentine@impextraco.com

through flow cytometry while liver impairments were checked through the quantification of serum concentrations of aspartate aminotransferase (AST) and alkaline phosphatase (AP). Early negative effects on intestinal integrity were evaluated through goblet cell counting in jejunum by immunohistochemistry.

RESULTS

T-2 toxin appeared to impair immunological parameters of the T-2 treatment birds compared to C, as indicated by significantly higher amount of suppressor macrophages (T-2, 0.32%; C, 0.04% peripheral blood mononuclear cells (PBMC)) and T-helper lymphocytes (T-2, 6.27%; C, 1.88% PBMC) after 14 days of exposure. The increased amount of macrophages is an indication of reduced phagocytosis efficiency and the higher T-helper cell count implicates an elevated metabolic cost to sustain homoeostasis, as no infective challenge was present and the control showed significantly less cells. Elitox® protected animals exposed to the toxin, as it modulated the immune response of ET-2 treatment birds, resulting in T-helper lymphocytes and suppressor macrophages being significantly lower than in the T-2 treatment (T-2, 3.61% PBMC; ET-2, 0.105% PBMC). Intestinal integrity as well as liver function were altered by the action of T-2 toxin after 28 days of exposure. T-2 birds showed significantly increased levels compared to C for AST (C, 196; T-2, 224.85 units/l), AP (C, 2.5×10^{-3}; T-2, 3.15×10^{-3} units/l) and goblet cells (C, 10.58; T-2, 23.55 cells/field), indicating liver and unspecific intestinal damage. In Elitox® treated groups, these negative effects appeared significantly reduced, evidencing the efficacy of the product in neutralizing the toxin: AST and AP for ET-2 birds were 196.82 units/l and 2.93×10^{-3} units/l, respectively, and the amount of goblet cells was 17.25 cells/field.

CONCLUSION

The addition of a mycotoxin eliminator (Elitox®) to broiler feed could prevent the early damages induced by T2-toxin.

Effects of Wheat Pentosan Content on Hepatic Antioxidants in Growing Broilers

V. Pirgozliev,[1]* F. Karadas,[2] M. Karakeçili,[2] S.P. Rose,[1] I. Whiting,[1] P.R. Shewry,[3] A. Lovegrove,[3] T. Pellny[3] and A. Amerah[4]

[1]*NIPH, Harper Adams University, Shropshire, UK;* [2]*Yuzuncu Yil University, Van, Turkey;* [3]*Rothamsted Research, Harpenden, UK;* [4]*Danisco Animal Nutrition, Wiltshire, UK*

INTRODUCTION AND AIMS

Wheat is the most widely used cereal in UK poultry diets. The nutritive quality of wheat for broilers is variable and may be affected by the non-starch polysaccharides (NSPs; primarily pentosans). Wheat pentosans can be a reason for increased digesta viscosity, microbial proliferation in the gut and reduced performance. The increased microbial proliferation may lead to an increase in toxin production in the gut that may impair the antioxidant status of the birds. Although much research has been conducted on the effect of wheat pentosans on performance and nutrient availability, there is a lack of information on the effect of pentosans on the antioxidant status of poultry. The objective of this study was to examine the effect of two wheat samples with different total pentosan contents on growth performance, and concentration of antioxidants such as total vitamin E, coenzyme Q_{10}, and carotenoids in the liver of growing broilers.

MATERIAL AND METHODS

Low (LP) and high (HP) pentosan wheat samples, containing 81 and 94 mg/g total pentosans, respectively, were used in the study. These were produced by Rothamsted Research from Yumai 34 wheat crosses selected for different pentosan contents but with similar proximate nutrient compositions. The determined

*Corresponding author: vpirgozliev@harper-adams.ac.uk

in vitro viscosity of the LP and HP were 1.81 versus 3.73 cP, respectively. The contents of total carotenoids and vitamin E in the LP and HP wheats were 1.24 versus 1.47 mg/kg, and 33.3 versus 35.3 mg/kg, respectively. Two diets were formulated to contain 206 g crude protein/kg, and 12.67 MJ/kg apparent metabolizable energy (AME). The main ingredients of the diets were 670 g/kg of experimental wheat, 220 g/kg soybean meal, 50 g/kg full fat soybean meal and 20 g/kg of vegetable oil. Eighty male 308 Ross broiler chickens were used in the study from 8 to 21 days of age. Each diet was replicated 20 times in a randomized block design, where two birds were allocated in each pen. Diets were fed as mash and feed and water were offered ad libitum. At the end of the study the birds were killed by cervical dislocation and their livers were collected for further analyses. The data were compared statistically by ANOVA.

RESULTS

Wheat pentosan content did not influence ($P > 0.05$) growth performance variables. However, birds fed LP wheat had lower ($P < 0.05$) *in vivo* viscosity compared to birds fed HP wheat, 4.5 versus 6.0 (SEM = 0.41), respectively. The hepatic concentration of carotenoids, coenzyme Q_{10} and total vitamin E of birds fed LP was higher compared to HP, 0.6 versus 0.5 mg/kg ($P < 0.05$; SEM = 0.02), 144 versus 126 mg/kg ($P < 0.1$; SEM = 7.5) and 38 versus 34 mg/kg ($P < 0.1$; SEM = 3.7), respectively. An increase in digesta viscosity may increase the retention time of nutrients in the gastrointestinal tract (GIT) providing relatively more substrate for utilization by microbes, thus increasing microbial proliferation. An increased microbial proliferation may be a reason for more toxins produced in the GIT of the birds, which will be absorbed across the intestinal mucosa and sent to the liver via portal circulation. Infectious diseases are often associated with reduction in tissue antioxidants, suggesting that higher concentrations of hepatic antioxidants may decrease the challenge provoked by infections.

CONCLUSION

The experiment has shown that feeding LP wheat improves the hepatic antioxidant content of growing broilers. In conclusion, the results of this study suggest that under commercial conditions, i.e. relatively high stocking density on floor pens, feeding LP wheat may have the potential to improve the birds' antioxidant status, which may help to reduce the impact of disease challenges.

POSTER 19
Ileal Amino Acid Digestibility of Wheat Dried Distillers Grain for Laying Hens

I. Whiting,[1]* V. Pirgozliev,[1] S.P. Rose,[1] A.M. Mackenzie[1] and A.M. Amerah[2]

[1]National Institute for Poultry Husbandry, Harper Adams University, Edmond, Newport, Shropshire, United Kingdom; [2]Danisco Animal Nutrition, Marlborough, Wiltshire, United Kingdom

INTRODUCTION AND AIMS

Wheat distillers dried grains with solubles (DDGS) is the predominant by-product of the bioethanol industry in the UK. It has the potential to be used as an alternative protein source for poultry diets. Although the amino acid digestibility of maize DDGS has been extensively studied for poultry, there is limited information on the amino acid digestibility of wheat DDGS when fed to laying hens. Research to date has found that bioavailability of amino acids in wheat DDGS is highly variable, particularly for lysine. The aim of the present experiment was to investigate ileal amino acid digestibility (IAAD) of wheat DDGS for laying hens.

MATERIAL AND METHODS

Eight wheat–soybean-based experimental diets were mixed containing either 150 g/kg or 300 g/kg from four DDGS samples produced by a single production plant (Ensus Limited, UK). The amino acid profiles of the four DDGS batches showed little variation. The contents of lysine, methionine and tryptophan ranged between 5.95 to 6.35 g/kg, 4.45 to 4.90 g/kg and 2.95 to 3.75 g/kg, respectively. A total of 144 22-week-old Hy-Line brown laying hens were randomly allocated to 48 layer cages (three birds per cage). Each diet was offered ad libitum to six cages for 8 days. Titanium dioxide was used as an indigestible marker. At the end of the study ileal digesta were collected and pooled into one pot per cage for determination of IAAD. Estimated values for IAAD of the four experimental DDGS samples were determined by extrapolation of regression lines using regression analysis.

*Corresponding author: iwhiting@harper-adams.ac.uk

RESULTS

The estimate of the mean for total amino acid digestibility was 0.510 (SE = 0.119). Amino acids with the lowest digestibility were lysine (0.198; SE = 0.184), threonine (0.345; SE = 0.147), cysteine (0.353; SE = 0.128) and aspartic acid (0.314; SE = 0.134). The estimates of the IAAD for the other limiting amino acids, methionine and tryptophan, were 0.569 (SE = 0.111) and 0.555 (SE = 0.133), respectively. The average of the indispensable and dispensable IAAD was 0.493 (SE = 0.125) and 0.532 (SE = 0.112), respectively.

CONCLUSION

The results from this experiment indicate that the bioavailability of amino acids in wheat DDGS is low with lysine being the least digestible. When formulating poultry diets containing wheat DDGS this should be taken into account and the use of a synthetic lysine should be considered.

INDEX

Page numbers in **bold** type refer to figures and tables.